Property of Stanley P. Huffman

MANUAL OF TROPICAL AND SUBTROPICAL FRUITS

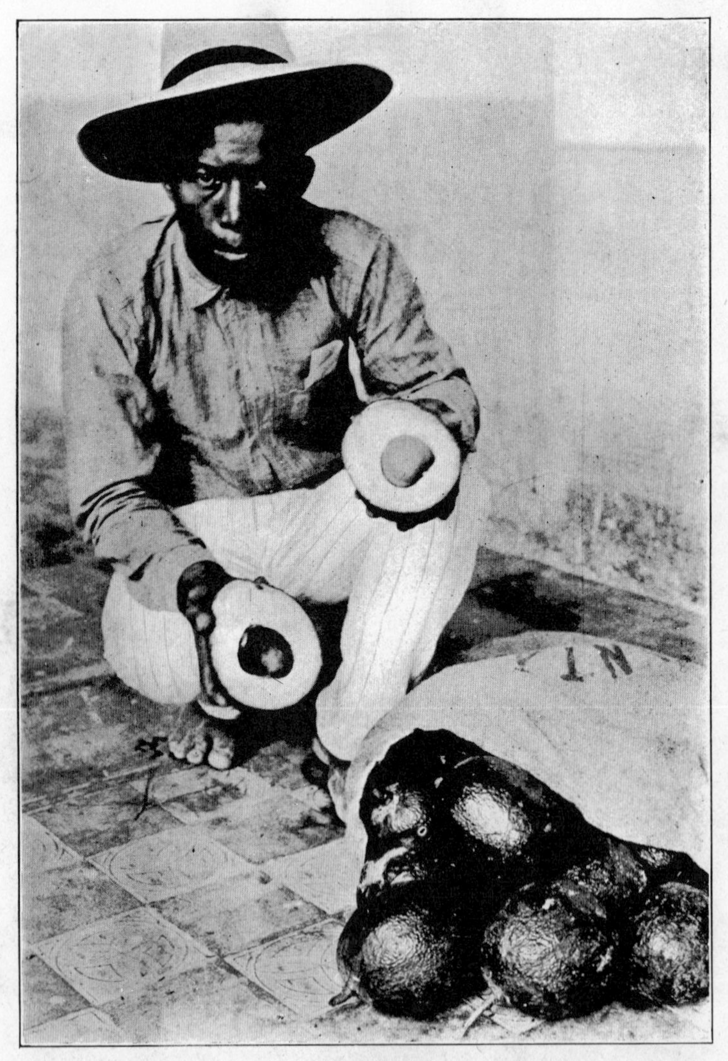

PLATE I. The Nimlioh avocado.

MANUAL OF TROPICAL AND SUBTROPICAL FRUITS

EXCLUDING THE BANANA, COCONUT, PINEAPPLE, CITRUS FRUITS, OLIVE AND FIG

A Facsimile of the 1920 Edition

BY

WILSON POPENOE

HAFNER PRESS
A Division of Macmillan Publishing Co., Inc.
New York
Collier-Macmillan Publishers
London

Originally Published 1920, by The Macmillan Company

This edition is reprinted by arrangement with Wilson Popenoe

HAFNER PRESS
A Division of Macmillan Publishing Co., Inc.
866 Third Avenue, New York, N.Y. 10022

Collier Macmillan Canada, Ltd.

Library of Congress Cataloging in Publication Data

Popenoe, Wilson, 1892–
Manual of tropical and subtropical fruits.

Facsimile of the 1920 ed. published by Macmillan, New York
1. Tropical fruit. 2. Fruit-culture—Tropics.
I. Title.
SB359.P55 1920 a 634.4 74-11080
ISBN 0-02-850280-9

Printed in the United States of America

printing number
3 4 5 6 7 8 9 10

INTRODUCTION
TO REPRINT EDITION

THIS work was first published more than fifty years ago after Liberty Hyde Bailey, the greatest of American horticulturists, became interested enough in tropical pomology to commission me to prepare a book on the subject for inclusion in his series of Rural Manuals being published by the house of Macmillan.

At the time the book was written, only a quarter of a century had passed since horticulturists in southern Florida had become interested in the commercial production of avocados and mangos, and their colleagues in southern California had begun to think of subtropical fruits other than the citrus. Because the seedlings of fruits improved by man through natural selection usually show too much variation in character and quality to make their commercial cultivation profitable, the dreams of these pioneers were realized through vegetative propagation, usually some form of grafting.

British horticulturists in India, and British and French horticulturists in the West Indies and their colleagues in other tropical regions had for many years been devoting attention to the development of mango culture. Theirs was not the first attempt. Early writings in Sanskrit make it seem possible that the improvement of the mango through vegetative propagation may have been practiced in very early times. For centuries, the Chinese had propagated their favorite fruit, the litchi or lychee, by air layering. Indeed, good bananas and pineapples had developed from wild forms because vegetative propagation was the natural means of reproduction.

By 1920 Liberty Hyde Bailey completed his monumental work, *The Standard Cyclopedia of Horticulture* which incorporated a goodly amount of material on tropical and subtropical fruits from other outstanding authorities. In writing my own book, I drew upon ten years' experience in various parts of the world and when the Manual was published in 1920 it enjoyed an enthusiastic reception in the horticultural press not only in the United States but also in Europe, the Far East, and such distant regions as South Africa and Australia. It was welcomed because it brought together, for the first time, comprehensive information on many species of tropical and subtropical fruits, and their horticultural varieties, if in fact they existed. It was especially esteemed because it contains a large amount of practical information on propagation and culture.

With the passing of the years, the book acquired a unique position among the many workers who have entered the field of tropical pomology. This fact, together with the difficulty in obtaining a copy today, has justified the Hafner Press in bringing out a reprint of the original edition.

In using the reprint, one must keep in mind that its real value to present day horticulturists lies in the fact that it is a unique source of background information. It goes without saying that during the fifty years since its publication much progress had been made. The horticultural varieties (cultivars, they are called today) of the avocado and mango which receive so much attention in this work have been superseded in nearly all cases by ones which produce better commercial characteristics, if not always better fruit. Indeed, readers should be reminded that the most popular commercial varieties of today may not be mentioned in a Manual of Tropical Fruits written in the year 2024. New varieties which originate as chance seedlings appear all the time, and the plant breeders are constantly seeking to produce new varieties capable of consistently bearing large crops, or resisting certain diseases, or extending the ripening season. Especially important has been the fight against the root-rot of avocados, *Phytophthora cinnamomi,* which has destroyed thousands of acres of trees in California, and has done much harm in other regions.

We have learned that if the flowers of the cherimoya are not hand pollinated (a laborious and expensive operation) the plants will only produce good crops when grown in an ideal climate—one that is neither too hot nor too dry. And we have been unable to control the erratic behavior of the litchi, which experts in Hawaii now say had best be considered a beautiful ornamental tree for the dooryard which every once in a while produces a fine crop of beautiful and delicious fruit. Finally we have not really been able to make the papaya behave like a civilized plant, nor have we eliminated from the otherwise delightful durian the smell of overripe Limburger.

JULY, 1974 WILSON POPENOE

Antigua, Guatamala

PREFACE

My intention in preparing the present work has been to bring together, for the guidance of those who live in the tropical and subtropical regions of the globe, the available information concerning the principal fruits cultivated, or which may be cultivated, in those regions. The banana, the coconut, the pineapple, the citrus fruits, the olive, and the fig are not included, however, since these have been fully treated by other writers. Nor have I attempted to describe all of the fruit-bearing plants of the tropics: rather has it been my aim to concentrate on those which most seem to merit extensive cultivation, the culture of many of which is as yet little understood. No work in the English language has attempted to cover this subject, and the few which have appeared in other languages do not contain the data concerning propagation and cultural practices which would make them useful to horticulturists. Unfortunately, as regards many of the less-known fruits, few data are available, but concerning the more important ones the researches of such workers as E. Bonavia, A. C. Hartless, and William Burns in India, H. A. Van Hermann, F. S. Earle, and C. F. Kinman in the West Indies, George B. Cellon, Edward Simmonds, W. J. Krome, P. H. Rolfs, and Reasoner Brothers in southern Florida, F. Franceschi (E. O. Fenzi) and Ira J. Condit in California, J. E. Higgins and his associates in Hawaii, P. J. Wester in the Philippines, and L. Trabut in the Mediterranean region, have brought to light much valuable information. The work of such men as G. N. Collins, O. F. Cook, David Fairchild, W. E. Safford, and Walter T. Swingle, of the Bureau of Plant Industry,

United States Department of Agriculture, has also added materially to our knowledge of the subject.

References throughout the book indicate the extent of my indebtedness to these and other investigators. In order that the work may reflect as fully as possible the total knowledge at present available on any topic, I have drawn freely from all sources, exercising, at the same time, all possible care to avoid perpetuating the more than numerous errors with which the literature of tropical fruits is burdened.

For the past seven years, during a large part of which time I have traveled as Agricultural Explorer for the United States Department of Agriculture, I have had exceptional opportunities for gathering, at first hand, information for this work. In the course of my journeys I have visited Hawaii, Japan, the Philippines, the Straits Settlements, India, Arabia, North Africa, Mexico, Guatemala, the West Indies, and Brazil. This field work has alternated with and been supplemented by practical experience with the cultural problems of tropical and subtropical fruit-growing in California and Florida. To those familiar with the thorough and exhaustive treatises which have been published on the northern fruits, however, the present work will no doubt appear superficial in character. Necessarily it is so. Present knowledge of the greater number of tropical fruits is superficial, and many years must pass before it will be possible for a thoroughly comprehensive treatise to be offered on any one of the species here considered, excepting possibly the date.

I have been assisted and encouraged in the preparation of this work by many persons. It is a particular pleasure to acknowledge my indebtedness to Charles Fuller Baker, now Dean of the College of Agriculture, University of the Philippines, under whose guidance I first took up work in tropical pomology, and whose boundless enthusiasm for tropical plants has been a constant inspiration to me; to F. Franceschi, formerly of Santa Barbara, California, who was one of the pioneers in the introduction and cultivation of tropical fruits in California; and above all, to my present chief, David Fairchild, and my colleagues in the Office of Foreign Seed and Plant Introduction of the Bureau of Plant Industry. To Dr. Fairchild America is indebted for many choice varieties of the mango, the date, and other tropical fruits which are now cultivated in the United States, and for his assistance and encouragement in my own investigations I owe him a debt of gratitude which I can never pay.

W. J. Krome of Homestead, Florida, has criticized the chapters on the avocado and mango, and added many notes of interest and value to the former. W. E. Safford of the Bureau of Plant Industry has revised the chapter on the annonaceous fruits, and Henry Pittier of the same Bureau that on the sapotaceous fruits. To my brother, Paul Popenoe, I am indebted for most of the chapter on the date. H. H. Hume of Florida has criticized the chapter on the kaki. J. N. Rose of the United States National Museum has furnished most of the data on the tuna and pitaya. Sidney F. Blake of the Bureau of Plant Industry has been of much assistance on matters of botanical nomenclature. J. Smeaton Chase of Palm Springs, California, has rendered valuable aid in the preparation of the manuscript. To all of these men I express my sincere appreciation of their help.

The line drawings with which this work is illustrated have been made by Mrs. R. E. Gamble of the Bureau of Plant Industry. Nearly all of them, as well as most of the half-tone plates, are from my own photographs; a few are from photographs by P. H. Dorsett of the Bureau of Plant Industry.

WILSON POPENOE

WASHINGTON, D. C.,
October 1, 1919.

TABLE OF CONTENTS

CHAPTER I

CHAPTER II

CHAPTER III

CHAPTER IV

CHAPTER V

CHAPTER VI

CHAPTER VII

CHAPTER VIII

CHAPTER IX

CHAPTER X

CHAPTER XI

PLATES

MANUAL OF TROPICAL AND SUBTROPICAL FRUITS

CHAPTER I

THE OUTLOOK FOR TROPICAL FRUITS

THE thickly peopled countries of the Temperate Zone must look more and more to the tropics to supplement their own food resources, whether by direct supplies, made possible in increasing measure by ever-improving means of transportation, or by furnishing plants which may be cultivated in mild-wintered regions such as California and Florida. Both forms of contribution will be largely in the item of fruits. As examples of the first class, the banana, because of its immense yield and quick production, has already been exploited on a large scale, and the coconut, through its product copra, has become an economic factor of prime importance; in the second (or rather, in both) the avocado, still a novelty but of very great possibilities as adaptable to growth in our own country, is on the verge of taking a high place among the food crops contributed by the tropics.

Many other fruits of the Torrid Zone, not all of them so important, yet all valuable in degree in the dietary of the race, must be grown in ever-increasing quantities, not only to supply northern markets, but also, — and even more important, — to enable the native populations of the tropics as well as settlers from the North to obtain abundantly and cheaply this most wholesome source of human energy.

For, strange as it may seem to many who have never lived or traveled in the hot belts of the earth, those lands come far short of conforming to that conventional idea of the tropics, as regions where luscious fruits grow wild upon every tree and the languorous native has only to stretch forth his hand to obtain his dinner. It is a well-attested fact that the inhabitants of many tropical countries suffer for want of sufficient fresh fruit; and it is also true that much real starvation in densely populated hot regions, India for example, could be averted by planting on a wholesale scale fruit-trees such as the avocado, whose product has a relatively high food value.

The reason for this scarcity of fruits in precisely those regions where, by climatic indications, one would expect them to be most abundant, is not to be found in any single fact, but is, perhaps, largely the result of three causes: first, the enervating effect of heat, which discourages man from undertaking work which can be avoided; second, the one-sided exploitation of many tropical regions for the production of materials such as rubber and cotton, without sufficient regard to supplying wholesome foodstuffs for those who labor in producing these articles; and third, the long time required by tree-fruits to yield returns, as compared with the annual crops such as corn, beans, and squashes. This last factor is particularly disastrous where primitive races of people are concerned, for such almost invariably devote their attention in the main to crops which give quick returns, — the very crops which must depend absolutely on the season's rainfall.

It is, indeed, only as scattered, often neglected, specimens in dooryards and around cultivated fields that many of the tropical fruit-trees exist. Others, such as the mango and the breadfruit, are given more attention, yet they rarely receive more than a fraction of the solicitous care which northerners lavish on their apples, peaches, and pears.

With the exception of a few species, such as the banana

and the coconut, the tropical fruits have received scientific attention only when their culture has been brought northward to the extreme limit of their zone, as, in the case of certain of them, it has been in California and Florida. Even here their study and improvement have only been undertaken in very recent years; many species, in fact, are still in the condition of wild plants, so that it is no wonder their fruits are sometimes looked on by northern horticulturists as almost without value. The case is well put by Hartwig, who writes, in his work "The Tropical World":

"It may easily be imagined that the tropical sun, which distills so many costly juices and fiery spices in indescribable multiplicity and abundance, must also produce a variety of fruits. But man has yet done little to improve by care and art these gifts of Nature, and, with rare exceptions, the delicious flavor for which our native fruits are indebted to centuries of cultivation, is found wanting in those of the torrid zone. In our gardens Pomona appears in the refined garb of civilization, while in the tropics she still shows herself as a savage beauty, requiring the aid of culture for the full development of her attractions."

The exceptions to this condition, however, are notable, and scarcely so rare as Hartwig and others have believed. The mango, in its finer Indian varieties, offers an example of improvement through selection and vegetative propagation which equals that of the peach, if indeed the advance from wild to cultivated forms has not been greater in the former than in the latter fruit. Those who have tasted the luscious Pairi mango of Bombay, or the Mulgoba as now grown in Florida, will recognize the probable accuracy of this statement.

Many other tropical fruits might be mentioned which compare favorably with the best products of high cultivation in the Temperate Zone. Who, that has had the opportunity of judging, has not felt, as he lifted the snowy segments of the mangosteen from their cup of royal purple, that here was a fruit not excelled by any other in the world? The cherimoya of tropical America leaves little to be desired, while the litchi is preferred in China, not without reason, to the finest orange or peach. American residents in Hawaii consider the papaya the most delicious of breakfast-fruits, surpassing in their estimation the cantaloupe or muskmelon. To the Japanese taste there is no better fruit than the kaki, while to the Arab the date is the quintessence of richness and flavor.

The ignorance, or tardiness of adoption, of the art of grafting has, in many tropical countries, prevented the development of superior fruits. The superb apples and pears of the Temperate Zone, and the splendid mangos of India, could not be grown without grafting, since improved varieties of nearly all tree-fruits tend to revert to the wild type when propagated by seed. The finest fruits are, in fact, artificial productions which can only be maintained by artificial means; under free competition of natural selection they would disappear.

Because of this rare occurrence, among tropical fruits, of fine horticultural varieties as compared with the profusion of semi-wild seedlings, much criticism has been ignorantly directed at these fruits in general. C. F. Baker, who has done much to advance the science of tropical pomology, graphically states the case as follows:

"On hearing some aspersions cast upon the caimite (*Chrysophyllum Cainito*), a valuable and delicious fruit at its best, a Cuban was heard to remark, 'There are caimites, and there are caimites!' A similar remark might be made of most tropical fruits. The methods of seed selection, of breeding, and of vegetative propagation have rarely been brought to bear on any of these things. As for systematic search for the better forms now existing, and the rapid building up of really comprehensive experimental plantations of them in the tropical botanic gardens and experiment stations, we have yet a field of highly useful, most remunerative, and intensely interesting work before us."

It is to this field that attention must be devoted, if the agricultural development of the tropics is not to become even more one-sided than it is to-day. British horticulturists in India and Ceylon, French in the Oceanic colonies, and American in the subtropical parts of California and Florida, as well as in the West Indies, Hawaii, and the Philippines, have done notable work during the past quarter of a century; yet when their achievements are considered alongside the possibilities, it is evident that hardly has a beginning been made with this promising field.

"*Botanicus verus,*" said the great Linnæus, "*desudabit in augendo amabilem scientiam,*" — "The true botanist will sweat in advancing his beloved science." Even so must the investigator who undertakes to further the progress of tropical pomology expect to find hard work, at times under trying climatic conditions, — to sweat indeed, — unless his lot is cast in the delightful climate of the tropical highlands, or in subtropical regions such as California and Florida. But the subject is one which offers such a wealth of fascinating problems and gives promise of such valuable results, that for a long time to come it can hardly fail to attract the needful few among the many whose tastes incline them toward pomological pursuits.

It is indeed fortunate for our country that its boundaries include areas where certain of the most valuable tropical fruits can be cultivated. Of these areas, the warmer parts of Florida and California seem destined, by reason of their favorable situation with respect to the great centers of our population, to take the lead in the production of such fruits for supplying the northern markets. The advantageous climate of these states as regards living and working conditions, as compared with the tropics, makes it probable also that they will be the field of more activity along lines of horticultural investigation than will the strictly tropical countries where the fruits are native. Of course, it is not possible to cultivate within the boundaries of the continental United States all of the fruits discussed in this work. Many of them are uncompromisingly tropical in character and refuse to accommodate themselves to regions where the temperature ever falls as low as the freezing point. It is a noteworthy and hopeful circumstance, however, that certain of the tropical fruits attain their greatest perfection when grown at the extreme northern or southern limit of their zone, when pushed, so to speak, right up against the frost-line. For example, the citrus fruits have been brought in California and Florida to a higher degree of excellence than has been reached by them in strictly tropical regions.

It has been thought in the past that it might be possible, by means of a process of acclimatization, to adapt even the more tender species of tropical plants to conditions in California and Florida, and ultimately to cultivate them on a commercial scale in those states. In the light of present knowledge, however, it seems probable that ability to withstand frost is not greatly increased by submitting a plant to lower temperatures than those to which it has been accustomed, even when this is carried through several consecutive generations, and the chances of acclimatizing in California such fruits as the strictly tropical annonas are not great.

Many of the tropical fruits have as yet scarcely been brought under cultivation, and systematic cultivation of the more important ones, such as the avocado and mango, is of such recent origin that cultural practices have not yet become standardized. New developments are constantly taking place. It is, therefore, inevitable that many of the practices herein described will be obsolete a few years hence.

Regarding the use of the terms tropical and subtropical a few words of explanation are necessary. Plants which will not grow where the temperature falls much below 40° (where temperatures are mentioned in this work, they refer to the Fahrenheit scale) are here termed strictly tropical; by tropical plants are meant (following P. H. Rolfs) those of the zone in which the coconut can be grown; and by subtropical plants, those of the zone of the orange. The next region, in point of minimum temperatures, should be termed the semi-tropical, but this term is frequently confused with subtropical and had better be avoided by stretching the use of the word subtropical to cover the region in which the loquat, the pomegranate, and the date can be grown. It must be borne in mind, however, that knowledge regarding the frost-resistance of plants is still meager. Because a certain species has safely passed through a temperature of 25° above zero in a particular instance in California, it need not follow that the plant will withstand the same temperature in another region, nor even that the same individual specimen in California would withstand again 25° if in different physiological condition.

With a few exceptions, the common names for the fruits are those recommended by the American Pomological Society (Proceedings 1917). The pomological nomenclature (names of fruit varieties) also follows, so far as is possible, the Code of Nomenclature of that Society. In spelling names which have come into the English from the Arabic or some other oriental alphabet, the system has been followed elaborated by the International Congress of Orientalists at Geneva in 1894, and now generally adopted by those having to do with the transliteration of oriental names; which is, that vowels should take the value they possess in Spanish and other Latin languages, and consonants the value they possess in English. The names in this work most affected by the application of this principle are those of varieties of the mango, date, and pomegranate, and the common names of a few minor fruits. Current spellings rejected as incorrect are given in the synonymy of varieties.

The botanical nomenclature is intended to conform to the International Rules, better known as the Vienna Rules. These are the ones followed by European, as well as many American, botanists. In the botanical synonymy all names are included which commonly appear in the publications of the United States Department of Agriculture, when they differ from those adopted under the Vienna Rules.

CHAPTER II

THE AVOCADO

Plates I–IV

North American horticulturists are accustomed to view the avocado as one of the greatest undeveloped sources of food which the tropics offer at the present day. From their standpoint they are correct, but the inhabitants of Mexico and Central America would consider it more logical to assert that the Irish potato is a new crop deserving of extensive cultivation. North Americans view the avocado as a possibility, but to the aboriginal inhabitants of tropical America it is a realized possibility.

"Four or five tortillas [corn cakes], an avocado, and a cup of coffee, — this is a good meal," say the Indians of Guatemala.

It is precisely this condition, — the importance of the avocado as a food in those parts of tropical America where it has been grown since immemorial times, — that has led students of this fruit in the United States to predict that avocado culture will some day become more important than citrus culture in California and Florida.

To a certain extent, the avocado takes the place of meat in the dietary of the Central Americans. It is appetizing, it is nourishing, it is cheap, and it is available throughout most of the year. When these last two conditions have been reproduced in the United States, will not the avocado become a staple article of diet with millions of people?

There is every reason to believe that eventually the avocado will be as familiar to American housewives as the banana is to-day. The increasing scarcity of meat, and the fact that an acre of land will yield a larger amount of food when planted to avocados than it will in any other tree crop known at present, assures the future importance of the avocado industry in this country.

Horticulturally speaking, the avocado is a new fruit. In Central America it has been grown mainly as a dooryard tree, and no care has been given to its propagation or culture. During the last fifteen years the horticulturists of California and Florida have devoted systematic attention to vegetative propagation, to cultural methods, and to the development of superior varieties. In these two states the avocado has been regarded as a fruit of great commercial possibilities. Cuba, Porto Rico, and several other countries are also giving serious consideration to commercial avocado culture.

During summer and autumn the avocado is regularly offered in the markets of New York, Boston, Philadelphia, and other large cities. Many persons who ten years ago were not even familiar with its name have now learned to appreciate the merits of this unique fruit. However, production is not yet great enough to place the avocado in the position which ultimately it must occupy, — that of a staple foodstuff, rather than a luxury or a salad-fruit.

The avocado orchards of California, Florida, Cuba, and Porto Rico now have a total area approaching one thousand acres. As with every young horticultural industry, the problems of propagation, culture, and marketing have been numerous, and many of them remain to be solved. The avocado growers of California have formed a coöperative organization for the purpose of attacking these problems more efficiently. Especially important is the question of varieties, which must, in many cases, be settled individually for each locality. Experience of the last fifteen years has brought to light many of the fundamental requirements of the avocado tree and has suggested cultural practices and methods which are producing satisfactory results. In addition, problems of budding and grafting have been mastered, and these means of propagation are practiced successfully by nurserymen, with the result that trees of the best varieties are obtainable in quantities which permit of extensive commercial plantings. A large number of varieties is being tested, and experience in handling and marketing the fruit is being gained rapidly.

Botanical Description

The genus Persea, to which the avocado belongs, is a member of the laurel family (Lauraceæ); hence it is related to the cinnamon tree, camphor, and sassafras. The avocados cultivated in the United States usually have been considered to represent a single species, *Persea americana,* but careful study shows that they are derived from two species, as follows:

P. americana, Mill. (*P. gratissima,* Gaertn.). All of the varieties classified horticulturally as belonging to the West Indian and Guatemalan races are of this species. It is the common avocado of the tropical American lowlands, and the one which has been most widely disseminated throughout the tropics.

P. drymifolia, Cham. & Schlecht. (*P. americana* var. *drymifolia,* Mez). This includes the small avocados of the Mexican highlands, now grown in California, Chile, and to a very limited extent in southern France, Italy, and Algeria. Horticulturists in the United States use the term "Mexican race" to indicate avocados of this species.

In addition to these two species, a third is well known in southern Mexico and Guatemala, and has recently been introduced into the United States. This is the coyó or chinini, *P. Schiedeana,* Nees. The yás of Costa Rica (probably *P. Pittieri,*

Mez) is another species which is likely, when known in this country, to be classed popularly as an avocado.

The two species from which the cultivated avocados are derived are closely alike in many respects. It is easy to distinguish them by the smell of the crushed leaves; those of *P. drymifolia* possess an aromatic odor, resembling that of anise or sassafras, which those of *P. americana* entirely lack. The flowers of *P. drymifolia* are typically more pubescent, and the under surfaces of the leaves more glaucous, than those of *P. americana.* The fruits also are distinct, having a thin, almost membranous skin in the former species, and a thick leathery or brittle skin in the latter. The horticultural differences are of more interest here than the botanical; they will be referred to later, in the discussion of the horticultural races.

Seedling avocados of both species vary in habit of growth, being sometimes short and spreading, but more commonly erect, even slender. On shallow soils they may not reach more than 30 feet in height, while on deep moist clay-loams they sometimes reach 60 feet. Budded trees are usually more compact in habit than seedlings, and probably will not attain such great ultimate dimensions.

While the avocado is classed as an evergreen, trees of some varieties cast their foliage at the time of flowering, the new leaves making their appearance almost immediately. The leaf-blades are multiform, some of the commonest shapes being lanceolate, elliptic-lanceolate, elliptic, oblong-elliptic, oval, ovate, and obovate. The apex differs from almost blunt to acuminate, while the base is usually acute or truncate. The length of the blades ranges between 3 or 4 inches and as much as 16 inches. *P. drymifolia* usually has smaller leaves than *P. americana,* both species exhibiting a wide diversity in leaf form.

In the United States the flowers appear from November to

May, according to locality and variety. Occasionally some of the Mexican avocados (*P. drymifolia*) bloom in November, while the Guatemalan varieties (*P. americana*) may not begin flowering until March or April. The flowers (Fig. 1) are produced in racemes near the ends of the branches, and are furnished with both stamens and pistils, all of them being inherently capable of developing into fruits. From their immense number, however, it is easy to see that only a minute percentage can actually do so. They are small and pale green or yellowish green in color. At first glance they appear to have six lanceolate or ovate petals, but on closer examination these are seen to be perianth-lobes; the usual differentiation into two whorls or series, calyx and corolla, does not occur in the avocado. The perianth-lobes are of nearly equal length in most varieties, the inner three occasionally being longer than the outer; they are more or less pubescent, heavily so in *P. drymifolia*, sometimes almost glabrous in *P. americana*. The nine stamens are arranged in three series; the anthers are 4-celled, the cells opening by small valves hinged at the upper end. At the base of each stamen of the inner series are two large orange-colored glands which secrete nectar, presumably for the attraction of insects. Inside the stamens are three staminodes or vestigial stamens. The ovary is 1-celled, and contains a single ovule; the style is slender, usually hairy, with a simple stigma.

Fig. 1. Flowers of Fuerte avocado. ($\times\frac{1}{2}$)

The fruit is exceedingly variable in both species. The smallest fruits of *P. drymifolia* are no larger than plums, while the largest of *P. americana* weigh more than three pounds. The form in both species is commonly pear-shaped, oval, or obovoid, but ranges from round and oblate at one extreme to long and slender, almost the shape of a cucumber, at the other. The color varies from yellow-green or almost yellow through many shades of green to crimson, maroon, brown, purple, and almost black. The skin is as thin as that of an apple in many varieties of *P. drymifolia;* in *P. americana* it is occasionally a quarter of an inch thick, and hard and woody in texture. The fleshy edible part which lies between the skin and the seed is of buttery consistency, yellow or greenish yellow in color, of a peculiarly rich nutty flavor in the best varieties, and contains a high percentage of oil. The flesh is traversed from the stem to the base of the seed by streaks or fine fibers (invisible in the ripe fruit of many varieties) which represent the vascular system. The single large seed is oblate, spherical, conical, or slender, inverted so that the young shoot develops from the end which lies toward the stem of the fruit. It is covered by two seed-coats, varying in thickness, often adhering closely to one another. The cotyledons are normally two, occasionally three in *P. drymifolia,* white or greenish white in color, smooth or roughened on the surface.

History and Distribution

The native home of the avocado is on the mainland of tropical America. *Persea drymifolia* is abundant in the wild state on the lower slopes of the volcano Orizaba, in southern Mexico, as well as in other parts of that country. The extent of its distribution is not precisely known. The native home of *P. americana* has not been determined with certainty, since the tree has been so long in cultivation and few efforts have as yet been made to locate the region in which it is truly indigenous.

Jacques Huber, in the Boletim do Museu Goeldi, says: "Everything indicates that the avocado, originally indigenous to Mexico, has been cultivated since immemorial times, and that it very early spread through Central America to Peru; then into the Antilles, where its introduction is mentioned by Jacquin; and much later into Brazil." He also remarks that its presence in Peru in pre-Colombian days is indicated by the indigenous name, *palta,* and the finding of fruits in the graves of the Incas. W. E. Safford, however, says that no vestiges of the avocado are found in the prehistoric graves of the Peruvian coast, nor is it represented in the casts of fruits and vegetables discovered among the terra cotta funeral vases so abundant in the vicinity of Trujillo and Chimbote.

While it is probable that the avocado is of relatively recent introduction into Brazil, and that its presence in Peru in pre-Colombian days may be open to question, the existence of native names for it in many different languages, as well as references by the early voyagers, indicate that at the time of the Discovery it was cultivated, if not indigenous, in extreme northern South America and from there through Central America into Mexico.

The first written account of the avocado, so far as known, is contained in the report of Gonzalo Hernandez de Oviedo (1526), who saw the tree in Colombia, near the Isthmus of Panama.

Pedro de Cieza de Leon, who traveled in tropical America between 1532 and 1550, mentions the avocado as one of the fruits used by the Spaniards who had settled in the Isthmus of Panama, and as being an article of food among the natives of Arma and Cali, in Colombia.

Francisco Cervantes Salazar, one of the earliest chroniclers of Mexico, gives evidence that the avocado was well known in the markets of Mexico City as early as 1554, which was very soon after the Conquest. In a later work, the "Crónica de Nueva España," written about the year 1575, he described the fruit. Both in this work and in his earlier one, "Mexico en 1554," he uses the name *aguacate.*

Sahagun, another early chronicler of Mexico, who wrote some time previous to 1569, briefly describes the Mexican avocado (*Persea drymifolia*) under the Aztec name, which he spelled *aoacatl.*

Acosta, writing in 1590, distinguished clearly between the Mexican form and that grown in Peru. He used the Peruvian name *palta,* in place of the Mexican *ahuacatl* or any of its corruptions.

Garcilasso de la Vega, writing in 1605, states that the name *palta* was applied to this fruit by the Incas, who brought the tree from the province of Palta to the valley of Cuzco.

One of the most valuable accounts written in the early days is that of Hernandez, as edited and published by the friar Francisco Ximenez in 1615. Hernandez, who was a physician sent by the King of Spain to study the medicinal plants of Mexico, was evidently familiar only with the Mexican avocado (*P. drymifolia*); at least, if he had seen the lowland species he makes no mention of it.

Another excellent account was written in 1653 by Bernabe Cobo, a priest who had traveled widely in tropical America. He was the first, so far as known, to mention the Guatemalan avocados. After describing at some length the West Indian race, as it is now called, mentioning in particular the varieties grown in Yucatan and those of certain sections of Peru, he says:

"There are three distinct kinds of paltas. The second kind is a large, round one which is produced in the province of Guatemala, and which does not have as smooth a skin as the first. The third is a small palta which is found in Mexico which in size, color and form resembles a Breva fig; some are round and others elongate, and the skin is as thin and smooth as that of a plum."

Thus it is seen that the three groups of cultivated avocados, recognized at the present day by horticulturists under the names of West Indian, Guatemalan, and Mexican, were distinguished as early as 1653 by Padre Cobo.

Hughes, in his important work "The American Physician" published in 1672, devotes a short chapter to "The Spanish Pear." His reference to its having been planted in Jamaica by the Spaniards is in agreement with other accounts, all of which indicate that the avocado was not cultivated in the West Indies previous to the Discovery.

Sir Hans Sloane, in his catalog of the plants of Jamaica, published in 1696, briefly describes the avocado, cites numerous works in which it is mentioned, and gives as its common name "The avocado or alligator pear-tree." This is the first time that either of these names appears in print, so far as has been discovered.

It is useless to enter into a discussion of all the common names which have appeared in the literature of this fruit. G. N. Collins [1] lists forty-three, but many of them are of limited use, and others are the clumsy efforts of early writers to spell the names they had heard.

The correct name of this fruit in English is at present recognized to be avocado. This is undoubtedly a corruption of the Spanish *ahuacate* or *aguacate,* which in turn is an adaptation of the Aztec *ahuacatl.* The Spaniards, who probably introduced the avocado into Jamaica, brought with it the Mexican name. When Jamaica was taken by the British this name began to undergo a process of corruption, during which such forms as *albecata, avigato,* and *avocato* were developed. Frequently the term "pear" was added to these, in conformity with the tendency of the early English colonists to apply familiar names to the fruits which they found in America. We have many other evidences of this tendency, *e.g.,* star-apple, custard-apple, hog-plum, Spanish-plum.

The name avocado or avocado-pear was one of the numerous corruptions which found its way into print, first appearing, so far as known, in 1696 (see above). For some reason it has

[1] Bull. 77, Bureau of Plant Industry.

outlived many other corruptions. Since it is reasonably euphonious, well adapted to the English language, and widely used, it has been officially adopted by the California Avocado Association and is used in the publications of the United States Department of Agriculture, as well as by horticultural societies and horticulturists generally. The name alligator-pear, which seems to have appeared in the same way and about the same time as the term avocado, is considered decidedly objectionable, and a vigorous effort is being made to eliminate it from popular usage.

Ahuacate (more commonly but less correctly spelled *aguacate*) is the name at present used in Mexico, Central America as far south as Costa Rica, and the Spanish-speaking islands of the West Indies, as well as in a few other parts of the world. The original form *ahuacatl* is still employed in those sections of Mexico where the Aztec or Mexican language has not been replaced by Spanish. The avocado tree is *ahuacaquahuitl*, a combination of *ahuacatl* and *quahuitl* (tree). There were at least two towns in ancient Mexico named Ahuacatlan. This word was expressed in the picture writing of the Aztecs by means of the sign of the avocado tree and the locative suffix *-tlan*, indicated by teeth set in the trunk of the tree (Fig. 2). The picture thus read *ahuacatlan*, or "place where the ahuacate abounds." The word *ahuacatl* has two meanings; one, the fruit of the avocado tree, and the other, testicle.

FIG. 2. Sign of the avocado tree used by the Aztecs.

The name *pahua* (from the Aztec *pauatl*, fruit) is applied in certain parts of Mexico to avocados of the Guatemalan and West Indian races, distinguishing them from the thinner skinned and smaller *ahuacates* of the Mexican race.

In southern Costa Rica the common name is *cura*, while in the western part of South America the Peruvian name *palta* is current. The latter occurs in the Quichua language, and is of unknown derivation.

The names current in various European languages are mainly adaptations or corruptions of the Spanish *ahuacate* or *aguacate*. The Portuguese name, used principally in Brazil, is *abacate;* the French generally call the fruit *avocat;* while the German name is *advogado* or *avocato*.

In all probability the avocado was brought to Florida by the Spaniards, but the first introduction of which a record has been found was in 1833, when Henry Perrine sent trees from Mexico to his grant of land below Miami.

The first successful introduction into California is believed to have been in 1871, when R. B. Ord brought three trees from Mexico and planted them at Santa Barbara. It seems strange that so valuable a fruit should not have been introduced into California by the Franciscan padres, who came from Mexico in the latter part of the eighteenth century and to whom credit is due for the introduction of the orange, the olive, and the vine.

According to Higgins, Hunn, and Holt,[1] the avocado was grown in Hawaii as early as 1825, although it did not become common until after 1853.

The avocado is now cultivated to a very limited extent in Algeria, southern Spain, and France, and has even fruited in the open at Rome. Naturally, only the hardiest varieties succeed in the Mediterranean region. In India and other parts of the Orient it has never become common, although it may have been introduced as early as the middle of the eighteenth century. In Réunion and Madagascar it seems to be more abundant. In Polynesia it has become well established, considerable quantities of the fruit having been shipped from the French island of Tahiti to San Francisco. It is gaining a foothold in northern Australia, and is grown in Natal, Mauritius, Madeira,

[1] Bull. 25, Hawaii Agr. Exp. Sta.

and the Canary Islands. In the Philippines its culture has been established since the American occupation, many varieties having been introduced by the Bureau of Agriculture.

While it will thus be seen that the avocado has spread from its native home entirely around the globe, it is still most abundant, and of the greatest importance as a food, in tropical America. Throughout Mexico, Central America, and the West Indies seedlings are common in dooryards, thriving with practically no attention and yielding generously of their delicious and nourishing fruits. Rarely in these countries, however, has the avocado been developed as an orchard crop; but this is not surprising in view of the fact that orchards of fruit-trees are almost unknown in the tropics.

Composition and Uses of the Fruit

Due to the investigations of M. E. Jaffa and his associates at the University of California, much light has been thrown on the food value of the avocado in recent years. The following table shows the composition of several well-known varieties, one of each of the recognized horticultural races, and the hybrid Fuerte. In presenting this table, which is based on the work of Jaffa, it is necessary to explain that the proportions of the constituents have been found to change in each variety according to the degree of maturity of the fruits. They may fluctuate also in different years. Variation is particularly noticeable in regard to the fat-content. For example, in specimens of the Chappelow examined at different times, the percentage of fat ranged from approximately 14 to 30, while in specimens of the Challenge it ran from 3 to 17. Fruits showing the lowest percentages were immature at the time the analyses were made, but they were no more so, probably, than many which are put on the market. Up to a certain point, the fat-content increases with the maturity of the fruit; after this point is reached, there is quite often no further increase, no matter how long the fruit may remain on the tree.

The total dry matter in the edible portion of the avocado is greater than in any other fresh fruit, the one nearest approaching it being the banana, which contains about 25 per cent. An average of twenty-eight analyses showed the avocado to contain about 30 per cent.

TABLE I. COMPOSITION OF AVOCADO VARIETIES

VARIETY	WATER	PROTEIN	FAT	CARBO-HYDRATES	ASH
	%	%	%	%	%
Trapp . . . . (West Indian)	78.66	1.61	9.80	9.08	0.85
Sharpless . . . (Guatemalan)	71.21	1.70	20.54	5.43	1.12
Puebla . . . . (Mexican)	63.32	1.80	26.68	6.64	1.56
Fuerte . . . . (Hybrid)	60.86	1.25	29.14	7.40	1.35

The protein-content, which has been found to average about 2 per cent, is higher than that of any other fresh fruit.

The percentage of carbohydrates is not high compared with that of many other fruits, because the avocado contains almost no sugar. F. B. La Forge of the Bureau of Chemistry at Washington has found in the avocado a new sugar, called D-Mannoketoheptose, which is believed to be present in amounts varying from 0.5 to 1 per cent.

The amount of mineral matter is much greater than is found in other fresh fruits. Soda, potash, magnesium, and lime compose more than one-half the ash or mineral matter, which places the avocado among the foods which yield an excess of the base-forming elements, as opposed to nuts, which furnish acid-forming elements in excess.

Jaffa [1] says: "So far as protein and ash in fresh fruits are concerned, the avocado stands at the head of the list, and with reference to the carbohydrates, contains on an average fully 50 per cent of that found in many fresh fruits. These facts alone would warrant due consideration being given to the value of the avocado as a fresh fruit. Its chief value as a food, however, is due to its high content of fat. This varies, as shown by the analysis, from a minimum of 9.8 per cent to a maximum of 29.1 per cent, with an average of 20.1 per cent. The only fruit comparable with the avocado in this respect is the olive."

Experiments carried on at the University of California have shown that the digestibility of avocado fat is equal to that of butter-fat, and not below that of beef fat.

As to the caloric or energy-producing value of the avocado in twenty-eight varieties examined, one pound of the flesh represents an average of 1000 calories. The maximum and minimum were 1325 and 597 respectively. The maximum "corresponds to about 75 per cent of the fuel value of the cereals and is not far from twice that noted for average lean meat."

In the following table the avocado is compared, in caloric value, with several common foodstuffs. For this comparison a pound of avocado flesh has been considered to represent 1000 calories; this is not showing the avocado at its best, for, as just stated, in some varieties a pound represents over 1300 calories:

	Calories
100 grams (about 3½ oz.) boiled rice	322
100 grams white bread	246
100 grams avocado	218
100 grams egg	166
100 grams lean beef	100

It must not be assumed from the figures that the avocado has a total food value greater than that of lean beef. It is

[1] Bull. 254, Calif. Agr. Exp. Sta.

only the caloric or energy-producing values that are shown, and much of the value of meat as a food lies, of course, not in the energy which it produces, but in its ability to build up and repair the tissues of the body.

In the United States the avocado is commonly used in the form of a salad, either alone or combined with lettuce, onions, or other vegetables. Up to the present, no satisfactory ways of cooking or preserving this fruit have been developed. Experiments in extracting a table- or cooking-oil have been encouraging, but as yet the production of avocados in this country is not great enough to permit the commercial development of this field. In the tropics, the fruit is added to soups at the time of serving; mashed with onions and lemon juice to form the delectable *guacamole* of Cuba and Mexico; or eaten as a vegetable, without the addition of any other seasoning than a little salt. In Brazil it is looked on more as a dessert than as a staple foodstuff, and is made into a delicious ice-cream. Numerous recipes appear in cook-books which have been published in Cuba, Florida, California, and Hawaii.

Climate and Soil

It is impossible to define in few words the climatic conditions most favorable to the avocado, since the different races do not always succeed under the same conditions. The subject must, therefore, be considered from the standpoint of races.

The West Indian race, which comes from the moist lowlands and seacoasts of tropical America, is more susceptible to frost than the others. Hence, when grown near the northern limit of the subtropical zone, it requires more protection from possible severe frosts than the Guatemalan race, which comes from the highlands of southern Mexico and Guatemala, or the hardy avocados from central and northern Mexico which constitute the Mexican race. Not a few losses have already resulted from attempts to grow West Indian avocados in locations in California subject to occasional severe frosts. In this state, the Guatemalan and Mexican races are the only ones to plant. The same is true of central and northern Florida, where the West Indian race has nearly always succumbed to cold.

In Florida, the region in which avocado culture is at present conducted commercially lies south of Palm Beach on the east coast and south of Tampa Bay on the west. Of the orchards which are now in bearing, the largest are situated close to Miami and Homestead. On the west coast the most important plantings are near Fort Myers. Most of the orchards in Florida are planted to Trapp, a variety of the West Indian race. The planting of hardy Guatemalan kinds will probably extend the commercial culture of this fruit many miles to the northward of the present limits of the zone. In addition, it will make avocado growing safer in all regions by lessening the possibility of frost injury. The Mexican race is known to have fruited as far north in Florida as Gainesville and Waldo.

In California, most of the young orchards, as well as the old seedlings which have fruited for some years, are in the vicinity of Los Angeles, Orange, and Santa Barbara. The coastal belt between Santa Barbara and San Diego, including the foothill region some distance from the coast, has been tested sufficiently to show that planting may proceed with confidence. In the interior valleys comparatively few trees have been planted, and these mainly in recent years. Much less is known, therefore, regarding the adaptability of the avocado in these situations. Old seedlings are to be seen at Visalia, San Luis Obispo, Berkeley, Los Gatos, and Napa, indicating that some varieties may be grown successfully as far north as the Sacramento Valley. Sections of the San Joaquin Valley which have proved suitable for citrus culture, such as the Porterville district, should prove safe for the hardier varieties of

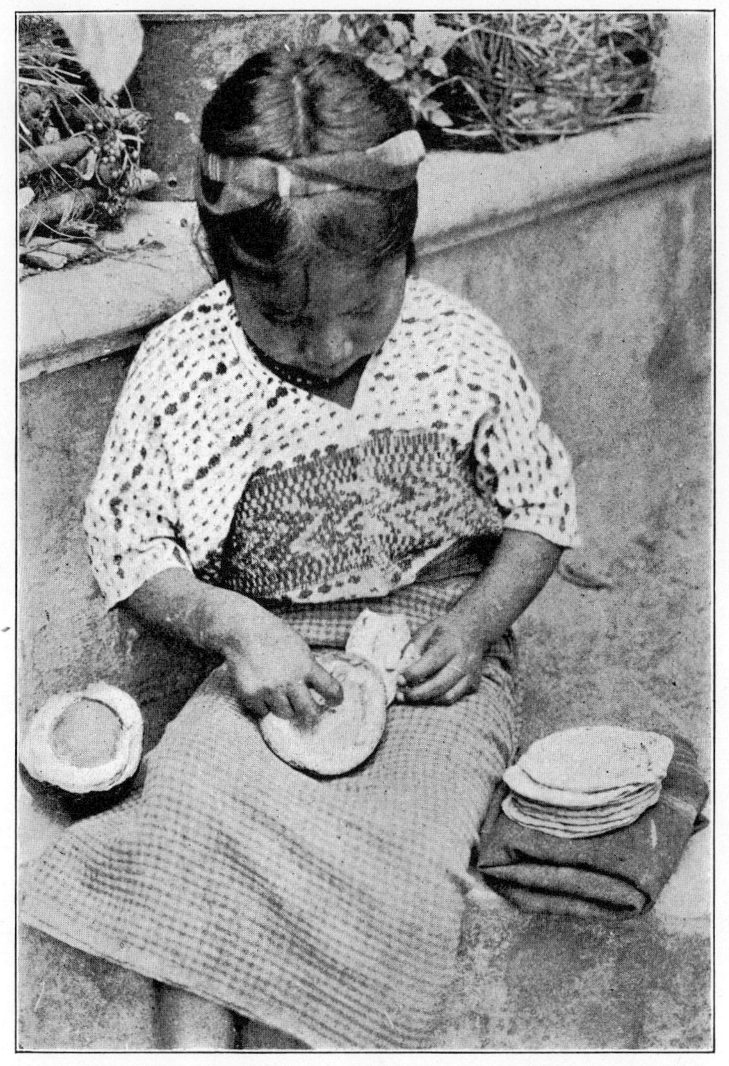

PLATE II. Four or five tortillas (corn cakes) and a good-sized avocado are considered a good meal by the Guatemala Indians.

avocados as well. Experimental plantings in the Imperial and Coachella Valleys have up to the present served only to indicate that the atmosphere of these regions is too dry. The leaves turn brown and fall off, no matter how much water is applied at the root.

As a guide to planters in regions where the avocado has not been tested, it may be said that experience has shown the Guatemalan race to be about as hardy as the lemon. Certain kinds probably are hardier than that, while others are known to be more tender. The Mexican race, in its hardier varieties, withstands a little more frost than the orange. The West Indian race is distinctly more tender than either the Guatemalan or Mexican.

W. J. Krome's experience at Homestead, Florida, leads him to say: "As a general rule West Indian avocados, beyond one year old, will not be damaged by a temperature of 32° unless that temperature holds for a longer period than two or three hours. When four or five years old they will stand 26° or 27° without injury, except to tender growth, but below that temperature there is likely to be considerable damage. At 22° five-year-old Trapp trees were killed back to wood one inch in diameter. At a temperature somewhere between 22° and 24° Guatemalans have, with a few exceptions, shown almost no damage beyond a slight singeing of the leaves."

In the Report of the California Avocado Association for 1917, H. J. Webber publishes the following summary of the effect of different temperatures on avocado trees in California in the cold weather of the winter of 1916–1917:

"30° F. Nothing injured so far as could be observed.

"29° F. No injury of account; only traces on most tender growth of West Indian and Guatemalan varieties.

"28° F. New foliage scorched on Guatemalan types; West Indian varieties showing considerable damage.

"27° F. Mexican varieties with new tips slightly scorched; Guatemalan with almost all new foliage injured; West Indian badly damaged.

"25° to 26° F. Mexican varieties with new foliage injured but some dormant trees uninjured; all Guatemalan sorts with new foliage badly injured and some old foliage scorched.

"24° F. Some dormant Mexicans uninjured; Guatemalan varieties badly injured, small limbs frozen back.

"21° F. All Guatemalan types killed to bud; a few of the hardiest Mexicans, such as Knowles and San Sebastian, with young leaves only injured."

The observations reported to Webber showed that young trees were injured at higher temperatures than older ones, when the variety was the same in both cases. It was also observed that trees in rapid growth were more severely injured than those which were in semi-dormant condition. Krome of Florida reports an opposite state of affairs. He says: "At the time of the January 1918 freeze, Trapp trees which had borne heavy crops and were in a hard, completely dormant state suffered a great deal more injury than trees which, owing to light crops the preceding season, were in full growth." Possibly the trees were weakened by over-production of fruit, and thus more susceptible to frost-injury. The subject demands further investigation.

Webber further says: "Trees which needed irrigation when the freeze came suffered rather severely, as did also trees that had been irrigated three to five days before the freeze and were thus gorged with water. The least injury seemed to be on trees that had been thoroughly irrigated two or three weeks before the freeze, and had water supposedly in what might be termed the optimum amount."

In regions subject to frosts, it is necessary to protect the trees during the first two or three winters with shelters of burlap, or by placing palm-leaves, pine boughs, or other material around them. Frost-fighting with orchard heaters is sometimes practiced where heavy frosts are expected.

A sharp frost at the time the tree is in flower may result in a crop failure, although the danger from this source is probably not great either in California or Florida, and has been over-estimated in the past. In Florida, the West Indian race usually blooms late enough to escape the coldest weather, while in California the Mexican race, though it blooms in winter, is sufficiently hardy to withstand ordinary frosts, and the Guatemalan race does not bloom until April or May. The latter race is, therefore, the safest in this respect.

In California, avocado culture is not dependent on rainfall, since irrigation is commonly practiced. In Florida, on the other hand, very few crops are irrigated, and up to the present it has been the general custom not to irrigate avocado trees, except during the first two or three summers. It is coming to be recognized, however, that a wet spring is followed by a good avocado crop and a dry one by a poor crop (a condition exactly reversed with the mango). As a result of this observation, irrigation is beginning to be practiced in southern Florida, especially in seasons when the rainfall is below normal.

The necessary soil-moisture can be supplied easily and satisfactorily, but the relative humidity of the atmosphere cannot be altered artificially; hence in regions where the humidity is exceedingly low the avocado suffers in the dry portion of the year. In Florida no attention need be paid to this subject, since the humidity closely approaches that of the West Indies and other regions where the avocado is at home. Humidity may prove, however, to be the limiting factor in parts of California. Tests in the Imperial and Coachella Valleys indicate that the trees are seriously injured by the dryness of the atmosphere. Experience shows that the Mexican race is less susceptible than the West Indian. None of the varieties so far tested, however, has proved to be so resistant to atmospheric dryness as the orange or grapefruit. Shading may help to limit the injury from this source. It has been found very beneficial in the coastal belt of California, where young avocado trees are often injured during the dry summer months by sunburn.

Another climatic factor which deserves consideration is the danger from high winds. The lower east coast of Florida is occasionally visited by a West Indian hurricane which defoliates trees, strips them of their crops, or even breaks them down. Certain parts of California are also subject to occasional high winds, less severe than the hurricane but nevertheless capable of doing much damage. To minimize the danger from this source, it is advisable to keep the trees as low as possible through pruning, since they are then much less liable to injury. The low tree has an additional advantage in that it permits of picking the fruit without the use of tall ladders, and keeps the branches more readily accessible for pruning, spraying, or thinning the fruit.

In regard to soil the avocado seems to be unusually adaptable, succeeding on the sandy lands of southern Florida, the volcanic loams of Guatemala and Mexico, the red clays of Cuba and Guatemala, the granite soils of California, and even on heavy adobe, provided the drainage is good. This question is less important, therefore, than many others connected with avocado culture. The chief requisite is good drainage.

Most of the avocado groves of southeastern Florida are situated upon limestone of the kind shown as Miami oölite. This formation comprises a narrow strip of land extending from above Fort Lauderdale on the north to some miles below Homestead on the south, being widest near the latter place, and nowhere more than thirty feet above sea level. In many parts of this region the rock comes to the surface; toward the northern end it is commonly overlaid with six inches to two feet of loose light-colored quartz sand, while below Miami the surface soil becomes very scanty, but heavier in nature, containing some clay in certain localities, and being strongly impregnated with iron, giving it a reddish color. The rock itself is soft and porous, and in the process of erosion has broken down unequally, leaving a jagged surface or the characteristic pot-hole formation. When first grubbed it crumbles and is readily worked, but on exposure to the air it gradually hardens, owing to the deposition of carbonate of lime following evaporation of the moisture held in the interstices.

The growth made by the avocado upon this rocky land is rather remarkable; it seems, in fact, that young orchards have done better around Homestead, where the rock comes to the surface, than they have in those areas north of Miami where there are six to eighteen inches of sand on top of the rock. The reason for this may lie in the moisture-retaining properties of this soft limestone; the roots, which are always close to the surface, here probably are kept more uniformly supplied with moisture during a period of dry weather than on light sandy soils which dry out rapidly.

The heavier Florida soils seem to be much more favorable to the growth of the tree than light sands. A yellowish or brownish subsoil in many parts of Florida indicates good avocado land. The avocado prefers a moist heavy loam, and the closer this can be approached the better will be the results.

The soils of California are probably more nearly ideal for avocado culture than any of those in southern Florida. Sandy loam, which is abundant in the southern part of the state, produces excellent growth and is giving good results. Adobe does not seem so desirable, yet good trees have been grown upon it at Orange.

Red clay has been satisfactory in Cuba and Central America, while heavy clay where well drained has produced good trees in Porto Rico.

Many problems connected with avocado culture remain to be solved. One of the most important is the adaptability of the tree to low wet lands in southern Florida. It has been the general opinion that avocados should not be planted on land where the water-table is less than three feet below the surface. Krome has observed groves on low rock-land which have been killed or badly injured by overflows, even where the water came scarcely as high as the crown roots and remained there only a few days. In several plantings on marl prairie, however, experience has been quite different. Trees on this type of land have been submerged twenty-four hours without damage to them. On the low islands along the western coast of Florida, salt water sometimes floods the groves, and this has proved fatal to many trees. It is probable, also, that the failure of one or two plantings on this coast can be attributed to the fact that the water which stands about two feet below the surface of the land is saline in character.

Until more experience has been gained regarding the adaptability of the avocado to low flat ground, occasionally subject to overflow, orchard plantings should be limited to lands where the water-table is three feet or more below the surface.

In California, the best site for the orchard is a gently sloping hillside, or level ground adjacent to a slope. If of this character, and well drained but naturally retentive of moisture, the situation may be considered excellent. In regions subject to heavy winds, it is well to select a piece of ground which is sheltered by surrounding elevations.

Cultivation

Regarding the best time to plant avocados in southern Florida, Krome says:

"I have planted at least a few avocados every year since 1905 and these plantings have been made during every month of the year. When I have the land prepared and the trees available I do not hesitate to plant at any season but I endeavor to make my arrangements so that all of my main settings will be between the 15th of September and the 20th of October, *i.e.* during the last month of the regular rainy season, after the hottest weather of the year is past. When avocados are planted in the spring in Florida they have immediately ahead of them our most trying months of drought, March, April and usually most of May. During this period the plants must be watered with the greatest regularity or they will suffer. Following the dry weather of our spring months the trees have the benefit of the rainy season but in Florida our rains are quite frequently uncertain during July and August and there will be need for watering any trees planted during the preceding three or four months. For the past month (June) we have averaged at least two applications of water per week to avocados planted during March, April and May. These spring-planted trees must also withstand the sun's rays during our season of greatest heat and shading is usually a necessity if sun-scald is to be prevented. There are no good reasons why trees thus planted should not be brought through to fall in good condition but it requires a great deal of additional work and expense as compared with trees planted during the latter part of September or first half of October, when rains are of almost daily occurrence and the plants after setting need very little further attention. Furthermore, spring-planted trees very seldom make sufficient growth over those planted in the fall to acquire any considerably greater degree of resistance to cold the following winter. I have always found that trees planted in March fare just about as badly as those planted in September when we have severe cold the next winter."

In California it is not desirable to plant earlier than March, because of danger from late frosts. April and May are good months, and November planting has been successful. Planting in midsummer is to be avoided, but it may be done successfully if the trees are carefully shaded and watered until they have become established.

Avocados are sometimes interplanted with other fruit-trees, such as grapefruit and mangos. This is scarcely to be recommended, since avocados require different cultural treatment.

In Florida, budded avocados are planted 20 by 20 feet (108 to the acre) to 26 by 26 feet (64 to the acre), some growers preferring to have the trees close together so that they will soon shade the ground, others desiring to give more room for ultimate development. On light sandy soil the trees are usually set closer than on heavy soils, 20 by 20 feet being a suitable distance in the first case, 24 by 24 in the second. In

California they should not be spaced closer than 24 by 24 feet, making 75 to the acre; 30 by 30 feet (48 to the acre) is preferable.

Holes for planting should be prepared a month in advance, with a small quantity of fertilizer incorporated in each. Barnyard manure is commonly used for this purpose in California, while South American goat manure and pulverized sheep manure, 2 or 3 pounds to each hole, have proved satisfactory in Florida.

In planting, the tree should be set so that the point of union between the bud and the seedling stock is slightly above the surface. Deeper planting may not be objectionable in California, but in Florida shallow planting seems to be best. A liberal watering should be given immediately after planting.

Tillage, mulching, and cover-crops.

The ground around the young trees should be kept liberally mulched with weeds, straw, barnyard litter, seaweed, or any coarse material which is not injurious and will not pack and form a layer impervious to air and water. Through the winter a mulch is not necessary in California, but in Florida it has been found desirable, in some sections at least, to maintain one throughout the year. In Porto Rico, G. N. Collins observed that the avocado tree was seldom, if ever, found in perfectly open places, with the bare ground around the roots exposed to the sun. While this principle applies more particularly to Florida and other regions distinctly tropical in character, it may be proved to hold good in California as well. Definite knowledge on this point is still lacking. Up to the present it is the practice of many California orchardists to cultivate the soil regularly after each irrigation, as with citrus fruits. Deep cultivation seems to produce no harmful results in California, where the roots go far down into the soil, but in southern Florida it must be practiced with caution. In this region the feeding roots extend practically to the surface, and deep cultivation destroys many of them, thus cutting off a large part of the tree's food supply. On shallow soils the most healthy and vigorous trees are those which are mulched. The mulch should extend at least two feet in each direction from the trunk of the young tree, and as the latter increases in size and its roots reach out on all sides, the mulch must be enlarged to be always a little wider than the diameter of the crown.

Mulching serves two purposes: it prevents the soil from drying out rapidly, and it protects the delicate feeding roots from injury due to excessive heating of the soil. This protection is of particular importance in Florida, where in many places the land is sandy and becomes exceedingly hot if exposed to the sun.

When the trees are of mature size, the shade furnished by their own foliage, together with the fallen leaves which carpet the ground, aids materially in maintaining the soil in good condition; but additional loose material, especially during the summer, is highly desirable.

The use of green cover-crops between the rows is decidedly beneficial, but they must not be brought close enough to the trees to rob them of their food. In Florida, cowpeas and velvet beans have been used for this purpose, cowpeas being preferred. A clump of pigeon peas (*Cajanus indicus*) planted four feet to the south of each young tree will provide shade during the first summer or two, serve as a protection from wind, and aid in enriching the soil. In California, purple vetch (*Vicia atropurpurea*), common vetch, and the other cover-crops used in citrus culture will probably prove satisfactory. Up to the present time they have not been extensively tried in connection with avocado culture.

Fertilizer.

Little systematic attention has yet been given to this subject. Not only is the question difficult, but it is also one of the most important in connection with avocado culture in Florida. The following extracts from a paper by Krome, published in the 1916 Report of the California Avocado Association, present the results of several years of experimentation:

"The nature of the plant food required by the avocado has not been very satisfactorily determined, but it has become evident that a scheme of fertilization must be worked out differing considerably from that which has been generally adopted for citrus. Broadly speaking the application of commercial fertilizers deriving their elements of plant food from wholly chemical sources has not proved successful. In many instances, through lack of more definite information, growers have given their avocados the same fertilizers which they have used on their citrus trees. Where the formulæ have been those most frequently applied to citrus, with nitrogen derived from sulphate of ammonia or nitrate of soda, potash from sulphate of potash, and phosphoric acid from acid phosphate, the results with the avocado have been generally unsatisfactory. However, when the formula used has been of the type known as 'young tree' fertilizer, carrying a proportionately higher percentage of ammonia largely derived from organic sources, better effects have been obtained.

"It has become fairly well established as a fact that of two avocado trees of the same variety, one which is well nourished and kept in growing condition during the entire summer and fall will produce larger and finer appearing fruit than one which is permitted to become more or less dormant through lack of fertilizer, but it is quite certain that the semi-dormant tree will carry its fruit without dropping for a considerably longer time. There is therefore a rather delicate adjustment to be made in order to bring the tree into condition such that it will hold its crop until late in the season and at the same time will not 'go back' to an extent that will be seriously detrimental to its further development or jeopardize the crop for the following season.

"Following such applications of fertilizer as are made to restore the tree to good condition after it has passed through the period of bloom and fruit setting there should certainly be at least one further fertilizing during the summer or early fall to provide the nourishment necessary for the production of the crop. And it may be added here that the drain on an avocado tree in bringing its fruit to maturity seems to be vastly greater in proportion than the same effort on the part of a citrus tree. The writer cannot vouch for the soundness of the theory, but it has been thought that this is probably due to the different character of the fruit. In the case of any citrus, water constitutes a large percentage of the fruit either by weight or volume, while with the avocado the proportion of oils is much higher and it would seem reasonable that to supply these components would be a heavier draft upon the tree. At any rate the fact is certain that an avocado tree must be furnished with a sufficiency of plant food if it is to be expected to produce full and regular crops.

"Avocados of the West Indian type begin to ripen in Florida about the middle of July and the heaviest portion of the seedling crop matures between August 20th and October 10th. At that period the crop from Cuba and other West Indian islands is likewise being shipped and the large quantity of fruit thus thrown on the market, together with the fact that during the summer and early fall the avocado must compete with northern-grown fruits and vegetables, tend to force prices so low, that at times it is difficult to dispose of the Florida seedlings with any margin of profit. After the middle of October the price of avocados begins to climb and during November and December very satisfactory figures are usually obtained. For this reason the large plantings of budded trees which have been made during the past few years have practically all been of late maturing varieties such as the Trapp and Waldin. These varieties mature their fruit so that it may be picked early in October if desired, but under proper conditions will carry at least a portion of their crop into December and in some cases until well along in January.

"Just how late in the season an application of fertilizer can be made without bringing about a tendency for the tree to mature and drop its fruit at too early a date depends somewhat on weather conditions. Fertilizer applied to Trapp trees about the middle of August of the season just passed, apparently had no detrimental effect as to the fruit holding well, while an application of fertilizer given the same trees about the first of September of the preceding year was followed, within a few weeks, by heavy dropping of fully matured fruit. The application made in August of the present year was at the beginning of several weeks of dry weather, while that of the previous season was followed by heavy rains and these differences in moisture probably had considerable to do with the effects of the fertilizer.

"This second problem is one of great importance to the Florida avocado grower as between December 1 and December 15 the value of his product not infrequently more than doubles and the premium to be gained by being able to carry his fruit until the latest possible date is well worth his very best efforts.

"It is our plan at Medora Grove to give the trees a heavy fertilizing immediately after the crop has been picked and a light application about the first of February, which brings them to their blooming stage in good condition, quite thoroughly recuperated from their fast during the fall.

"This program provides for five or six applications of fertilizer during the year, which is probably one or two more than is given by most growers, the difference being in the method of carrying the trees through the spring period. The quantity of fertilizer used at each application varies of course with the size of the tree, quantity of fruit it is carrying and the analysis of the fertilizer. For ten year old trees as high as 25 pounds at a single application has been used with good results. For four year old trees, bearing their first crop, four applications of from three to four pounds each, one of four and one-half and one of five pounds have brought the trees through the year in fine shape. As materials from which fertilizers suitable for avocados may be compounded, cottonseed meal, castor pomace, tankage, ground tobacco stems and ground bone are to be recommended, with a certain amount of nitrate of soda used as a source of nitrogen when quick results are sought as in the case of trees which have 'started back.' Previous to the war scarcity of potash, it was thought advisable to use formulæ giving from four to six per cent of that element, but the enforced limitations to the percentage of potash obtainable during the past two years has had no apparent ill effects upon the trees or fruit and seemingly a range of from zero to four per cent will provide all the potash that an avocado tree requires under Florida conditions. A formula that has given good results is built up of cottonseed meal, castor pomace, tankage, and ground tobacco stems, analyzing 4 per cent to 5 per cent ammonia, 6 per cent to 7 per cent phosphoric acid and 2 per cent potash.

"The trees were usually cultivated by hoeing three times each year and a heavy mulching of dead grass or weeds during the dry winter season. If instead of the dead grass a mulching of compost or well rotted stable manure is used the results are even more satisfactory and the February application of fertilizer may then be omitted entirely."

In California, stable manure has been practically the only fertilizer used up to the present. The necessary nitrogen can be obtained from this source, and the organic matter added to the soil is also of benefit.

Irrigation.

An abundance of water is especially important during the first two or three years after the tree is planted, if rapid healthy growth is to be maintained. In Florida, particularly in sections where the soil is deep, many young groves have in the past suffered for lack of water. One of the most experienced growers near Miami states that trees which have had abundant irrigation are as large at four years of age as non-irrigated trees at six years. Their larger size enables them to yield commercial crops earlier than non-irrigated trees.

In California it is the general practice to irrigate avocados in the same manner as citrus fruits. The amount of water necessary for maximum development varies considerably on different soils, but during the first few years a thorough irrigation every ten days during the dry season is not too much.

The importance of an abundance of moisture in the soil at the time the fruit is setting has already been mentioned in the discussion of the climatic requirements of the avocado. Several crop failures in Florida have been blamed on unusually dry weather during this period. A drought probably does little harm if it occurs when the trees are just beginning to bloom, but if it continues the flowers are likely to drop and the crop to be a failure. This has been the experience with Trapps when grown on deep sand; on heavy soils, which are more retentive of moisture, the danger is less.

In order to avoid crop failures from this cause, the grower should certainly be prepared to irrigate at the time the fruit is setting. In southern Florida this is usually in March and April. When a prolonged dry spell occurs just at this time, as is sometimes the case, two or three thorough irrigations, a week apart, may suffice to save a considerable amount of fruit.

In California, if the soil is allowed to become too dry during the hot summer months, young trees are frequently given a setback from which they are slow to recover. This has been observed in Florida as well, particularly on deep sandy soils.

The method of applying water varies in different regions. In California the basin system is commonly used, especially when the number of trees to be irrigated is small. Basins should be filled with coarse strawy manure to serve as a mulch. In many orchards the trees are irrigated by the furrow system which is used with citrus fruits, the soil being cultivated after each irrigation. In southern Florida other methods are made necessary by the fact that water cannot be run in furrows over the sandy soil. Revolving sprinklers, placed at the proper distance so that all the ground will be covered by their spray, are sometimes employed. Where economy of water is a factor, these are less desirable than the basin system. Taken in all, it seems that the best method of irrigating is to form around the tree a basin as wide as the spread of the branches (or wider during the first two years), to fill it with weeds, straw, manure, seaweed, or other loose mulch, and then to apply water at least once in two weeks when the rainfall is not sufficient to maintain the tree in good growing condition.

Pruning.

The amount of pruning required by the avocado depends largely on the variety. Some make short stocky growths and form shapely trees without the assistance of the pruning-shears, while others take long straggling shapes and do not branch sufficiently to form a good crown. These latter must be cut back heavily. Trapp, and other varieties of the West Indian race in general, usually make low stocky trees, branching abundantly and forming plenty of fruiting wood. With such forms, pruning is reduced to the minimum, consisting principally in removing fruit-spurs which die back after the crop has been harvested, and in the occasional cutting back of a branch to produce a crown of symmetrical form and good proportions. Beyond this very little pruning is done in Florida orchards.

With the Guatemalan race, more training is often necessary to produce a tree of ideal proportions, since some varieties tend to make long unbranched growths. In others the lateral branches are very weak and scarcely able to bear their own weight if allowed to develop unhindered. With these, careful attention should be given during the first few years to producing a well balanced tree capable of carrying good crops of fruit.

The Mexican race usually shows a tendency to grow more stiffly erect than the others, and make stout rigid branches which are capable of bearing heavy crops. In order to keep some of these varieties from becoming too tall and slender, it is necessary to top them when young, perhaps pinching out the buds of the main branches later on to induce branching.

It is not desirable to have the crown so dense that light will not reach all parts freely. When the crown is too thick, fruit is produced only on its outer surface, and much of the fruit-bearing capacity of the tree is thus wasted.

Thus it can be seen that no specific rules for pruning, covering all varieties, can be laid down, other than that the object should be to produce a tree having a broad, strong, well-branched crown of good proportions and great fruiting capacity, preferably headed low (about 30 inches above the ground), in order to shade the soil beneath it. After the tree has reached maturity little pruning is required, provided it has had the benefit of careful training during the first few years. Experience along this line is meager, however, and the future will bring out many new points of importance.

In top-working old seedlings, it is often necessary to cut off large limbs. The stubs should be smoothed off and covered with a coating of grafting-wax. The same rule applies to cuts made in the course of ordinary pruning with young as well as old trees. When secondary branches are removed, they should be cut as close to their junction with the main branch as possible, and the cut should be parallel with the main branch. The cut surface should be treated with a coating of grafting-wax. Paint is sometimes used for this purpose, but in Florida it has been found injurious, especially to young trees. If the stubs are not waxed, they often allow fungi to start and destroy the wood. The entrance of such fungi is facilitated by the fact that the pith sinks in the cut ends of large limbs, leaving a small cavity to collect water and maintain the moist conditions which are so favorable to fungous growth.

Opinions differ as to the best time for pruning. In Florida late fall and winter, November to February, have proved suitable. In California the best growers seem to favor spring or fall. According to Krome, pruning in hot weather often results in serious injury. The most favorable times seem to be early spring, before growth has commenced and before the heat of summer, and autumn after hot weather is past.

Propagation

Avocados do not come true from seed; that is, a tree grown from a seed of the Trapp variety will not produce Trapp fruits, although it may produce fruits similar in character. For commercial purposes it is necessary to propagate the trees by budding or grafting, in order to insure good fruit of uniform quality and to eliminate sparse bearers, or trees otherwise undesirable.

Seedling avocados are often grown, especially in the tropics. While named varieties cannot be propagated in this way, if the seed is taken from good fruit the tree which it produces is likely to bear such fruit. But occasionally seedling trees do not bear, and some have other undesirable qualities, so that it is always best to plant a budded tree. Seedlings can only be recommended, in fact, where a tree is desired for the dooryard merely, in which case the ornamental appearance of the avocado makes it eminently satisfactory. If such trees do not bear well no special loss is entailed.

Since 1901, when George B. Cellon first budded the avocado commercially, several methods of vegetative propagation have been applied to this plant by nurserymen. While all of these have been successful in the hands of certain propagators, shield-budding, which was originally used by Cellon, has proved the most generally dependable, and is now employed by most nurserymen in California and Florida. It is, therefore, given major consideration here, while methods of grafting are described in less detail.

Stock plants.

In Florida it has been the custom to bud or graft West Indian varieties on seedlings of the same race. In California the Guatemalan race has usually been budded on the Mexican, in the belief that the superior hardiness of the latter would make the budded tree less susceptible to cold and also because seeds of the Mexican race are more easily obtainable. Recently in Florida the Guatemalan has been budded on the West Indian, the West Indian on the Mexican, and so on; and these experiments, although not extensive, have served to indicate that the question of stocks is of great importance, and demands further investigation. Not only does it appear that the hardiness of the tree may in a measure depend on the nature of the root stock, but the congeniality of the various races, when budded on each other, seems to differ. Attempts to bud the West Indian on the Mexican have produced rather indifferent results in Florida, the buds making a poor union and growing very slowly. The Mexican race has not been tried on the West Indian extensively, but this practice appears to succeed better than the reverse. The Guatemalan buds well on the West Indian, but is perhaps preferable on Guatemalan roots.

Seeds are usually obtainable most abundantly in August and September in Florida, a month or two later in California, having reference to the West Indian race in the former state and the Mexican in the latter. These two races are those generally used for seedling stocks. The seeds should be planted soon after removal from the fruit, although they are viable for several weeks if kept cool and dry. Seeds of the Mexican race have even been kept for three or four months in good condition, in the dry climate of California.

Previous to the issuing of Quarantine Order No. 12 of the Federal Horticultural Board, prohibiting the importation of avocado seeds from Mexico, many thousands were imported annually to California from that country. In shipping these, the best results were obtained when the seeds were removed from the fruit, washed immediately, dried in the shade, and packed loosely in wooden boxes without the addition of moistened sawdust, charcoal, sphagnum moss, or other material. The percentage of loss with such seeds was insignificant. When shipping seeds from moist tropical regions, greater difficulty is experienced, decay being more troublesome. Good results are sometimes secured by shipping in slightly dampened charcoal, but where the distance is not too great the best method seems to be to wash and dry the seeds and then pack them loosely in wooden boxes, as above described.

Seeds are planted in pots, boxes, flats, or in the open ground. For nursery work on a large scale, planting in flats and seed-beds has given excellent results. The seedlings are transplanted almost as soon as they have sprouted. In California seeds planted in the seed-bed during autumn, October to December, will make plants six to twelve inches high by March or April, when they may be planted out in the field in nursery rows.

While seedlings are sometimes budded in pots or boxes, field budding is more satisfactory, as it is difficult to bring pot-grown trees into the vigorous growth essential to success in budding.

Planting in the field should be done in California as soon as danger from frost and cold weather is past. Nursery rows should be 3 to 4 feet apart, with the plants 18 inches apart in the row (or about 12 inches in Florida). Partial shade should always be given the young plants for a few days after they are set in the open, especially if they have been sprouted, as they should be, under a lath- or slat-house. In Florida, seeds planted in August may be set out in the field in November, and should make trees ready to bud by January or February, which is the proper season for budding in that state.

For germinating seeds, a light, loose, sandy loam is preferable, pure sand sometimes being used in California if the seedlings are to be transplanted as soon as they have germinated. Four-inch pots are large enough for seeds of the Mexican race, but frequently a five- or six-inch pot is necessary to accommodate the West Indian. In Florida, wooden boxes about 6 inches in each dimension are often used, while in California tin cans are employed, but the latter are much less desirable than clay pots. When planted in flats or seed-beds, the seeds may be placed close together. The pointed end of the seed, — or in the case of round seeds, the end which has been toward the stem in the fruit, — should be uppermost, and it is usually allowed to project above the surface of the soil, not more than four-fifths of the seed being below the surface. If the seed-coats are loose and come off easily, it is well to remove them before planting.

The soil should be kept moist while the seeds are germinating. The time required for germination varies greatly, sprouts sometimes appearing within two weeks from planting, while in other instances they may be two or three months in starting. A month is the average time in warm weather.

Essential features of bud propagation.

Shield-budding is most successful when the stocks are small and full of vigor. If the plants are once allowed to cease the rapid thrifty growth with which they spring from the seed, the wood hardens, sap is less abundant, and if the bud unites at all there is great difficulty in forcing it into growth. Those who do not devote their undivided attention to the propagation of the avocado sometimes allow the seedlings to get into this condition before they attempt to bud them, and as a result failures are numerous. It must be stated unreservedly that shield-budding of the avocado, to be successful, must be made the subject of careful and intelligent study on the part of the nurseryman, who must exercise constant vigilance to keep the stock plants in perfect condition. If this is done, and budwood is intelligently selected, success is within reach, but the number of failures from neglect or ignorance of these two points might well be enough to discourage the beginner from attempting to bud the avocado. It is only through the closest application to minute details that real success in budding avocados can be achieved, and it may truthfully be said that those who have produced budded trees in quantity have invariably been men who have devoted their best efforts to the work and made it a painstaking study.

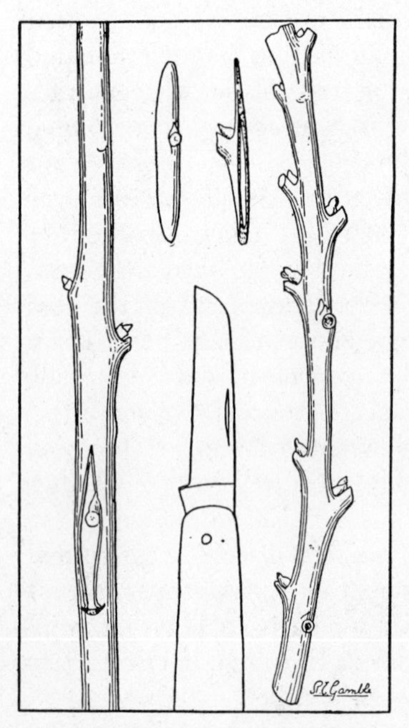

FIG. 3. Shield-budding the avocado. On the left, a bud properly inserted; above the knife blade, two buds of proper size and shape; and on the right, budwood with good "eyes." The method of wrapping the inserted bud is shown in Fig. 11.

Budding (Fig. 3).

As soon as the stock plants are large enough to receive the bud conveniently they should be budded, provided the season is favorable. In southern Florida the best months for budding the West Indian race are November, December, January, and February. Budding can be continued into March with success, but after warm weather commences the percentage of failures is too high to make the undertaking profitable. In California the best time is as soon as the sap has begun to flow freely. This usually occurs late in April or early in May, at which season there is a period of three or four weeks when budding is more successful than at any other time of the year. After this short period, however, avocados are in active growth and the proper sort of budwood is difficult to obtain, hence it is best to wait until the growth has hardened sufficiently to make good budwood. This will usually be late in June or in July, when budding can be recommenced and continued until autumn. October and November are good months, although not quite so favorable as the first-named period in the spring. Buds inserted in autumn frequently push out within five or six weeks and must be protected carefully during the ensuing winter. Unless the work is done very late in the autumn, the buds cannot be held dormant until spring.

Selection of the proper type of budwood requires more experience and judgment than any other feature of avocado propagation, since the character of the buds differs widely among varieties of the same race. Some kinds make such poor budwood that not more than 50 per cent of the buds will grow even for the most skillful propagator; in other varieties, such as Taft and Fuerte, 95 per cent of the buds can frequently be made to develop into trees. In general, it may be said that the budwood should be of recent growth, not soft enough to snap on bending but beginning to mature. In early spring, budwood must be obtained from mature growth of the previous fall and early winter. In summer it must be obtained from the current season's growth. In some sorts, such as Fuerte, very young budwood can be used successfully, but that which has commenced to mature is usually better. Buds can sometimes be cut from the tips of the branchlets and from 6 to 12 inches from the tip, according to the variety and the condition of the wood. Buds which have broken into growth should be avoided, in the case of most varieties, at least; so should those from which the outer bud-scales have dropped, as this is indicative of old wood, and such buds, when inserted, will frequently "drop their eyes" and leave a blind shield from which a tree cannot develop.

To insert the buds, an incision is made in the stock, as close to the ground as convenient, either in the form of a T or an inverted T. No particular advantage seems to be derived from either form of incision, both being used quite successfully. The bark should not be opened by using the ivory end of the budding-knife, as this injures the delicate tissues below; if the bark does not separate from the wood readily enough to allow the bud to be pushed in easily, the stock is too dry to be budded. The propagator should always aim to have the stock plants in such vigorous condition that he can force the bud into the incision with very slight pressure and without loosening the bark with his knife. The most skillful budders, when making the horizontal cut of the incision, turn the knife blade forward dexterously, forcing the bark away from the stock and leaving a sufficient opening in which to insert the point of the bud. The latter is then pushed in very gently and wrapped immediately with a strip of waxed cloth, raffia, soft cotton twine, or plain tape. This should be wound firmly around the stock, from the bottom upward, and fastened securely at the upper end, above the incision, by slipping the end through the last loop and drawing it down tightly.

In cutting the buds, an extremely thin-bladed, sharp-edged budding-knife should be used, and it should never be allowed to become the least bit dull. A razor-strop is usually worn by budders, attached to the belt; after ten or fifteen buds have been cut, the knife is given a few strokes on the strop to keep it in perfect condition. It should be the aim of the budder to cut the bud with one sliding stroke of the knife, keeping the blade as nearly parallel with the budstick as possible, so

Plate III. Avocado-growing in the Mexican highlands.

that the cut surface will be flat and not rounded at the ends. Buds which are gouged out do not fit snugly on the stock. It is well to cut the buds somewhat larger than citrus buds, 1 inch being the minimum length, and 1½ inches the ideal for most varieties. This must vary, of course, with the size of the stock and budwood, large stocks sometimes taking a bud 2 inches long.

Opinions differ as to the best material for wrapping, some preferring waxed cloth, while others have found plain cloth tape equally good, and still others use raffia successfully. Waxed cloth is doubtless the safest, but the objection to it has been that in hot weather the wax melts and works its way into the bud, sometimes killing it. This can be avoided by using a compound of 1 pound beeswax and ¼ pound rosin. The cloth, preferably a cheap grade of bleached muslin, should be torn in strips 6 inches wide, made into rolls 1 inch in diameter, and boiled for fifteen minutes in this mixture. It may then be kept until needed, when it is torn into narrow strips of the proper width and length for tying buds.

Three weeks after insertion the buds should have united with the stock, and the wraps must be loosened or they will soon bind the stock, if growth is active. They should not be removed until the end of six weeks or two months. In order to force the bud into growth, the tree should be topped at the time the wrap is first loosened, 3 or 4 inches being removed from the tip. The axillary buds along the stem will then break into growth; some of these should be allowed to develop for a while, to keep up an active flow of sap. In another four or five weeks the top should be cut back farther, but a few axillary buds still left on the seedling to grow and maintain the flow of sap. If the stock is cut back too heavily the first time, the eye may fall from the bud, leaving a blind shield. Lopping, as practiced with many other fruits, is not altogether successful with the avocado.

As soon as the bud has made a growth of 3 or 4 inches, it should be tied back to the stem of the seedling with raffia. Later it must be stake-trained, and when it has reached a height of 24 to 30 inches it should be forced to branch and form a shapely top. The stub which remains from the seedling stock should not be cut off until the bud has developed to the height of one foot. In California it is usually considered best to remove the stub in winter; it should be cut off just above the bud, and the cut surface covered with grafting-wax, or shellac made with alcohol and a little rosin. Common paint should not be used for this purpose.

Field-grown trees, after they have reached the proper size, are either lifted and put into pots or boxes, where they are held until established and then planted in the field; or they are balled at any time after they have gone dormant in late winter, and heeled-in under a plant-shed, where they can be kept until spring and then planted out. In Florida, field-grown plants are usually lifted and set in wooden boxes 5 × 5 × 12 inches in size. As soon as they are placed in these boxes, they must be set in partial shade and watered copiously. When they have become established, which will be within a month or six weeks, they can be transplanted to the orchard.

Transplanting with bare roots has not proved generally satisfactory in California. Regarding his experience with it in Florida, Krome says:

"This may become one of the recognized methods of planting and under certain conditions it has many advantages over setting either boxed or balled plants. Two years ago I moved about four hundred seedlings with semi-bare roots and lost only three trees in the process. The trees were two year stocks averaging four feet in height grown in a 'red-flat' at my own grove. We began transplanting during July but most of the trees were moved in September. We waited until the trees had reached a dormant state between flushes and then defoliated them and pruned back the most tender growth. We moved them only after three o'clock in the afternoon when the greatest heat of the day was over, digging only as many trees as could be carefully planted during the remainder of that day. Before digging we wet down the surrounding soil until it puddled easily. The trees were dug with as much of the root systems as could well be handled and the roots were immediately wrapped in wet burlap and the trees placed in the shade. We did our defoliating and pruning back considerably ahead of the digging and found that trees which had been cut back for a week or more and had just started a new growth could be moved as successfully, and in fact grew off better, than those which had been more recently defoliated.

"Since then we have carried on experiments in this line at our nursery, using trees with roots entirely bare, and have had a very low percentage of loss. Upon our recommendation a number of avocado growers in South Dade have tried the method with a limited number of trees and without exception have expressed themselves as intending to make all their plantings hereafter with bare-root trees.

"The two essentials seem to be getting the tree into proper condition before moving from its original position and plenty of water after transplanting."

Grafting.

One method of grafting has been employed extensively for the production of nursery stock in Florida, and another has been used on a limited scale for top-working old trees.

The system extensively used is a modified form of the side-graft employed with other plants. The seeds are germinated in a seed-bed; when the sprouts have reached a height of 5 or 6 inches the plants are dug and laid on the bench. A cut an inch long is made on one side of the sprout, just above the seed, and a thin section of the stem removed, exposing the tissues. The cion is then taken from the tip of a very small branchlet, preferably one which has not fully matured. It should be about 1 inch long, and provided with one or two axillary buds as well as the terminal. It is trimmed on one side to a tapering point at the lower end, and this cut surface is placed against the cut on the stock, after which it is bound carefully in place. The plant is then potted, placed under partial shade, and carefully watered from day to day. After a union is effected, the top of the seedling is removed and the cion allowed to develop.

Top-working old trees.

Large numbers of seedling avocados have been planted in Florida and California. Many of these produce fruits inferior in quality to the best budded varieties, while quite a number do not produce at all. It is often desired, therefore, to convert such avocados into budded trees of choice varieties, and this can easily be done.

Several methods of top-working are employed, the most satisfactory one being shield-budding. When trees are to be top-worked by this means, they should be cut back in November or December in Florida, February or March in California, removing three-fourths of the main limbs a foot or two from their union with the trunk, the remainder being left to keep the tree in vigorous condition. The limbs should be cut off with a sharp saw, to avoid splitting or tearing on the lower side. The stubs should be covered with a good coating of grafting-wax.

When growth has commenced, in early spring, numerous sprouts will appear around the upper ends of the stubs. Only three or four of the strongest should be allowed to remain on each stub, and when these have reached the diameter of one's little finger, they may be budded in the same manner as seedlings, with a large bud, preferably from growth which is not mature. The exceedingly vigorous growth of these sprouts makes success much more certain than in budding seedlings in the nursery. Because of the rapid growth, it is necessary to loosen the wraps frequently to keep them from binding. They should not be removed entirely before the buds have developed to a length of 6 or 8 inches. The sprouts rising from the upper side of the stub form stronger unions with the latter than do those from the lower side.

Cleft-grafting, another method employed in top-working old

trees, is most successful with seedlings two to four years old, but can also be used on older trees. While it has not been practiced extensively, it has given good results in the grove of W. J. Krome, at Homestead, Florida. Krome has worked out the method here described.

The trees to be grafted should be sawed off 2 to 4 feet from the ground, according to size, this work being done during November and December in Florida, though it has been successful as late as March. With two-year-old seedlings the trunk itself is sawed off; on larger trees it is well to go above the trunk and saw off the main branches a foot from their union with the trunk. A cleft is then prepared in the stump, not by splitting it with a grafting tool as is usually done with fruit-trees in the North, but by using a saw. After sawing to a depth of 4 to 8 inches, depending on the size of the stub, the saw is removed and a soft wooden wedge is inserted in the top of the cleft and driven down until the lower end of the cleft begins to split. This produces the steady pressure necessary to hold the cion firmly in place.

Cions are cut from wood of larger size and more mature growth than is used for budding, branches about ½ inch in diameter being preferable. The cion, which should be 6 to 9 inches long, is trimmed on two sides throughout the lower half to a slender tapering point at the bottom. It is then placed in position in the cleft and forced downward until the upper end of the cut surface is flush with the top of the stub. One cion is placed in the cleft at each side of the stub, nearly even with the surface of the bark on the outside. The wedge which has been used to keep the cleft open is now partly withdrawn until the cions are clamped firmly by the pressure of the two halves of the stub, when it is sawed off flush with the top of the stub and allowed to remain in place so that the pressure on the cions will not become too great.

After the cions are properly placed, the cleft is filled with plastic grafting-wax so that air is excluded. Wax is also rubbed over the outside of the cion where it fits into the stub. The stub is then firmly wound with strips of waxed cloth, covering the top as well as the sides. A collar made of builder's paper is then tied around the stub, extending an inch above the tops of the cions. This collar is filled with sand. Particular attention must be given to insuring a layer of sand between the cions and the side of the collar, since otherwise the latter transmits heat from the outside and kills the cions. Vent holes should be made in the paper near the top of the stub to drain off the water which collects within the cup.

Nothing more remains to be done until the cions have had time to unite with the stock. Two or three months after growth has commenced the sand may be removed and the collar taken off. As a rule, only the stronger of the two cions develops. Both may start to grow but one eventually outstrips the other in most cases, and the weaker one succumbs.

This method appears to produce vigorous trees. Its use has been attended by excellent results at Homestead.

The Crop

The age at which budded avocado trees come into bearing varies with the different races, and also among the varieties of the same race. Furthermore, experience indicates that many kinds will bear at an earlier age on the sandy soils of southern Florida than on the heavier lands of California. In the latter state, budded trees of the Mexican race frequently come into bearing the second or third year after they are planted in the orchard; the Guatemalan race shows greater range among the numerous varieties, some, for example the Lyon, commencing to bear within eighteen months or two years from the time of budding, while others, for example Taft, have not borne earlier than the fourth or fifth year. Trapp and several other

West Indian varieties have been grown for four or five years in southern California without bearing fruit. They are sometimes injured by cold, but, allowing for setbacks from this cause, the West Indian race does not fruit so early in California as in Florida. The Mexican race usually fruits at an early age in both regions.

As a rule, budded trees of the West Indian race are precocious in Florida. Trapp is remarkable in this respect; and in addition it has a strong tendency toward over-production which must be checked during the first few years by thinning the fruit. Trapp trees will often produce a few fruits the year after they are planted in the orchard, and at three years from planting may begin to yield commercial crops. If grown under irrigation, so that their development has been rapid, the trees may be allowed to carry thirty or forty fruits the third year after planting, but during the first year it is best to remove all fruits, and the second year not more than half a dozen should be allowed to mature. When grown without irrigation, the tree is rarely large enough at three years of age to carry more than twelve or eighteen fruits without injury to itself, unless soil conditions have been very favorable. The mistake is often made of allowing Trapps to over-bear when young, with the result that they die back following the fruiting season.

Seedlings vary even more than budded trees in the age at which they begin fruiting. The Mexican race often fruits at two or three years from seed. The Guatemalan race, in California, has occasionally fruited at three or four years, but more commonly comes into bearing at six or seven years. The West Indian race, in Florida, does not usually come into bearing earlier than five or six years from seed.

In California, no figures showing the yield of a budded orchard have as yet been obtained, but in Florida, where the avocado industry is older, interesting data are available. While the figures given may not apply to both regions and will certainly vary greatly with different sorts, they serve at least to show what may be expected from one variety under certain conditions.

According to George B. Cellon, a Trapp tree seven to ten years old will yield, under good cultural treatment, between five and ten crates of fruit, counting forty fruits to the crate, which is about the average pack. The returns from one of the largest groves near Miami for two seasons, however, show an average of only one and one-half crates to a tree. This is a low yield, and should certainly be exceeded. Krome, who has kept careful crop records, finds that his Trapp trees at five years of age yield one to four crates a tree, two and a half crates being the average. Charles Montgomery of Buena Vista, Florida, has obtained yields of about the same amount, his estimate being that a mature Trapp grove should produce 500 crates to the acre.

The yield of other varieties in Florida is not so well known, since none except Pollock has been planted to any extent, and even this variety is grown in comparatively small numbers. In regularity of bearing Trapp excels Pollock, the latter showing a tendency to fruit in alternate years.

In Guatemala and Mexico, many seedling trees of the Guatemalan race tend to produce good crops only in alternate years. The feature is not so marked in trees of the West Indian race which have been observed, nor in those of the Mexican; nor is it true that all Guatemalans possess it. It is possible that over-production one season results in a crop failure the following one, and it is probable that unfavorable cultural conditions have something to do with the matter.

Season

The season during which avocados are obtainable in southern Florida has been, until very recently, from July until January. A few Trapps may hang on until February or even as late as

March, but the fruit is so scarce after the early part of January that it need scarcely be reckoned with. The earliest varieties of the West Indian race begin to ripen in July, while the bulk of the seedling crop matures in August and September. During this season avocados are cheap, and the markets of the North are receiving shipments from Cuba, but there is a certain demand for high-class fruit even during the summer, and such varieties as Pollock are profitably grown in a small way. It has always been recognized, however, that the most profitable avocados are those which can be marketed in winter, for not only is the cheap seedling fruit out of the way at that time, but the markets of the North are not filled to overflowing with peaches, plums, grapes, and other standard fruits.

It is, therefore, the late Trapps which have been the most profitable in Florida, and the constant search has been for even later varieties which would make it possible to supply the markets during late winter and early spring. Such have not been found among those of the West Indian race, but the Guatemalan meets this demand, and varieties of this race will, in all probability, soon be planted extensively in Florida. The Guatemalan kinds which have already fruited at Miami and elsewhere have served to indicate that the season during which this race will ripen is, roughly speaking, November to May.

In California a given variety of the Guatemalan race ripens one to two months later than in Florida, so far as present experience goes. The season of this race in California extends from January or February, when the earliest sorts appear in the market, to autumn. Following the Guatemalans, the Mexican varieties mature, their season in general being October to January, although there are some kinds which mature a few fruits in spring. Thus it can be said that there is never a day when ripe avocados are not obtainable in California.

While the Mexican race has received little attention in Florida, it seems likely to become of considerable value for the cooler sections of the state, now that varieties of good size and quality are obtainable. Chappelow has been in bearing at Miami for some years, maturing there in June and July, which is considerably earlier than in California.

In Cuba it is said that trees growing on dry soils will hold their fruits longer than those growing on low moist land. Occasional seedling trees (West Indian race) are found throughout Cuba which have the reputation of carrying their fruits until Christmas or even later. Such trees are, of course, highly profitable to their owners, since avocados are in great demand in Habana during the winter months, and the supply at present is limited.

Picking, Packing, and Marketing

Avocados are picked best with orange clippers. The stem is usually swollen just above the point of attachment with the fruit; it should be severed with the clippers immediately above this swollen portion. In order to supply the early markets, avocados are sometimes picked before they are fully mature, a custom which should be discouraged. Immature fruits are certain to be inferior in flavor, and should they fall into the hands of those who were trying the avocado for the first time they would be certain to give a bad impression. Trapps are usually left on the tree as long as possible, in order to obtain the high prices which late fruit commands; when they begin to change from bright green to yellowish green they must be picked or they will drop. If they are picked only a day or two before they would drop, they are sure to ripen in transit and reach the market in an over-ripe condition. To prevent this, Cellon advises that questionable fruits be laid aside for twenty-four hours; if at the end of this time they are still firm, they may safely be packed for shipment.

The standard package for avocados in southern Florida is the tomato crate, which measures about 12 × 12 × 24 inches. It is sometimes used with a partition in the center, sometimes without. Excelsior is placed above and below each layer of fruits as a cushion, and is stuffed around them freely to hold them in place and prevent bruising. Some growers wrap each fruit in tissue-paper, but the wisdom of this practice is doubtful. The fruits heat more quickly when wrapped, and as heating greatly hastens the ripening process it should be avoided as much as possible. Avocados must not be packed under such great pressure as oranges, more care being necessary in nailing on the top of the crate to avoid crushing the fruits.

The number of fruits to a crate varies from twenty-three to fifty-four with Trapp, the average being about forty. Pollocks run from eighteen to thirty-six to a crate, while seedlings run from twenty-eight to ninety. Quotations, f. o. b. southern Florida, are sometimes made by crate, sometimes by dozen fruits. The following figures on Trapps are those quoted by one of the principal shippers at Miami during the past several years:

First week in October, 54s (that is, fruits which pack 54 to the crate), 75 cents a dozen; 50s, 85 cents; 46s, $1; 36s, $1.30; 28s, $1.75; 23s, $2. After November first the price is increased on all sizes, as follows: 50s, $1.50 a dozen; 46s, $2; 36s, $3. At Thanksgiving the prices vary from $3 to $4 a dozen for 24s, 36s, and 46s, and about Christmas they advance to $4 to $6 a dozen.

Pollocks are quoted during August as follows: 36s, 75 cents a dozen; 28s, $1; 24s, $1.50; 18s, $2. The quotations on high-grade seedling fruits at the same time are as follows: 50s to 60s, 60 cents a dozen; 46s, 75 cents; 36s, $1; 28s, $1.50.

Prices on Trapps a crate vary from about $2 in early October to as high as $36 for the last few crates at the end of the season in February; these figures are f. o. b. southern Florida. From one of the principal groves near Miami the entire crop has been marketed for several years at an average net price of $5.25 a crate averaging forty fruits. The average return from 1400 crates shipped from another grove was $5.50 a crate.

Trapps have been shipped from southern Florida to all parts of the United States. A few years ago one grower sent small consignments every day during a large part of the season to Seattle, Washington, and did not receive a complaint of a crate received in bad order. These shipments were on the road eight days, and were not sent in cold storage. It is the general practice to ship from Florida by express. The shipping qualities of Trapp are much better than those of the average seedling.

At present most of the Florida Trapp crop goes to the markets of the eastern United States, Washington, Philadelphia, New York, and Boston each taking a good share. Some growers have shipped heavily to Chicago and other points in the Middle West, and small shipments go to the Pacific Coast each year.

The production in California has not yet become great enough to permit of commercial shipments to eastern markets, the crop being consumed locally. Since most of the returns up to the present time are based on the crop from the parent seedling tree of each variety, they are of little value to show the probable profits from a budded orchard of the same sort. The most remarkable record which has been made by a commercial planting of budded trees is that of J. T. Whedon at Yorba Linda. Whedon's planting of the Fuerte variety, containing fifty trees (less than one acre), produced a crop of fruit when five years old which sold for $1700.

Pests and Diseases

In the early stages of many horticultural industries insect pests and fungous diseases are not troublesome, but as the industry develops its enemies become more numerous. So it has been with the avocado. During the first few years in which this fruit was planted commercially in Florida little injury was caused by parasites, but recently it has been necessary to combat vigorously the insects which prey on the tree, and also several fungous diseases.

In California the avocado has, up to the present, been comparatively free from the attacks of insect and fungous pests; yet several insects have made their appearance in the orchards and must be watched carefully lest they become so numerous as to cause serious harm.

Thrips and red-spider are the most common insects which attack the avocado in Florida. Red-banded thrips (*Heliothrips rubrocinctus* Giard.) and the greenhouse thrips (*Heliothrips hæmorrhoidalis* Bouché) feed on the foliage, sometimes causing much damage. Both these species are exceedingly small, soft-bodied, fringed-winged insects, with piercing mouth-parts by means of which they puncture the epidermis and extract the juices from the leaves. They are most destructive in early spring, their numbers being greatly reduced when the summer rains commence. Spraying with nicotine solutions has been quite effective in controlling them.

The red-spider (*Tetranychus mytilaspidis* Riley) also does considerable damage during the spring months. This insect, which is scarcely larger than a pin point, can be detected on the foliage without the aid of a magnifying glass because of its bright red color. It feeds on the avocado by piercing the leaf tissues and extracting the plant juices. Often it becomes so abundant as to cause the leaves to assume a brownish, sickly appearance. It occurs commonly in California as well as in Florida, but has not yet been reported as attacking avocados in California. Lime-sulfur mixtures have been used successfully in combating this insect. For citrus trees, H. L. Quayle recommends commercial lime-sulfur, dry sulfur and hydrated lime, and distillate emulsion. These may all prove to be effective with the avocado as well.

Among the scale insects which commonly attack the avocado, the most important are the black scale (*Saissetia oleæ* Bern.), and a soft white scale (*Pulvinaria pyriformis* Ckll.), the latter being a serious pest in Florida. Severe infestations of the black scale are occasionally found on old seedling trees in California, but this insect has not yet become a pest in the young avocado groves of that state. The wax scale (*Ceroplastes floridensis* Comst.) is occasionally found on avocados in Florida, but rarely requires combative measures. All of these scale insects, as well as a white fly (*Trialeurodes floridensis* Quaint.), which has become troublesome on some of the Florida Keys, can probably be controlled by the use of oil sprays.

The citrus mealy-bug (*Pseudococcus citri* Risso) has been reported on the avocado in Ventura County, California, but it is not known to have caused extensive damage. The avocado mealy-bug (*Pseudococcus nipæ* Mask.), which is a serious pest in Hawaii, has been found in southern Florida groves. It sometimes becomes very troublesome. D. F. Fullaway of Hawaii recommends that it be controlled by spraying with oil-emulsions.

The presence of the avocado weevil (*Heilipus lauri* Boh.) in California, where it was probably introduced from Mexico in avocado seeds, caused the Federal Horticultural Board to prohibit the importation of seeds of the Mexican race from Mexico and Central America. This insect is a small black beetle which tunnels in the seeds, and is said to do considerable damage.

Other seed weevils attack the avocado in various parts of the tropics. H. S. Barber describes the more important ones, so far as they are known, in the Proceedings of the Entomological Society of Washington, March, 1919. *Heilipus pittieri* Barber, from Costa Rica, is similar to *H. lauri*. *Conotrachelus perseæ*

Barber does great damage to avocados in Guatemala. Its larvæ have been found in avocado seeds sent to the United States, but it is believed the species has not become established in this country. Once thoroughly established, the seed weevils are difficult to exterminate, hence it is to be hoped that they will not gain a foothold in this country.

In Guatemala, *Trioza koebelei* Kirkaldy (and perhaps other species) produces leaf-galls on the avocado, often in such great numbers as seriously to affect the health of the tree.

In addition to these insects, a number of others have been reported as attacking the avocado in various parts of the tropics. These include numerous scale insects, both armored and unarmored, several borers, and the well-known Mediterranean fruit-fly (*Ceratitis capitata* Wied.); the better-known species are listed in the Manual of Dangerous Insects published by the Department of Agriculture.

In the dry climate of California, fungous parasites give the avocado grower comparatively little trouble, but in Florida and in many parts of the tropics they may require stringent combative measures.

The following extracts from a paper by H. E. Stevens, published in the Proceedings of the Florida State Horticultural Society for 1918, cover the situation as regards fungous pests in Florida as it exists at the present time:

"Leaves and frequently fruits of the avocado are attacked by a fungus which is probably a species of Gloeosporium. The affected leaf is usually attacked at the tip, and the disease gradually spreads until the greater part of the blade is involved, when the leaf falls. Severe attacks may cause considerable defoliation of trees and result in the death of young terminal twigs. Fruits may be attacked when small, in which case severe shedding may follow. If the more mature fruits are attacked, a brown spotting is produced and the skin may crack.

"Another common type of injury, frequently noted on the fruits, is referred to as anthracnose by some of the growers. This type of injury is very similar to melanose of citrus fruits in general appearance. It is superficial and appears in the form of dark reddish brown caked masses on the surface of affected fruits. The markings are hard, compact, and the surface is cracked or broken. The injury may cover only a part or the whole surface of the fruit. It makes an unsightly fruit, but apparently does not affect the quality. The disease is apparently caused by a fungus, perhaps a Gloeosporium or a closely related species.

"Another fungus, a species of Colletotrichum, is often observed in diseased spots on leaves and fruits. This fungus is closely related to Gloeosporium and the injuries with which it is associated resemble those caused by the latter fungus. It is probably the cause of some of the injuries that are classed as anthracnose.

"In the control of these leaf and fruit spots, Bordeaux mixture has given satisfactory results where applied in time. As soon as the injuries begin to appear, spraying should be made and continued until the disease is checked. Two or three applications may be necessary, made at intervals of two or three weeks. If the fruit is near maturity, it is advisable to substitute ammoniacal solution of copper carbonate for the Bordeaux mixture, to prevent any disagreeable stain that may result from the use of the latter. Aside from spraying, all dead wood should be kept out of the trees, as this is likely to harbor these fungi from one season to the next.

"Avocado scab is of more than ordinary interest, owing to its close connection with citrus scab, and the fact that it has come into existence within the past three or four years. It is in all respects a new disease that has had its beginning in Florida.

"Scab is chiefly a disease of the tender growth, and at present it is found more abundantly in the nurseries, where it is particularly severe on seedling plants. It also attacks budded varieties in the nursery. The disease has been found on young and old bearing trees in the groves, affecting the leaves, and in a few cases the injury was observed on fruits. At present it is more common in the nurseries, but it may soon prove a serious pest in the groves.

"Scab forms definite spots or patches on the young, tender leaves and shoots, and severe attacks may cause the foliage to curl or become distorted. The more mature leaf tissue is not affected, but old leaves will be found bearing spots that were formed when the tissue was young. The spots are usually small, raised, circular to irregular, purplish brown to dark in color, and may vary from a sixteenth to an eighth of an inch in diameter. They may appear scattered over the surface, or several may grow together, forming irregular patches. The spots penetrate the leaf tissue, and they are visible on both sides. They are usually more prominent on the upper surface of the leaf, in which case the under surface of the spot will be slightly bulged and marked by a discolored area. The centers of the spots are composed of dead cells, more or less spongy in character and brownish in color.

In the earlier stages the surfaces may show a fuzzy, whitish growth — the fruiting parts of the fungus. The surfaces of older spots are darker in color and frequently covered with a dark webby fungous growth. On young shoots and twigs the spots appear more elevated, small, oval, dark purplish brown to black, and have comparatively smooth surfaces. This same type of spot is observed on the fruits.

"It is plainly evident that the avocado scab fungus is none other than *Cladosporium citri*, which causes citrus scab. The two fungi agree in structure and growth habits, and both are parasitic on citrus.

"Only tentative control measures for avocado scab can be suggested at the present time. Spraying with Bordeaux mixture for the disease in the nursery has given good results in some cases, in others less satisfactory. If the new growth can be protected while it is putting out, the disease may largely be avoided. The sprayings should be made when the foliage begins to put out, and continued until the leaves are nearly developed. The 4–4–50 Bordeaux mixture may be applied at intervals of ten days or two weeks, or often enough to keep the young foliage well protected. The fungus develops more rapidly during cool weather where moist conditions are provided. Shade and a crowded condition of the trees also seem to favor the development of the scab."

Many growers in southern Florida who have planted the Trapp avocado have been troubled by the trees dying back following the production of a heavy crop of fruit. Krome of Homestead has given this subject much study, and writes as follows regarding it in the 1916 Report of the California Avocado Association:

"Avocado trees of the West Indian race, when in good condition of growth, are prone to put on a tremendous bloom from which a setting of fruit is apt to result so heavy as to be entirely beyond the carrying capacity of the tree. Following this abnormal effort there is often a period of apparent exhaustion during which the tree seems to realize that it has 'bitten off more than it can chew,' and to be seeking the best method to recoup from its over exertion. This is a critical time in the life history of the tree and calls for intelligent handling on the part of the grower. If left to its own devices the tree will endeavor to carry the over crop, draining upon its reserves until its vitality has been seriously impaired. Evidences of this condition are usually very apparent. The tree drops a large portion of its leaves, the younger branches change in color from a dark green to a saffron yellow and no new growth is put on. Lack of sufficient foliage to provide proper shade often results in serious sunburning of the more tender branches, and the low state of vitality lays the tree particularly liable to the inroads of disease, especially of the anthracnose fungus which seldom loses such an opportunity for making an attack. Finally the tree is compelled to drop practically its entire crop of fruit and is left in a condition which means, at the very best, a set-back of two seasons in its development and not infrequently results in its actual death.

"To obviate overblooming, particularly in the case of young trees, is very difficult, for the better the cultural condition of the tree, the more likely this is to occur. The usual procedure has been to thin the over crop of fruit and this method of handling works quite satisfactorily provided the set-back to the tree has not already been brought about through the excessive bloom. However, the avocado requires a longer period than most fruits between the first appearance of the bloom and the setting of the fruit and it often happens that the damage to the tree has made considerable advance before relief by stripping can be obtained. In this event removal of the entire crop and further careful attention is necessary.

"In an effort to overcome this difficulty, I have during the past two seasons resorted to frequent applications of fertilizer, in order to offset the heavy drain upon the vitality of the trees during the blooming period. In the spring of 1916, following a season favorable to growth, the avocado trees at Medora Grove began to bloom about the middle of March. Immediately afterward a light application of fertilizer, carrying ammoniates from readily available sources was made. The bloom was the heaviest known in a number of years and persisted until about the middle of April. Between April 15th and 20th, another light application of the same fertilizer was made and this was followed by a third application the latter part of May, when a fertilizer somewhat higher in phosphoric acid, largely derived from low grade tankage, was used. As a result of this treatment a full crop of fruit was set and in most cases carried through to maturity without damage to the trees. When an over crop was set at first, as a rule dropping took place without a reduction in vitality, until the proper carrying capacity had been reached, and the remainder of the crop was matured. In a few cases stripping was necessary, but among nearly two thousand trees of varying ages, not more than eight or ten showed any appreciable damage."

In both California and Florida, avocados sometimes crack open while hanging on the tree. This has occurred in varieties of the Guatemalan and Mexican races, but is most common in the latter. The cracks are usually situated towards the apex of the fruit, and are often very extensive. W. R. Horne, H. S. Fawcett, and others have noted the presence of several fungi in the cracks and the flesh beneath them, but up to the present it is believed that these fungi are secondary, and not the cause of cracking.

Races and Varieties

The avocados cultivated in the United States are classified horticulturally in three races: the West Indian, the Guatemalan, and the Mexican. The West Indian and Guatemalan races, so far as can be judged at present, are two expressions of one botanical species, *Persea americana,* while the Mexican race represents a distinct species, *Persea drymifolia.*

Horticultural varieties of the avocado, when propagated from seed, do not reproduce the parent fruit in every detail. Seedlings from a round green fruit of the West Indian race may produce fruits oblong or pyriform in shape, and red or purple in color, varying from the parent in numerous other ways as well. But these seedlings will always be like their parents in certain respects, because they belong to the same race and will reproduce the racial even though not the individual characteristics.

To use the definition of H. J. Webber,[1] "Races are groups of cultivated plants that have well-marked differentiating characters, and propagate true to seed except for simple fluctuating variations." Technically speaking, the Mexican avocados should not be called a race, since they really represent a species; the West Indian and the Guatemalan, however, do not appear to differ from each other except in minor characters.

The classification of avocados into these three races has been useful, inasmuch as it brings together all those varieties which have several characteristics in common. In fact, the mere

[1] In the Standard Cyclopedia of Horticulture.

statement that an avocado belongs to the West Indian, Guatemalan, or Mexican race gives one an idea of the relative hardiness, season of ripening, and commercial character of the fruit.

The botanical standing of the cultivated races, as at present understood, and the characters which serve to distinguish them horticulturally, are shown in the following key:

1. Leaves anise-scented; skin of fruit thin (rarely more than $\frac{1}{32}$ inch in thickness) *Persea drymifolia*
   MEXICAN RACE of horticulture
2. Leaves not anise-scented; skin of fruit thicker (from $\frac{1}{32}$ to $\frac{1}{4}$ inch in thickness) *Persea americana*
   - *a.* Fruit summer and fall ripening; skin usually not more than $\frac{1}{16}$ inch thick, leathery in texture.
     WEST INDIAN RACE
   - *b.* Fruit winter and spring ripening; skin $\frac{1}{16}$ to $\frac{1}{4}$ inch thick, woody in texture.
     GUATEMALAN RACE

One variety cultivated in the United States, the Fuerte, appears to be a hybrid between the Mexican and Guatemalan races. Others of similar origin are likely to appear at any time, hence it is desirable to establish a group to include hybrids.

The avocados of the West Indian race have been developed in the tropical lowlands; the Guatemalan race, on the other hand, is a product of the highlands. At intermediate elevations varieties appear which belong to neither of these races, but possess some of the characters of each. These intermediate forms cannot be classified with accuracy.

In selecting varieties for commercial planting, it must be borne in mind, first of all, that the tree must be vigorous and hardy enough to grow successfully in the particular location which the planter has in view. Secondly, it must in time produce sufficiently large crops of marketable fruit to make its culture commercially profitable. It is not necessary that it be very precocious; it is noticeable, in fact, that precocious varieties often fail to make vigorous trees. It is more desirable to have the tree devote itself during the first three years to the development of an extensive root-system and a well-branched crown capable of withstanding the drain imposed by the production of heavy crops of fruit than to have its growth limited and its vitality exhausted by premature fruiting. Thirdly, the fruit itself must be given consideration from a commercial standpoint. Attractiveness, flavor, shipping qualities, season, and other important characteristics should be considered in respect to the market it is proposed to supply. Naturally, good shipping quality can be sacrificed to some other point if the fruit is for local use, while it is essential if the fruit is destined for distant markets. The flavor and quality of the flesh should be as good as possible, and the seed should not be unduly large.

More than one hundred and fifty varieties have been propagated in the United States up to the present time. The larger part of these originated as seedlings in California and Florida; the remainder have been introduced from Mexico, Guatemala, Cuba, the Bahamas, Hawaii, and a few other regions.

Of this large number not more than a dozen are likely to be planted ten years hence. Indeed, most of them have already been discarded. New varieties are originating every year, however, and the introduction of promising sorts from foreign countries is receiving much attention. It is only by testing a large number of varieties from all of the important avocado regions of the tropics that the best available kinds for commercial cultivation can be obtained.

It is not desirable to burden such a work as this with descriptions of all the avocados which have been propagated. It is sufficient to include the more important ones which are at the present time being planted commercially. For descriptions of minor varieties, and for information regarding the behavior and value of new introductions, the reader is referred to the annual reports of the California Avocado Association. In 1917 this organization issued Circular No. 1, "Avocado Varieties Recommended for Planting in California," the suggestions contained in which have done much to eliminate from consideration numerous inferior sorts. The varieties recommended in this circular are as follows, the arrangement being according to season of ripening in California:

Spring varieties
Fuerte, Spinks, Blakeman, and Lyon
Summer varieties
Spinks, Blakeman, Lyon, Dickinson, and Taft
Fall varieties
Taft, Dickinson, and Sharpless
Winter varieties
Sharpless, Puebla, and Fuerte

Several of these varieties may be superseded within a short time by others which are now being tested in California. It is not to be expected that the industry can settle down to the cultivation of a few standard sorts until all of the promising ones have been tested, and this may require several years.

In Florida, the only variety which was extensively planted during the first fifteen years of the industry was Trapp. With the introduction of the Guatemalans, however, the question has become more complicated, and it will take some time to determine by actual trial which members of this race are most suitable for cultivation in different parts of the state.

It is probable that varieties will be obtained which will make it possible, both in California and Florida, to market avocados in every month of the year. Indeed, it is almost possible to do so at the present time. In other regions horticulturists should work toward this end by obtaining for trial varieties ripening at different seasons.

PLATE IV. *Left,* Puebla avocado tree producing its first crop at two years of age; *right,* the Fuerte avocado.

West Indian race.

This race is the predominant one in the West Indies and throughout the low-lying portions of the tropical American mainland. It is found as far north as Florida and the Bahama Islands, and as far south as central Brazil. From its home in America it has been carried to Madeira, the Canary Islands, parts of tropical Africa, Oceania, and the Indo-Malayan Archipelago. It is much more widely disseminated than either of the other races. The name South American race is sometimes applied to it, while P. H. Rolfs [1] termed it the West Indian-South American.

Practically all of the avocados cultivated in Florida previous to the introduction of the Guatemalan were of this race. In California it has never been extensively grown; only a few trees, in fact, are known to have fruited in that state. It is the most susceptible to frost of the three races, and is best suited to cultivation at low elevations in the tropics.

The foliage of the West Indian race lacks the anise-like scent which characterizes the Mexican; in general, it resembles the foliage of the Guatemalan closely, but often the young branchlets and the leaves are lighter in color. The fruits are produced on short stems; the smallest weigh 4 or 5 ounces, the largest 3 pounds or more. The surface is nearly always smooth, yellow-green to maroon in color, the skin rarely more than $\frac{1}{16}$ inch thick, pliable and leathery in texture. The seed is usually large in proportion to the size of the fruit, and often loose in the seed cavity. The cotyledons are often rough on the surface, with the two seed-coats frequently thick and separated, at least over the pointed end of the seed, one of the coats sometimes adhering to the cotyledons and the other to the wall of the seed cavity. The flowers are characterized by less pubescence than those of the Mexican race, but are very similar to those of the Guatemalan; sometimes they are almost

[1] Bull. 61, U. S. Dept. Agr.

devoid of pubescence. The flowering season is from February to March in Florida, the fruit maturing from July to November, in certain varieties sometimes remaining on the tree until December or January.

Pollock (Fig. 4). — Form obovate to oblong-pyriform; size very large to extremely large, weight commonly 25 to 35 ounces, but occasionally attaining to 50 ounces, length $6\frac{1}{2}$ to $7\frac{1}{2}$ inches, greatest breadth 4 to 5 inches; base narrow, flattened slightly, with the short stem inserted obliquely in a shallow, flaring, regular cavity; apex obliquely flattened or slightly depressed; surface smooth, light yellowish green in color, with numerous small greenish yellow or russet dots; skin less than $\frac{1}{16}$ inch thick, separating very readily from the flesh, tough and leathery; flesh firm, smooth and fine in texture, deep yellow changing to yellowish green close to the skin, almost without a trace of fiber discoloration; flavor rich, rather dry, very pleasant; quality excellent; seed conic, oblique at base, rather small, weighing about 4 ounces, usually fitting snugly in the cavity but sometimes loose, the seed-coats rather loose, more or less separate; season August and September at Miami, Florida.

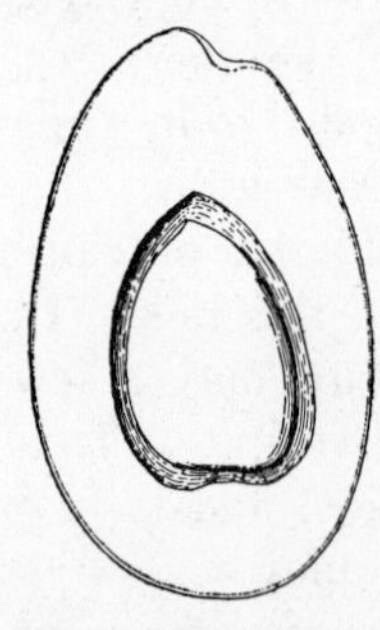

FIG. 4. The Pollock avocado. ($\times \frac{3}{14}$)

Originated at Miami, Florida; first propagated in 1901. It has been planted more extensively than any other West Indian variety except Trapp. It is remarkable for its large size and excellent quality.

Trapp (Fig. 5). — Form roundish oblate, obliquely flattened at the apex; size large to very large, weight 16 to 24 ounces, length 4 to $4\frac{1}{2}$ inches, greatest breadth $4\frac{1}{4}$ to $4\frac{3}{4}$ inches; base narrowing slightly, flattened around the deep, narrow, rounded, regular cavity in which the short stem is inserted; apex obliquely flattened; surface smooth to undulating or slightly pitted, pale yellow-green in color, with numerous small to medium sized, irregular, pale greenish yellow dots; skin $\frac{1}{16}$ inch thick, separating very readily from the flesh, firm, leathery and pliable; flesh firm, very smooth, rich cream-yellow, changing to pale green near the skin, fiber discoloration very slight; flavor

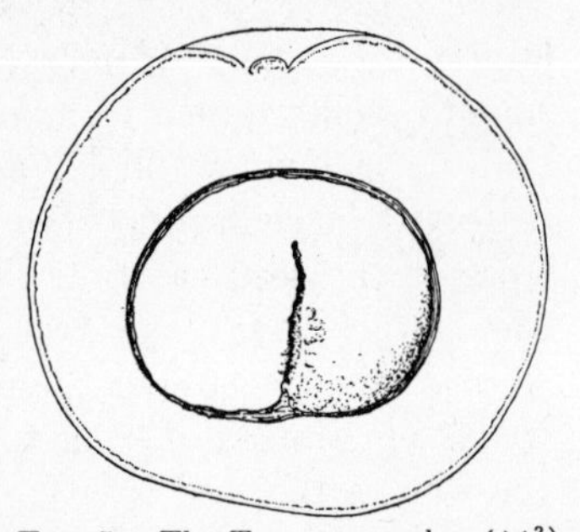

FIG. 5. The Trapp avocado. ($\times \frac{3}{8}$)

moderately rich, pleasant, quality good; seed broadly oblate, large, about 5 ounces in weight, nearly tight in the cavity, with the seed-coats adhering more or less closely to the cotyledons or sometimes to the lining of the cavity. Season commencing in late September or October at Miami, Florida, and extending until the end of December, with a few fruits hanging on until the end of February or March.

Originated at Coconut Grove, Florida; first propagated in 1901. An unusually late variety, and for this reason valuable. It was the only avocado planted extensively in Florida previous to the introduction of the Guatemalans. The tree is very productive, but is a weak grower and susceptible to frost.

Waldin. — Form oblong to oblong-pyriform; size large to very large, weight 18 to 28 ounces, length 5 to $6\frac{1}{2}$ inches, greatest breadth $3\frac{3}{4}$ to $4\frac{1}{2}$ inches; base somewhat narrowed with the rather short thick stem inserted squarely; apex slightly flattened; surface smooth, usually without markings; skin $\frac{1}{16}$ inch thick, separating readily from the flesh, tough and leathery in texture; flesh firm, deep yellow in color, smooth, with very little trace of fiber; flavor rich and pleasant; quality excellent; seed obovate, rather large, weighing about 5 ounces, usually tight in the cavity. Season October until early January at Homestead, Florida.

Originated near Homestead, Florida; first propagated in 1915. The tree is a strong grower, productive, and more resistant to cold and to fungous diseases than the average variety of its race. Valuable on account of its lateness in ripening, and the good quality of its fruits.

Guatemalan race.

Although planted in California as early as 1885, the Guatemalan race did not begin to attract attention until about 1910. With the increase of interest in avocado culture which had its inception in California about that time, a number of fruiting trees were brought to light, most of them grown from seed introduced about 1900 by John Murrieta of Los Angeles, although the first tree was planted by Jacob Miller at Hollywood. Because of the excellent commercial qualities of the fruits produced by these seedlings and the season at which they ripened, several of them were propagated and named as horticultural varieties. The number has now increased, both through the fruiting of seedlings locally and the introduction of selected varieties from southern Mexico and Guatemala, especially from the vicinity of Atlixco, Puebla, Mexico, which was the source of most of the seeds introduced by Murrieta and has since furnished budwood of many choice varieties.

In Florida this race came into notice even later than in California. Several trees grown from seeds sent from Guatemala by G. N. Collins about 1901 came into bearing at the Miami Plant Introduction Garden in 1911–1912, and their season of ripening, February to April, immediately stimulated interest in this race, since a winter-ripening avocado had been the greatest desideratum of Florida growers. Budwood of practically all the varieties growing in California was obtained, and the first offspring of these came into bearing at Miami in 1915. While it can thus be seen that the Guatemalan race is new to Florida, it promises to become of great commercial value, and it has the decided advantage that its culture will be possible farther north than that of the West Indian race. Up to the present the trees are successful under Florida conditions. The varieties that have so far fruited ripen from October to May.

In other countries the distribution of this race is limited. It was introduced into Hawaii in 1885, and has recently begun to attract attention in that territory. Lately it has been planted in Cuba, where it promises to be successful. It has also been introduced into Porto Rico and a few other regions, but only within the last few years.

The foliage of the Guatemalan race, as of the West Indian, lacks the anise-like odor which characterizes the Mexican. It is commonly deeper colored than the West Indian, the new growth often being deep bronze-red. The fruits, weighing 4 ounces to more than 3 pounds (commonly 12 to 20 ounces), and borne on long stems, are light green to purplish black in color. The surface is often rough or warty, especially toward the stem end of the fruit. The skin is usually over $\frac{1}{16}$ inch, sometimes $\frac{1}{4}$ inch, thick. This characteristic, together with the texture of the surface, is variable, occasional forms being found which have the skin scarcely thicker or rougher than in the West Indian race. It is usually harder, however, and more coarsely granular in character. The seed completely fills the cavity. The cotyledons are nearly or quite smooth, the seed-coats thin, closely united, and adherent to the cotyledons throughout. The flowers, more finely pubescent than in the Mexican race, are similar in character to those of the West Indian. They appear much later than those of the Mexican race, usually beginning to open in late spring, about the time those of the West Indian race (in Florida) are setting fruits. Unlike both the other races, the fruit does not ripen in the ensuing summer, but is carried over into the following autumn, winter, or spring; while in California, fruits which develop from flowers appearing in June may remain on the tree until a year from the following October. The ripening season in general is winter and spring in Florida, somewhat later in California, where the earliest varieties at present cultivated begin to ripen late in January or in February, and the latest ones hang on the tree until October.

Blakeman. — Form broad pyriform to obconic, oblique, broad at the basal end; size above medium to very large, weight 14 to 20 ounces, length 4 to $4\frac{3}{4}$ inches, greatest breadth $3\frac{1}{4}$ to $3\frac{3}{4}$ inches; base rounded, the long stem inserted obliquely in a very shallow cavity; apex broadly rounded, obliquely flattened or slightly depressed on one side, with the stigmatic point raised; surface slightly undulating to roughened, but not so rough as in many other Guatemalan varieties, dark green with numerous large yellowish or reddish brown dots; skin thick and woody, separating readily from the flesh, brittle, granular; flesh fine-grained, firm, deep cream-yellow in color, tinged with green near the skin, free from fiber or discoloration; flavor rich, pleasant; quality very good; seed broadly conic, medium sized, fitting tightly in the cavity with both seed-coats adhering closely. Season April to August at Hollywood, California.

Originated at Hollywood, California; first propagated in 1912, under the provisional names *Habersham* and *Dickey No. 2.*

Dickinson (Fig. 6). — Form oval to obovate, sometimes almost pyriform; size small to medium, weight 9 to 14 ounces, length 3½ inches, greatest breadth 2¾ inches; base not noticeably flattened, the long stem inserted in a very small and shallow cavity; apex rounded; surface very rough, verrucose or tuberculate around the base, dark purple in color with large, irregular, maroon dots; skin very thick, especially near the base, separating fairly readily from the flesh, coarsely granular, woody, brittle; flesh buttery, pale greenish yellow, free from fiber, of pleasant flavor; quality good; seed roundish oblate, medium sized, tight in the cavity, with both seed-coats adhering closely. Season June to October at Los Angeles, California.

Originated at Los Angeles, California; first propagated in 1912. Vigorous in growth and precocious in fruiting.

Lyon. — Form broad pyriform, indistinctly necked, and sometimes oblique at the apex; size above medium to large, weight 14 to 18 ounces, length about 5½ inches, greatest breadth 3½ inches; base narrow, the long stout stem inserted obliquely almost without depression; surface undulating to rough, bright green in color, with numerous small yellowish or russet dots; skin moderately thick, separating very readily from the flesh, coarsely granular, brittle; flesh smooth, firm, deep cream colored, tinged with green toward the skin, free from fiber discoloration, the flavor very rich and pleasant; quality very good; seed broad conic, medium small to medium in size, fitting tightly in the cavity with both seed-coats adhering closely. Season April to August at Hollywood, California.

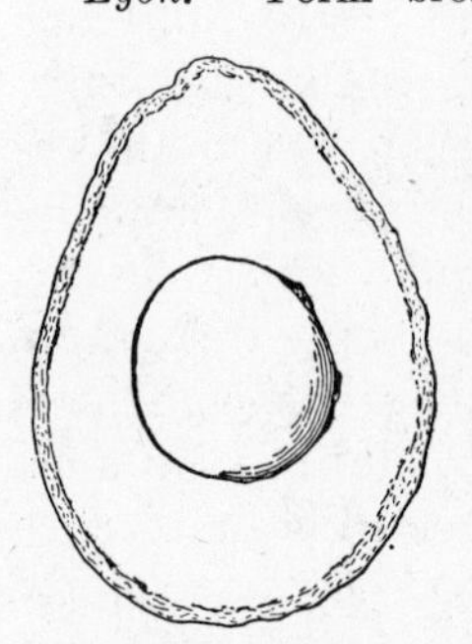

Fig. 6. The Dickinson avocado. (× ⅜)

Originated at Hollywood, California; first propagated in 1911. The tree is precocious in bearing, and the fruit is of excellent quality.

Sharpless. — Form slender pyriform to elongated pyriform with a long neck; size large to very large, weight 16 to 24 ounces, length 6 to 6½ inches, greatest breadth 3¼ inches; base very narrow, the long stem inserted obliquely without depression; apex rounded; surface slightly roughened or pitted, glossy, greenish purple to deep purple in color, with numerous yellowish dots; skin thick, separating readily from the flesh, granular and woody; flesh smooth, firm, cream colored, free from fiber discoloration, and of unusually rich pleasant flavor; quality excellent; seed oblate-oblique, small, weighing 1¼ ounces, fitting tightly in the cavity, with both seed-coats adhering closely. Season October to February at Santa Ana, California.

Originated near Santa Ana, California; first propagated in 1913. This is a fruit of fine quality, ripening very late in season.

Solano. — Form broadly obovate to oval; size above medium to large, weight 16 to 24 ounces, sometimes attaining to 28 ounces, length $5\frac{3}{4}$ inches, greatest breadth $3\frac{7}{8}$ inches; base rounded, with the long stem inserted obliquely without depression; apex oblique, slightly flattened; surface nearly smooth, somewhat glossy, bright green in color with numerous greenish yellow dots; skin moderately thick, separating readily from the flesh, granular; flesh firm, smooth, yellowish cream color, greenish near the skin, free from fiber discolorations and of mild pleasant flavor; quality fair; seed broadly conical to broadly ovate, small, fitting tightly in the cavity, with both seed-coats adhering closely. Season March to May at Los Angeles, California; October to November 15 at Miami, Florida.

Originated at Hollywood, California; first propagated in 1912. Productive, and a strong grower.

Spinks. — Form broadly obovate, or obconic; size extremely large, weighing from 18 to 34 ounces, length about 5 inches, greatest breadth about $4\frac{1}{2}$ inches; base narrow, rounded, with the rather short stout stem inserted almost squarely without depression; apex rounded; surface roughened, warty around the base, dark purple in color; skin thick, separating readily from the flesh, woody, granular, brittle; flesh firm, smooth, rich yellow in color, free from fiber, and of rich pleasant flavor; quality very good; seed nearly spherical, small, weighing 3 ounces, fitting tightly in the cavity with the seed-coats adhering closely. Season April to August at Duarte, California.

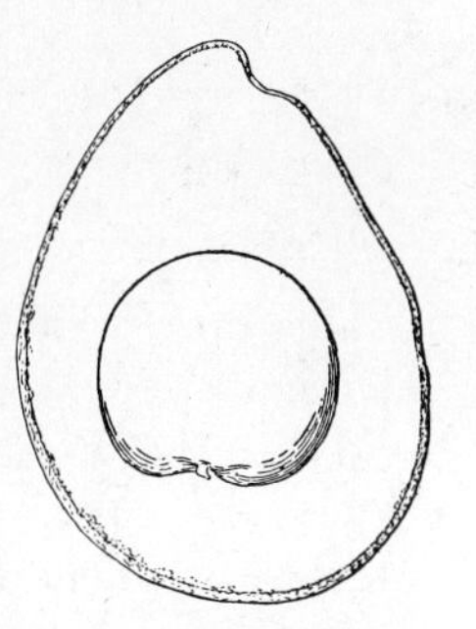

FIG. 7. The Taft avocado. ($\times \frac{1}{3}$)

Originated at Duarte, California; first propagated in 1915. The tree is vigorous and productive, and the fruit of excellent quality.

Taft (Fig. 7). — Form broad pyriform, slightly necked; size above medium to very large, weight 14 to 24 ounces, length 5 to $5\frac{1}{2}$ inches, greatest breadth $3\frac{3}{4}$ inches; base tapering, the long stem inserted obliquely without depression; apex rounded, with the stigmatic point raised; surface undulating to roughened around the base, deep green in color, with numerous yellowish dots; skin thick, separating very readily from the flesh, granular, rather pliable; flesh firm, smooth, light yellow in color with no trace of fiber discoloration; flavor unusually rich and pleasant; quality excellent; seed broadly conical, medium sized, fitting tightly in the cavity with both seed-coats adhering closely. Season May to October in southern California.

Originated at Orange, California; first propagated in 1912. The tree is a strong grower but has not proved very frost-resistant in Florida. Its bearing habits have not been satisfactory in California, but in Florida they promise to be better.

Taylor. — Form pyriform to obovate; size medium to large, weight 12 to 18 ounces, length 4 to $4\frac{1}{2}$ inches, greatest breadth $3\frac{1}{4}$ inches; base tapering, usually not distinctly necked, the long stem inserted obliquely almost without depression; apex rounded; surface undulating to rough, dull green in color, with numerous small yellowish dots; skin $\frac{1}{16}$ inch thick, separating readily from the flesh, granular and woody; flesh firm, smooth, yellowish cream color, pale green near the skin, free from fiber, and of fairly rich pleasant flavor; quality very good; seed conical, medium sized, tight in the cavity with both seed-coats adhering closely. Season January 15 to April 1, at Miami, Florida.

Originated at Miami, Florida; first propagated in 1914. This variety has been planted only in Florida, where it has proved to be vigorous and reasonably productive.

Mexican race.

This race, which embraces the hardiest avocados cultivated in the United States, is particularly valuable for regions too cold for the West Indian and Guatemalan varieties. It is extensively cultivated in the highlands of central and northern Mexico, whence seeds have been brought to California, resulting in numerous seedling trees scattered throughout the southern half of the state. In Florida it has never become popular, but good varieties have not been introduced until recently. Some of them promise to prove of value for the colder sections of that state.

From its native home in Mexico this race has spread to several other regions, most notably Chile, where it appears to be well known. It is the only race grown successfully in the Mediterranean region, trees having fruited at Algiers, in southern Spain, along the Riviera in southern France, and even in such a cold location as that of Rome. In tropical regions outside of Mexico it seems to be little cultivated.

The anise-like scent of the foliage and immature fruits is the most distinctive characteristic of the race and the one by which it is usually identified. The leaves are commonly smaller than those of the Guatemalan and West Indian races, and sharper at the apex. The fruit is small, 3 to 12 ounces in weight, rarely 15 or 16 ounces. The skin is thin, often no thicker than that of an apple, and usually smooth and glossy on the surface. The color varies from green to deep purple. The seed is commonly larger in proportion to the size of the fruit than in the Guatemalan race. The seed-coats are both thin, sometimes closely united and adhering to the cotyledons, sometimes separating as in the West Indian race. The flowers are heavily pubescent, and appear in winter or early spring, sometimes as early as November and usually not later than March. The fruit ripens in summer and autumn, commencing in June in Florida and August in California. Sometimes a second crop is produced from late flowers, ripening from March to May in California.

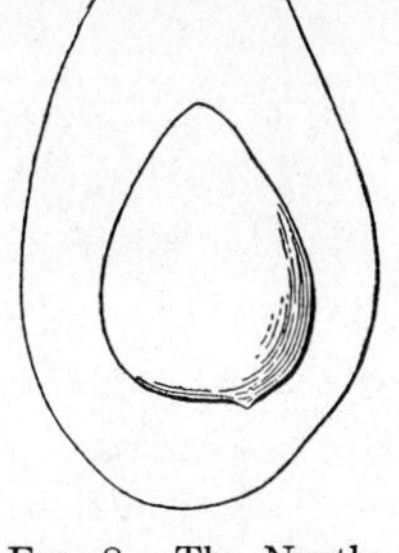

FIG. 8. The Northrop avocado. ($\times \frac{3}{7}$)

Northrop (Fig. 8). — Form obovate to pyriform, sometimes distinctly necked; size small, weight 5 to 8 ounces, length 4 inches, greatest breadth $2\frac{1}{2}$ inches; base narrow, the slender stem inserted squarely almost without depression; apex rounded; surface smooth, very glossy, deep purple in color, with a few small maroon dots; skin thin, adhering closely to the flesh, membranous; flesh buttery, cream yellow in color, practically free from fiber, and of rich flavor; quality good; seed oblong-conic, small, fitting tightly in the cavity with the seed-coats both adhering closely. Season October and November at Santa Ana, California, with a second crop maturing in April and May.

Originated near Santa Ana, California; first propagated in 1911 under the name *Eells*. The tree is vigorous, frost-resistant, and productive.

Puebla (Fig. 9). — Form obovoid, slightly oblique; size below medium to medium, weight 8 to 10 ounces, length $3\frac{1}{2}$ inches, greatest breadth $2\frac{7}{8}$ inches; base obliquely flattened, the stem inserted slightly to one side in a small shallow cavity; apex obliquely flattened but not prominently so; surface smooth, glossy, deep maroon-purple in color, with numerous reddish dots; skin less than $\frac{1}{32}$ inch thick, easily peeled from the flesh, firm in texture; flesh rich cream yellow near the seed, changing to pale green near the skin, buttery in texture, and of rich nutty flavor; quality very good; seed medium to large, tight in the cavity, with both seed-coats adhering closely to the cotyledons. Season December to February in southern California.

Originated at Atlixco, state of Puebla, Mexico; first propagated in 1911, in which year it was introduced into California. A vigorous and hardy variety, fruiting later in the season than most others of its race.

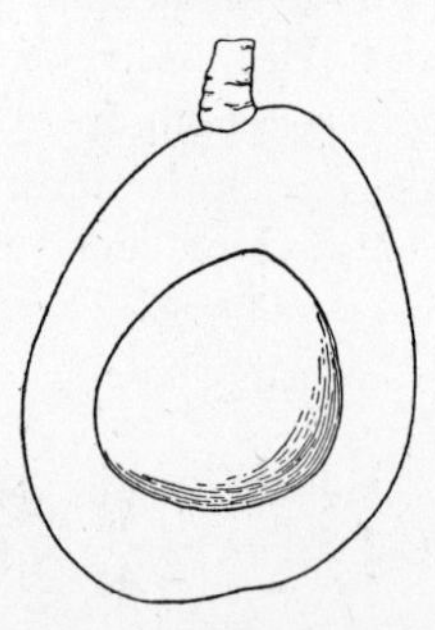

FIG. 9. The Puebla avocado. ($\times \frac{3}{8}$)

Hybrids.

This group has been established to include hybrids between *Persea drymifolia* (the Mexican race of horticulture) and *P. americana* (the Guatemalan and West Indian races). Fuerte is the only variety which at present falls within it, and even this is not definitely known to be a hybrid. It bears, however, many evidences of hybridity, and cannot rightly be classified either with the Mexican or the Guatemalan races.

Fuerte. — Form pyriform (not necked) to oblong; size below medium to above medium, weight 10 to 16 ounces, length 4 to $4\frac{1}{2}$ inches, greatest breadth $2\frac{1}{2}$ to $2\frac{7}{8}$ inches; base pointed; the stem inserted obliquely in a small shallow cavity; apex obliquely flattened, depressed around the stigmatic point; surface pebbled, sometimes slightly wrinkled around the stem, dull green, with numerous small yellow dots; skin about $\frac{1}{24}$ inch thick, separating readily from the flesh, pliable and leathery in texture; flesh rich cream yellow in color, greenish near the skin, of smooth buttery texture, and very rich flavor; quality excellent; seed small, tight in cavity, seed-coats closely surrounding cotyledons. Season January to August in southern California.

Originated at Atlixco, state of Puebla, Mexico; first propagated in 1911, in which year it was introduced into California. An unusual variety, apparently a hybrid between the Guatemalan race of *Persea americana* and the Mexican race (*P. drymifolia*). It is characterized by great vigor of growth, hardiness, good productiveness, and a long season of ripening differing from that of nearly all Guatemalan and Mexican varieties. The fruit contains as much as 30 per cent of fat, and is of very pleasant flavor.

CHAPTER III

THE MANGO

Plates V–VI

AKBAR, the Mughal emperor who reigned in northern India from 1556 to 1605, planted near Darbhanga the Lakh Bagh, an orchard of a hundred thousand mango trees. Nothing, perhaps, more eloquently attests the importance of this fruit and the esteem in which it has long been held than this immense planting, made at a time when large orchards of fruit-trees were almost unknown. Three hundred years after they were set out, the English horticulturist Charles Maries found some of these trees still in vigorous condition.

Few other fruits have the historic background of the mango, and few others are so inextricably connected with the folk-lore and religious ceremonies of a great people. Buddha himself was presented with a mango grove, that he might find repose beneath its grateful shade. The Turkoman poet Amir Khusrau, whose grass-covered tomb is still venerated at Delhi, wrote to this effect in Persian verse during the reign of Muhammad Tughlak Shah (1325–1351):

> The mango is the pride of the Garden,
> The choicest fruit of Hindustan.
> Other fruits we are content to eat when ripe,
> But the mango is good in all stages of growth.

In more recent times, British authors have not hesitated to lavish praise on this oriental King of Fruits. Fryer, in 1673, wrote regarding mangos that "The Apples of the Hesperides are but Fables to them; for Taste, the Nectarine, Peach, and Apricot fall short." Hamilton, who wrote in 1727, went even farther than this; he declares "The Goa mango is reckoned the largest and most delicious to the taste of any in the world, and I may add, the wholesomest and best tasted of any Fruit in the World."

These few quotations will suffice to show the long established prestige of the mango in its native home. After the development of trade between India and the outside world, its cultivation spread to other countries. At the present time the mango is a fruit of greater importance to millions throughout the tropics than is the apple to temperate North America.

In the past twenty years choice budded or grafted varieties have been planted in Florida and the West Indies, and the fruit has begun to appear in the markets of the North. The rich spicy flavor of the mango, its peculiarly tempting fragrance, and the beautiful shades of color which characterize many varieties, make it one of the most attractive dessert fruits on the American market.

In many instances travelers have made the acquaintance of this fruit through some of the fibrous seedlings which abound in all parts of the tropical world, and as a result may have formed an aversion for it difficult to overcome. It is only in the superb grafted varieties of the Orient, the product of centuries of improvement, that the mango exhibits its best qualities. There is more difference between an ordinary seedling and a grafted Alphonse than there is between a crab-apple and a Gravenstein.

Since the introduction of these choice varieties into tropical and subtropical America, mango culture has there taken on a new aspect. Previously limited to the production of seedling fruits usually of inferior quality though valuable for local consumption, the industry is now being developed with a view to supplying northern markets with fancy fruit.

While many of the common seedlings yield abundantly with no cultural attention, the production of fine grafted mangos is attended by certain cultural difficulties, some of which are yet to be overcome. Anthracnose, a fungous disease related to the wither-tip of citrus fruits, is a serious pest in many regions. The greatest difficulty, however, is the tendency of many of the choice Indian varieties to bear irregularly. In some cases good crops are produced not oftener than once in three or four years. Thorough investigation of cultural requirements together with experimental planting of many varieties is bringing to light the most productive kinds and the proper methods to be employed in their cultivation.

Botanical Description

The family Anacardiaceae, to which the mango belongs, includes a large number of plants found within the tropics and a few growing in the Mediterranean region, Japan, and temperate North America. The best known relatives of the mango are, probably, the cashew (*Anacardium occidentale*), widely cultivated in the tropics for its edible fruit; the pistachio nut (*Pistacia vera*) of the Mediterranean region; several species of Spondias which are grown for their edible fruits; the obnoxious poison ivy (*Rhus Toxicodendron*) of the United States; and the so-called peppertree, *Schinus molle,* familiar in the gardens and streets of southern California.

The cultivated mangos are usually considered as belonging to a single species, *Mangifera indica.* It has been pointed out by certain botanists, however, that probably other species have entered into the composition of cultivated forms. C. L. Blume[1] says that they have developed from many species scattered through tropical Asia, mainly in the Malay Archi-

[1] Mus. Lugd. Bat. 1, 190–191.

pelago. It is probable that some of the groups or races recognized as horticulturally distinct represent other species than *M. indica,* or hybrids. A species which has been regarded particularly as one of the ancestors of cultivated forms is *M. laurina.*

About forty species of the genus Mangifera are recognized by botanists, most of them coming from the Malayan region. Several are cultivated for their fruits, although on a limited scale. Some of them are perhaps not distinct from *M. indica,* as at present recognized. The following species merit consideration in connection with mango culture (the notes are based mainly on Hooker's Flora of British India and Blume's Museum Botanicum Lugduno-Batavum):

Mangifera altissima, Blanco. PAHUTAN. Indigenous to the Philippine Islands. Fruit large, closely resembling that of the mango, edible.

M. cæsia, Jack. BINJAI. Wild and cultivated in Malacca, Sumatra, and Java. Fruit oblong-obovate, reddish white in color, not of good quality.

M. fœtida, Lour. BACHANG. AMBATJANG. Distributed throughout the Malay Archipelago. Fruit variable in form, not compressed, green, with yellow flesh of disagreeable odor. Not esteemed, although sometimes eaten.

M. laurina, Blume. MANGA MONJET, MANGA PARI, etc. Wild and cultivated in the Malay Archipelago. Fruit elliptic-oblique, the size of a plum. Blume describes numerous varieties grown in Java and other islands. Certainly very close to *M. indica.*

M. odorata, Griff. KUWINI. BUMBUM. Wild in Malacca, cultivated in Java. Fruit oblong, yellowish green, the flesh yellow, sweet, with no turpentine flavor. "Often planted by the natives, who eat the fruit."

M. sylvatica, Roxb. Tropical Nipal, Sikkim Himalaya, and the Khasia mountains of India; Andaman Islands. The foliage is like that of the common mango; the fruit, ovoid, beaked, differs only slightly from that of *M. indica.*

M. verticillata, Rob. BAŬNO. Wild in the southern Philippine Islands. Fruit "very juicy, rich, subacid, quite aromatic, of excellent flavor."

M. zeylanica, Hook. f. Wild in Ceylon. Closely resembles *M. indica,* but is considered by Hooker to differ in habit and foliage, and in the character of the flowers. Fruit said to be small, edible.

The mango tree is evergreen. Seedlings on deep rich soils often reach immense size. One measured in Bahia, Brazil, had a spread of 125 feet and a trunk 25 feet in circumference. Trees believed to be more than a hundred years old are common in the Orient; not a few such are to be seen in tropical America, but the comparatively recent introduction of the mango into this hemisphere makes old trees less common than in India. Budded or grafted trees do not grow so large as do seedlings, and are probably shorter lived.

The crown is sometimes broad and round-topped; in other instances it is oval, giving the tree an erect or even slender form. The leaves are lanceolate, commonly to 12 inches in length, rigid, deep green, almost glossy, borne upon slender petioles 1 to 4 inches long. Growth is not continuous throughout a long season, but takes place in frequently recurring periods, each of which is followed by a period of inactivity. These periods of growth (commonly termed "flushes" by horticulturists) do not occur at fixed intervals, and in fact the whole tree does not always break out in new growth at the same time. It is a common occurrence for one side of the tree to be in active growth while the other side is dormant. The young leaves are usually reddish or coppery, and often hang limply from the ends of the branchlets. After the growth has begun to mature, they become turgid and soon lose their reddish color.

The small pinkish white flowers are borne in large panicles at the ends of the branchlets. In Florida and the West Indies the flowering season extends from December to April. Sometimes the trees bloom two or three times during the season. More than 4000 flowers have been counted on a single panicle, but not all of these are capable of developing into fruits, since the mango is "polygamous," that is, it produces two kinds of flowers: perfect ones having both stamens and pistils, and others which are unisexual. The unisexual flowers, which are staminate, commonly outnumber the perfect ones; usually, however, there is only one pollen-bearing stamen in each flower. The perfect blossoms are easily distinguished from the staminate by the presence in the former of the small greenish yellow ovary surmounting the white disk in the center.

The fruit varies greatly in size and character. The smallest kinds are no larger than good-sized plums, while the largest are 4 or 5 pounds in weight. The form is oval, heart-shaped, kidney-shaped, round, or long and slender. The skin is smooth, thicker than that of a peach, commonly yellow on the surface but varying greatly in color. Some varieties are delicately colored, deep yellow or apricot with a crimson blush on one cheek; others are an unattractive green even when ripe. The color depends to a certain extent on the climate in which the fruit is grown. The aroma is often spicy and alluring, indicative of the flavor of the fruit. The flesh is yellow or orange in color, juicy, often fibrous in seedlings and inferior budded varieties, but in the best sorts entirely free from fiber and of smooth melting texture. The seed is large and flattened, its tough, woody husk or outer covering inclosing a white kernel. The flavor of the mango has been likened to a combination of apricot and pineapple, yet it cannot be described accurately by any such comparison. It is rich and luscious in the best varieties, sweet, but with sufficient acidity and spiciness to prevent its cloying the palate.

History and Distribution

Alphonse DeCandolle considered it probable that the mango could be included among the fruits which have been cultivated by man for 4000 years. Its prominence in Hindu mythology and religious observance leaves no doubt as to its antiquity, while its economic importance in ancient times is suggested by one of the Sanskrit names, *am,* which has an alternative meaning of provisions or victuals.

Dymock, Warden, and Hooper (Pharmacographia Indica) give the following résumé of its position in the intellectual life of the Hindus:

"The mango, in Sanskrit *Amra*, *Chuta* and *Sahakara*, is said to be a transformation of Prajapati (lord of creatures), an epithet in the Veda originally applied to Savitri, Soma, Tvashtri, Hirangagarbha, Indra, and Agni, but afterwards the name of a separate god presiding over procreation. (Manu. xii, 121.) In more recent hymns and Brahmanas Prajapati is identified with the universe.

"The tree provides one of the pancha-pallava or aggregate of five sprigs used in Hindu ceremonial, and its flowers are used in Shiva worship on the Shivaratri. It is also a favorite of the Indian poets. The flower is invoked in the sixth act of Sakuntala as one of the five arrows of Kamadeva. In the travels of the Buddhist pilgrims Fah-hien and Sung-yun (translated by Beal) a mango grove (Amravana) is mentioned which was presented by Amradarika to Buddha in order that he might use it as a place of repose. This Amradarika, a kind of Buddhic Magdalen, was the daughter of the mango tree. In the Indian story of Surya Bai (see Cox, Myth. of the Arian Nations) the daughter of the sun is represented as persecuted by a sorceress, to escape from whom she became a golden Lotus. The king fell in love with the flower, which was then burnt by the sorceress. From its ashes grew a mango tree, and the king fell in love first with its flower, and then with its fruit; when ripe the fruit fell to the ground, and from it emerged the daughter of the sun (Surya Bai), who was recognized by the prince as his long lost wife."

When introduced into regions where climatic conditions are favorable, the mango rapidly becomes naturalized and takes on the appearance of a wild plant. This fact, together with the long period of time during which it has been cultivated throughout India, makes it difficult to determine the original home of the species.

Sir Joseph Hooker (Flora of British India) considered the mango to be indigenous in the tropical Himalayan region, from Kumaon to the Bhutan hills and the valleys of Behar, the Khasia mountains, Burma, Oudh, and the Western peninsula from Kandeish southwards. He adds, "It is difficult to say whether so common a tree is wild or not in a given locality, but there seems to be little doubt that it is indigenous in the localities enumerated." Dietrich Brandis (Indian Trees) says it is indigenous in Burma, the Western Ghats, in the Khasia hills, Sikkim, and in the ravines of the Satpuras. R. S. Hole, of the Imperial Forest Research Institute at Dehra Dun, considers that the so-called wild mangos which are found in many parts of India are mostly forms escaped from cultivation, as shown by the fact that they are always near streams or foot-paths in the jungle, where seeds have been thrown by passing natives.

Alphonse DeCandolle says: "It is impossible to doubt that it is a native of the south of Asia and of the Malay Archipelago, when we see the multitude of varieties cultivated in those countries, the number of ancient names, in particular a Sanskrit name, its abundance in the gardens of Bengal, of the Dekkan peninsula, and of Ceylon, even in Rheede's time. . . . The true mango is indicated by modern authors as wild in the forests of Ceylon, the regions at the base of the Himalayas, especially towards the east, in Arracan, Pegu, and the Andaman Isles. Miquel does not mention it as wild in any of the islands of the Malay Archipelago. In spite of its growing in Ceylon, and the indications, less positive certainly, of Sir Joseph Hooker in the Flora of British India, the species is probably rare or only naturalized in the Indian peninsula."

Most species of Mangifera are natives of the Malayan region. Sumatra in particular is the home of several. While it is known that the mango has been cultivated in western India since a remote day, and we find it to-day naturalized in many places, it seems probable that its native home is to be sought in eastern India, Assam, Burma, or possibly farther in the Malayan region.

The Chinese traveler Hwen T'sang, who visited Hindustan between 632 and 645 A.D., was the first person, so far as known, to bring the mango to the notice of the outside world. He speaks of it as *an-mo-lo,* which Yule and Burnell consider a phonetization of the Sanskrit name *amra.* Several centuries later, in 1328, Friar Jordanus, who had visited the Konkan and learned to appreciate the progenitors of the Goa and Bombay mangos, wrote, "There is another tree which bears a fruit the size of a large plum, which they call *aniba.*" He found it "sweet and pleasant." The common name which he used is a variation of the north Indian *am* or *amba.* Six years later (1334) Ibn Batuta wrote that "the mango tree ('anba) resembles an orange tree, but is larger and more leafy; no other tree gives so much shade." John de Marignolli, in 1349, says, "They also have another tree called *amburan,* having a fruit of excellent fragrance and flavor, somewhat like a peach." Varthema, in 1510, mentioned the mango briefly, using the name *amba.* Sultan Baber, who wrote in 1526, is the first to distinguish between choice and inferior varieties. He says, "Of the vegetable productions peculiar to Hindustan one is the mango, (*ambeh*). . . . Such mangos as are good are excellent."

The island of Ormuz, in the mouth of the Persian Gulf, was settled in early days by the Portuguese and became one of the great emporiums of the East. Garcia de Orta, a Portuguese from Goa, wrote in 1563 that the mangos of Ormuz were the finest in the Orient, surpassing those of India. It is probable, however, that the mangos known at Ormuz were not grown on the island itself, since it has very little arable land and water is exceedingly scarce. The Cronica dos Reys Dormuz (1569) says that mangos were brought to Ormuz from Arabia and Persia. Later, in 1622, P. della Valle speaks of the mangos grown on the Persian mainland at Minao, only a few miles from Ormuz.

The Ain-i-Akbari, an encyclopedic work written during the reign of Akbar (about 1590), contains a lengthy account of the mango. Akbar, it may be remembered, was the Mughal emperor who planted the Lakh Bagh at Darbhanga, and in other ways stimulated the cultivation of fruit-trees throughout northern India. Abu-l Fazl-i-'Allami, author of the Ain (translated by Blochmann), writes:

"The Persians call this fruit Naghzak, as appears from a verse of Khusrau. This fruit is unrivalled in color, smell, and taste; and some of the gourmands of Turan and Iran place it above muskmelons and grapes. In shape it resembles an apricot, or a quince, or a pear, or a melon, and weighs even one ser and upwards. There are green, yellow, red, variegated, sweet and subacid mangos. The tree looks well, especially when young; it is larger than a nut tree, and its leaves resemble those of a willow, but are larger. The new leaves appear soon after the fall of the old ones in the autumn, and look green and yellow, orange, peach-colored, and bright red. The flower, which opens in the spring, resembles that of the vine, has a good smell, and looks very curious. . . . The fruit is generally taken down when unripe, and kept in a particular manner. Mangos ripened in this manner are much finer. They commence mostly to ripen during summer and are fit to be eaten during the rains; others commence in the rainy season and are ripe in the beginning of winter; the latter are called Bhadiyyah. Some trees bloom and yield fruit the whole year; but this is rare. Others commence to ripen, although they look unripe; they must be quickly taken down, else the sweetness would produce worms. Mangos are to be found everywhere in India, especially in Bengal, Gujrat, Malwah, Khandesh, and the Dekhan. They are rarer in the Panjab, where their cultivation has, however, increased since his Majesty made Lahor his capital. A young tree will bear fruit after four years. They also put milk and treacle around the tree, which makes the fruits sweeter. Some trees yield in one year a rich harvest, and less in the next; others yield for one year no fruit at all. . . ."

The name mango, by which this fruit is known to English-speaking as well as Spanish-speaking peoples, is derived from the Portuguese *manga*. According to Yule and Burnell, the Tamil name *man-kay* or *man-gay* is the original of the word, the Portuguese having formed *manga* from this when they settled in western India. Skeat traces the origin of the name to the Malayan *manga*, but other writers consider the latter to have been introduced into the Malay Archipelago from India. The name *mango* is used in German and Italian, while the Dutch have adopted *manga* or *mangga*, and the French form is *mangue*.

In the Malay Archipelago and in many parts of Polynesia mangos are plentiful. W. E. Safford[1] writes, "The mango tree is not well established in Guam. There are few trees on the Island, but these produce fruit of the finest quality. Guam mangos are large, sweet, fleshy, juicy, and almost entirely free from the fiber and flavor which so often characterize the fruit." Excellent mangos were formerly shipped from the French island of Tahiti to San Francisco. Many choice varieties have been planted in the Hawaiian Islands. J. E. Higgins has written a bulletin on mango culture in this region.

On the tropical coast of Africa, extending south to the Cape of Good Hope, and in Madagascar, mangos are common. The French island of Réunion is the original home of several varieties now cultivated in the West Indies and Florida.

In Queensland, Australia, attention has been given to the asexual propagation of this fruit, and a limited number of choice Indian varieties have been introduced.

In the Mediterranean region the species is not entirely successful. Trees are reported to have produced fruit in several localities, but nowhere have they become commonly grown. In Madeira and the Canary Islands they are more at home; Captain Cook, when on his first voyage of discovery, reported in 1768 that mangos grew almost spontaneously in Madeira. C. H. Gable, who has recently worked on the island, says there are now only a few trees to be found, but that these bear profusely.

The Portuguese are given the credit for bringing the mango to America. It is believed to have been first planted at Bahia, Brazil, at an uncertain date probably not earlier than 1700. Captain Cook found in 1768 that the fruit was produced in great abundance at Rio de Janeiro. In the West Indies it was first introduced at Barbados in 1742 or thereabouts, the "tree or its seed" having been brought from Rio de Janeiro. It did

[1] Useful Plants of Guam.

not reach Jamaica until 1782. Its introduction into the latter island is described by Bryan Edwards:[1] "This plant, with several others, as well as different kinds of Seeds, were found on board a French ship (bound from the Isle de France for Hispaniola) taken by Captain Marshall of his Majesty's Ship *Flora,* one of Lord Rodney's Squadron, in June, 1782, and sent as a Prize to this island. By Captain Marshall, with Lord Rodney's approbation, the whole collection was deposited in Mr. East's garden, where they have been cultivated with great assiduity and success." Thirty-two years after its introduction, John Lunan stated that the mango had become one of the commonest fruit-trees of Jamaica.

It is said to have been introduced into Mexico at the same time as the coffee plant, early in the nineteenth century, the introducer having been D. Juan Antonio Gomez of Córdoba. It is evident that Mexico has received mangos from two sources; some from the West Indies, and others from the Philippines, brought by the Spanish galleons which traded in early times between Acapulco and Manila.

The cultivation of the mango under glass in Europe was attempted at an early day. A writer in Curtis' Botanical Magazine in 1850 says: "The mango is recorded to have been grown in the hothouses of this country at least 160 years ago, but it is only within the last twenty years that it has come into notice as a fruit capable of being brought to perfection in England. The first and, we believe, the most successful attempt was made by the late Earl of Powis, in his garden at Walcot, where he had a lofty hothouse 400 feet long and between 30 and 40 feet wide constructed for the cultivation of the mango and other rare and tropical fruits; but within these last few years we have known it to bear fruit in other gardens."

In the United States, cultivation of the mango is limited to southern Florida and southern California. It is believed the

[1] History of the West Indies, 1793.

PLATE V. *Left,* inflorescence of the Alphonse mango; *right,* a Cuban mango-vender.

species was first introduced into the former state by Henry Perrine, who sent plants from Mexico to his grant of land below Miami in 1833. These trees, however, perished from neglect after Perrine's death, and many years passed before another introduction was made. According to P. J. Wester, the second and successful introduction was in 1861 or 1862, by Fletcher of Miami. The trees introduced in these early years were seedlings. In 1885 Rev. D. G. Watt of Pinellas made an attempt to introduce the choice grafted varieties of India. According to P. N. Reasoner,[1] Watt obtained from Calcutta eight plants of the two best sorts, Bombay and Malda. "They were nearly three months on the passage, and when the case was opened five were dead; another died soon after, and the two remaining plants were starting nicely, when the freeze destroyed them entirely." In 1888 Herbert Beck of St. Petersburg obtained a shipment of thirty-five inarched trees from Calcutta. This shipment included the following varieties: "Bombay No. 23, Bombay No. 24, Chuckchokia, Arbuthnot, Gopàlbhog, Singapore, and Alphonse." In the latter part of 1889 Beck reported to the Department of Agriculture that all but seven of the trees had died. Further details regarding this importation are lacking, but it is not believed that any of the trees lived to produce fruit.

On November 1, 1889, the Division of Pomology at Washington received through Consul B. F. Farnham of Bombay, India, a shipment of six varieties, as follows: "Alphonse, Banchore, Banchore of Dhiren, Devarubria, Mulgoba, and Pirie." The trees were obtained from G. Marshall Woodrow, at Poona. After their arrival in this country they were forwarded to horticulturists on Lake Worth, Florida. Most of the trees succumbed to successive freezes, but in 1898 Elbridge Gale reported that one Alphonse sent to Brelsford Brothers was still alive, but was not doing well; and that of the five trees sent to himself only one, a Mulgoba, had survived. This

[1] Division of Pomology, Bull. 1.

tree began to bear in 1898, and is still productive, although it has not borne large crops in recent years. The superior quality of its fruit furnished the needed stimulus to the development of mango culture in this country, and considerable numbers of Mulgobas were soon propagated and planted along the lower east coast of Florida. Recently, numerous other Indian varieties have fruited in that state, some of them more valuable from a commercial standpoint than Mulgoba, so that the latter probably will not retain the prominent position which it has held.

As regards California, the exact date at which the mango was first introduced is not known, but it is believed by F. Franceschi that it was first planted at Santa Barbara, between 1880 and 1885.

Composition and Uses of the Fruit

The mango contains much sugar. The proportions of other constituents, such as acids and protein, are low in the ripe fruit. The following table, from analyses made in Hawaii by Alice R. Thompson, shows the composition of three well-known Indian varieties:

Table II. Composition of the Mango

Variety	Total Solids	Ash	Acids	Protein	Total Sugars	Fat
	%	%	%	%	%	%
Pairi	20.52	0.343	0.221	0.456	14.78	0.032
Alphonse	20.92	.469	.373	.919	14.64	.149
Totapari	15.27	.277	.578	.475	11.48	.065

In commenting on these and other data, Miss Thompson[1] says: "The total solids are high for the average fresh fruit;

[1] Hawaii Exp. Sta. Rept., 1914.

the total sugars vary from 11 to 20 per cent, according to the variety. In all samples the sucrose is the principal sugar present. The protein in several varieties is a little higher than is usual in fruits. The acidity varies and is as much as 0.5 per cent in one variety. Qualitative tests showed the presence of considerable amounts of tannin, but no starch was apparent."

The unripe fruit is characterized by the presence of malic and tartaric acids in considerable quantities. An analysis published in the Pharmacographia Indica shows the percentage of tartaric (with a trace of citric) to be 7.04, and the remaining free acid as malic, 12.66.

The Agricultural News (Barbados, September 27, 1913) published a comparison of the chemical composition of the apple with that of the Carabao mango, one of the principal Philippine sorts. It was found that "The former fruit contains 14.96 per cent solids, whereas the mango contains 17.2. In regard to sugar (total) the first-named fruit contains about 7.58 per cent, whereas the mango has 13.24. As regards protein (nitrogenous matter) the apple has about 0.22 per cent, and the mango 0.22 per cent also. The total acidity in the apple is 1.04 per cent, whereas in the mango it is only 0.14 per cent. In making these comparisons we have purposely taken one of the less nutritious varieties of mango, and it may safely be said that in regard to chemical composition the balance is on the side of the mango."

While the mango is most commonly eaten as a fresh fruit, it can be utilized in many different ways. Sir George Watt [1] says:

"Besides being eaten as a ripe fruit, numerous preparations are made of it. When green it is cut into slices, and after extraction of the stone, is put into curries, or made into pickles with other ingredients or into preserves and jellies. When young and green it is boiled, strained, mixed with milk and

[1] Commercial Products of India.

sugar, and thus prepared as the custard known as *mangophul*, or dried and made into the native *ambchur*. When very young it may be cut into small pieces and eaten in salad. So again, the ripe fruit is used in curries and salads, and the expressed juice when spread on plates and allowed to dry is formed into the thin cakes known as *ambsath*."

In the United States, mangos have up to the present been used chiefly as dessert fruits. To a less extent they have been made into chutney, — the spicy sauce well known to all those who have traveled in the Orient, — preserves, sauces, and pies. For these purposes the fruit is taken before fully ripe. The "mango pickles" sold in the northern United States are not made from the mango, but from a sweet pepper; the use of the name mango in this connection is unwarranted.

Mangos are canned in the same manner as peaches. Recently a firm at Muzaffarpur, India, has undertaken to develop an export trade in preserved mangos. About 18,000 cans were shipped to England in a single year. Consul General William H. Michael said of the product, "I have opened one can of the Bombay Extra mangos and find that they are carefully packed and retain their flavor as well as could be expected of this sort of fruit. In fact they are as well preserved and retain their flavor quite as well as do peaches canned in California."

Hindu and Muhammadan writers on Materia Medica discuss at length the medicinal virtues of the mango:

"Shortly, we may say that they consider the ripe fruit to be invigorating and refreshing, fattening, and slightly laxative and diuretic; but the rind and fiber, as well as the unripe fruit, to be astringent and acid. The latter when pickled is much used on account of its stomachic and appetizing qualities. Unripe mangos peeled and cut from the stone and dried in the sun form the well-known Amchur or Ambosí (Amrapesi, Sans.,) so largely used in India as an article of diet; as its acidity is chiefly due to the presence of citric acid, it is a valuable anti-scorbutic; it is also called Am-ki-chhitta and Am-khushk. The blossom, kernel, and bark are considered to be cold, dry and astringent, and are used in diarrhœa, etc. The smoke of the burning leaves is supposed to have a curative effect in some affections of the throat. According to the author of the Makhzan, the Hindus make a confection of the baked pulp of the unripe fruit mixed with sugar, which in time of plague or cholera they take internally and rub all over the body; it is also stated in the same work that the midribs of the leaves calcined are used to remove warts on the eyelids." (Dymock, Warden, and Hooper.)

Climate and Soil

While the mango grows in humid tropical regions subject to heavy rains throughout the year, it is not successfully cultivated for its fruit under these conditions. It requires the stimulus of a dry season to fruit abundantly. To a certain extent this stimulus can be given by artificial means, but there can be no doubt that the best regions for commercial mango culture are those in which there is a well-marked dry season occurring at the proper time of year.

This is illustrated by conditions in India. Lower Bengal is a humid region in which moisture-loving tropical plants are completely at home. Mango trees in this region are ragged in appearance, with foliage of an unhealthy color, and the fruit does not ripen well. In sharp contrast, the trees at Saharanpur, on the dry plains of northern India, are vigorous and stocky in habit, with abundant foliage of rich green color. They fruit more profusely than those in the moist lowlands, and the fruit ripens perfectly. Saharanpur lies at an elevation of 1000 feet, and has an annual rainfall of about thirty-five inches. During the season when mangos are ripening, no rain falls and the air is hot and dry. Temperature of 100° F., continued throughout day and night, are common. The monsoon, or rainy season, lasts but a few months.

The total amount of rainfall is not so important as the season during which it occurs. Where the dry season coincides with the normal flowering time of the mango, good crops of fruit can be expected, but it seems doubtful whether the finer grafted mangos can be cultivated successfully in regions where there is much precipitation during the flowering season. Some of the seedling races will fruit under these conditions, but the choice Indian varieties are more exacting in their climatic requirements.

On this point G. N. Collins[1] states: "The fact that the tree may thrive in a given locality and yet fail to produce fruit should always be kept in mind. It may be considered as proven that the mango will be prolific only in regions subjected to a considerable dry season. On the moist north side of Porto Rico the trees grow luxuriantly, but they are not nearly so prolific nor is the fruit of such good quality as on the dry south side, and in the very dry region about Yauco and at Cabo Rojo the fruit seemed at its best, while its abundance was attested by the fact that fine fruit was selling as low as 12 for a cent. In Guatemala and Mexico the mango was found at its best only in regions where severe dry seasons prevailed."

Fawcett and Harris[2] report similar conditions in Jamaica. They say: "Although the mango grows freely everywhere, it is not a fruitful tree in every district; in the southern plains and the low, dry limestone hills it produces enormous crops year after year, and very often two crops a year, the main crop from May to August, and the second crop later in the year. . . . In humid districts and along the northern coast the tree is not at all fruitful, except in very dry years, and in the wet districts like Castleton it rarely fruits."

In the Botanic Garden at Rio de Janeiro, Brazil, there is a magnificent avenue of mango trees planted by the emperor Dom João VI more than a century ago. So far as known these trees have never matured any fruits. They blossom, and occasionally set fruits, but the latter invariably drop off before reaching maturity. J. C. Willis, former director of the garden, attributes this to the fact that they are planted on low wet ground.

[1] Bull. 28, U. S. Dept. Agr.
[2] Bull. of the Bot. Dept., vol. 8, 1901.

Other mango trees in the immediate vicinity but on higher ground produce fruit regularly.

Mangos can be grown successfully on soils of several different types. In Porto Rico deep sandy loam has given excellent results. On this soil the tree makes rapid growth and attains great size. The sandy soils of southern Florida have proved satisfactory. Clay, provided it is well drained, seems to be good.

In India, some of the best mango districts are situated on the great Indo-Gangetic plain, where the soil is a deep, rich alluvial loam. This may perhaps be considered the best of all mango soils. An analysis of surface soil from the mango orchards in the Saharanpur Botanic Garden shows that it contains:

Lime (CaO)	1.20 %
Magnesia (MgO)	1.18 %
Potash (K_2O)	2.73 %
Phosphoric acid (P_2O_5)	0.18 %
Nitrogen	0.105%

C. F. Kinman[1] says:

"A shallow soil underlain with stone or hardpan, although sufficiently deep to produce shrubs or other low-growing wild vegetation, will not satisfy the needs of the deep rooted mango, whose growth in such ground will be slow and its yield poor, at least after the first few years. The application of fertilizers, however, will materially decrease the depth of the soil required. . . . Mango trees are often found on very light, unfertile sand, which may be a few feet in depth, and still produce flourishing growth if the subsoil is suitable. As the mango, like most other fruit trees, thrives best on a deep loose loam with good drainage and a high percentage of humus, those who intend planting it commercially should secure, if possible, this type of soil."

[1] Porto Rico Agr. Exp. Sta. Bull., 24.

Much more important than the mechanical or chemical composition, in most cases, is the drainage of the land. The mango avenue in the Botanic Garden at Rio de Janeiro illustrates this. If the subsoil is permanently wet or poorly drained, the tree cannot be expected to fruit profusely.

While the mango is more susceptible to frost than the hardier races of the avocado, mature trees have withstood temperatures below the freezing point without injury. In general it may be said that most varieties, if not in active growth at the time cold weather strikes them, will withstand 28° or 29° above zero, provided such temperatures are not of long duration. Young trees in vigorous growth may be injured seriously by a temperature of 32°. At Miami, Florida, five-year-old trees of one or two varieties were killed outright by a freeze of 26.5°. Old seedling trees have gone through temperatures lower than this without losing more than the smallest branches. The cultivated kinds show slight differences in hardiness. Observations have been made at Saharanpur and lists drawn up showing the relative susceptibility to frost of many varieties. The vagaries of the 1917 freeze in southern Florida, however, have resulted in an impression that such lists are not altogether dependable, and that much depends on local conditions, the physiological state of the tree, and other factors as yet not understood.

The mango resists heavy winds much better than does the avocado. The wood is tough, and ordinarily the tree (except in the Cambodiana group) assumes a low compact form if not crowded. It is not essential, therefore, that the young tree be trained with a view to making it of such form that it will be able to withstand a hurricane or cyclone.

Mango culture in California presents some unusual aspects. Although experience is limited, it is apparent that the great variations in temperature between night and day, coupled with the comparatively cold winters, have the effect of retarding the growth of the tree, as well as preventing the rapid development of the fruit. The dryness of the climate, on the other hand, makes the tree bear at an early age and yield very heavily. In certain situations near the sea, the summers are so cool that the fruit does not ripen properly. This has proved to be true of Santa Barbara, Hollywood, and San Diego. In the foothill regions, where the summers are warmer than near the sea, good mangos have been produced. It is necessary to protect the trees from frost while they are young; even large trees are sometimes injured by an unusually severe winter. All of the mangos which have fruited in California up to the present time have been seedlings or inferior budded varieties: only recently have budded trees of choice varieties been planted. Localities such as Glendora and Monrovia, which have warm summers and are comparatively free from winter frosts, are probably the most suitable for mango culture. The hot summer weather of such districts hastens the development of the fruit and brings it to maturity before the onset of cool weather in autumn.

Commercially, mango culture has never been considered promising in California. It should be possible to produce good fruit on a limited scale in a few of the most protected situations, but the greater number of mango trees which have been planted in the state have been killed by frost.

In Florida, commercial mango culture is successful from Palm Beach on the east coast and Punta Gorda on the west coast down to the southern end of the peninsula. There are a few trees as far north as New Smyrna on the east coast and Tarpon Springs on the west, but the hazards are great in any except the warmest parts of the state.

The largest commercial plantings have been made in the vicinity of Miami. There are a few small groves near Palm Beach and Fort Myers. At Oneco, near Bradentown, the Royal Palm Nurseries have one of the best variety collections in the state, but it is necessary to protect the trees during the winter. They are grown within a large shed whose top is made of thin muslin which can be removed in the summer.

In southern Florida the weather is normally dry during the flowering season. Sometimes there are light rains in this period, or many cloudy damp days. In such seasons many of the Indian mangos, notably Mulgoba, fail to bear good crops, although the seedling mangos which are found throughout this region fruit abundantly. Mangos differ in their ability to flower and fruit under adverse climatic conditions. Some of the Indian varieties will only flower after a period of three or four weeks of dry sunny weather; certain Cuban seedling races (and those of other countries as well), on the other hand, will insist on flowering even though the spring months are unusually wet; and if one crop of flowers is destroyed by the anthracnose fungus, as is often the case, they will flower a second and even a third time in an attempt to produce fruit. Methods of encouraging the Indian varieties to flower and fruit are discussed in a later paragraph.

The soils of the Fort Myers region produce larger trees than those of Miami. The latter, which are mainly light sands underlaid with oölitic limestone, are nevertheless satisfactory when properly fertilized. The mango requires much less fertilizer than the avocado or the citrus fruits, but it only reaches large size when grown upon reasonably deep soil.

Cuban soils are well suited to the mango. In commercial orchards near Habana, however, the anthracnose fungus has caused great damage and discouraged some of the growers. Methods of combating this pest are discussed under the heading diseases. In Porto Rico at least two orchards of considerable size have been planted with choice Indian varieties. Both of these are on the north side of the island, where the soil is excellent but the climate somewhat too moist for the best crop results.

Cultivation

The best site for the mango orchard is one which has good drainage together with soil of such nature that it will dry out thoroughly when no rain falls for a few weeks. In regions where the soil is deep and the trees consequently grow to large size, they should not be set closer than 35 by 35 feet. There are a few dwarf varieties, such as D'Or, which can be set much closer than this, but most of the Indian kinds ultimately make trees of good size. G. Marshall Woodrow recommends planting 20 by 20 feet, but in America this has not been found a good practice. Closer planting than 30 by 30 feet is undesirable except with dwarf varieties. Seedlings grow to larger size than budded or grafted trees, and need proportionately more space. On deep soils they will usually come to crowd each other in time if planted less than 40 or 45 feet apart.

April and May are considered the best months for planting in Florida. Midsummer planting is, however, much more successful than with the avocado. The principal point to be observed is the condition of the young tree at the time of planting. If it is not in active growth, it can be set at almost any season of the year, provided the weather is warm. In India it is recommended to plant at the beginning of the rainy season.

Holes 2 to 3 feet broad and deep should be prepared in advance of planting. Woodrow recommends that 20 pounds of fresh bones be placed in the bottom of each hole before filling in the soil. In Florida a small amount of commercial fertilizer is commonly used. The object in preparing the holes is the same as in planting other fruits, viz., to loosen the subsoil so that the roots can develop readily in all directions, and to place in the ground a supply of food for the young tree. It is sometimes recommended that stable manure be incorporated with the soil; this is a desirable practice, but it should be kept in mind that stable manure is not, generally speaking, suitable for bearing mango trees.

Well-grown budded or grafted trees, when shipped from the nursery, are eighteen inches to three feet in height, with stems one-half inch in thickness. They should be stocky and straight, with foliage of rich green color. Inarched trees are sometimes weak, crooked, and may have poor unions. While many inarched trees are produced and planted in certain parts of the world, notably in India, they seem much less desirable than the sturdy budded trees grown in the nurseries of Florida.

As soon as the young trees have been planted in the field, they should be shaded with a light framework covered with burlap or other cheap material. Palm leaves and pine boughs may be used for this purpose. The trees should, of course, be watered liberally as soon as they are planted, and in most regions the ground around the base of each should be mulched with straw or other loose material.

During the first four or five years, the trees should be encouraged to make vigorous rapid growth. After that the aim of the orchardist is to make them produce good crops of fruit. The object of early culture is, therefore, distinct from that of later years and somewhat different methods are required. The young growing tree can be given both water and fertilizer in liberal quantities; the mature tree, on the other hand, must be encouraged to flower and fruit by withholding water and fertilizer during certain portions of the year.

It must be admitted that the cultural requirements of the mango are not yet thoroughly understood. Varieties differ greatly in their reaction to the stimulus of tillage, irrigation, and manuring. A thorough study has not yet been made of the requirements even of a single variety. Horticulturists in India have devoted a limited amount of attention to the subject; but the mango seems to differ so markedly from other fruits which have been subjected to systematic cultivation that much further study will be needed before its habits are thoroughly understood.

The amount and character of tillage given to the orchard varies in different regions. In most parts of the tropics little attention is given to the mature tree. The soil beneath its spreading branches is often firmly packed down by the hoofs of domestic animals; or weeds may be allowed to grow unchecked. Needless to say, such treatment has little to recommend it. In Florida the land is sometimes given shallow cultivation during part of the year, and at other seasons leguminous cover-crops may be grown upon it, particularly if the orchard is not yet of bearing age. It is evident that the amount of nitrogenous fertilizer required by bearing groves is small. Over-stimulation results in vigorous development of foliage but no fruit.

Growers of grafted mangos in India resort to various expedients to check the vegetative activity of the tree and encourage the development of fruit. Thomas Firminger[1] says: "The mango, like all other fruit trees, is much benefited by having the earth around it removed, and the roots left exposed for a space of two or three weeks. This should be done in November, and in December the roots should be well supplied with manure, and then covered in again with entirely fresh earth, and not that which had been previously removed." Woodrow notes that "the mango growers near Mazagon, Bombay, who produced such famous fruit before the land was occupied with cotton mills, applied ten pounds of salt to each tree at the end of September; this would arrest growth in October and November, and encourage the formation of flower buds. In a moist climate, and the intervening ground occupied with irrigated crops, this system is highly commendable, but with a dry climate it is unnecessary."

The failure of many varieties to fruit abundantly is often

[1] Manual of Gardening for India.

attributed to imperfect pollination, attacks of insect pests, and other causes which are discussed in a later paragraph. It seems probable that too much emphasis has in the past been placed on these factors, and that the problem is largely a physiological one, connected with the nutrition of the tree. It is for this reason that the two quotations above are illuminative. They show that the nutritional problem has been recognized by early students of mango culture; yet no one has taken up the subject in sufficient detail to master it.

The mango requires less water than the avocado, although young trees are benefited by frequent irrigations. In Florida, old mango trees will be found growing and fruiting in fence corners and abandoned gardens where they have to depend entirely on rainfall. They are much more successful under such conditions than the avocado. Orchards of budded or grafted trees are rarely irrigated after the trees have attained a few years' growth. In other regions treatment must be different. In California, for example, irrigation should be practiced as with citrus fruits. J. E. Higgins remarks concerning Hawaii: "Liberal moisture must be supplied to the roots, from 50 to 70 inches per year being required, according to the retentive power of the soil and the rate of evaporation. In the case of bearing trees the heaviest irrigation should be given from the time when the flower buds are about to open until several weeks after the fruiting is over, withholding large amounts of water during two or three months preceding the flowering season." Regarding India, Woodrow says: "When fruiting age is attained there need be no necessity for irrigation from the time the rain ceases in September till after the flowers have 'set,' that is, till the young fruit appears; thereafter, irrigation over the area covered by the branches once in fifteen days or so is desirable while the fruit is increasing in size, but may be discontinued when ripening approaches."

All writers point out the necessity of applying a check to

vegetative growth previous to the flowering season. Ringing and hacking the trunk are two of the commonest practices, while root-pruning is occasionally performed in India. Recent experiments indicate that a liberal application of potash is extremely beneficial. Mulgoba trees at Miami, Florida, and Guanajay, Cuba, which were heavily fertilized with potash, produced much larger crops than those fertilized in the ordinary way. A standard commercial fertilizer especially prepared in Florida for use on mango trees contains:

Ammonia	5 to 6 %
Phosphoric acid	7 to 9 %
Potash	9 to 11%

These elements are derived from ground bone, nitrate of soda, dried blood, dissolved bone black, and high-grade potash salts.

Woodrow recommends for India that young trees be fertilized liberally with barnyard manure; but he adds that as soon as they come into bearing the application of manure must be stopped, and leguminous cover-crops planted between the rows. These crops can be plowed under, thereby enriching the soil in the necessary degree and at the same time keeping down weeds. The best legumes for this purpose, according to Woodrow, are *Crotalaria juncea*, *Cicer arietinum*, *Phaseolus aconitifolius*, and *Phaseolus Mungo*. P. J. Wester says, "The velvet bean (*Stizolobium Deeringianum*), Lyon bean (*Mucuna Lyoni*), the cowpea (*Vigna Catjang*) and related species may be used with good success in the Philippines. Of these the Lyon bean is preferable in the Philippines, since here it produces a greater amount of growth per acre than any other legume." In Florida velvet beans, cowpeas, and the bonavist bean (*Dolichos Lablab*) have been used. Growers should plant a number of different legumes experimentally to determine which are the best for their particular localities.

Numerous experiments to test the effectiveness of girdling

and root-pruning have been made at the Porto Rico Agricultural Experiment Station. C. F. Kinman reports of them:

"Girdling, branch pruning, and root pruning are common practices, but they should be used with caution and moderation, as a tree may easily be so severely injured as to prevent its bearing for one or more seasons. Pruning back the ends of the branches to induce blossoming has been practiced with good results at the station. In the operation, from a few inches to a foot of the end of the branch was removed, depending upon the stage of maturity of the wood, leaving a few nodes from which the leaves had not fallen. From these nodes blossoms developed profusely, no blossoms appearing on untreated branches. To secure best results, the pruning should be done in the late summer or fall, several months before the blossoming time. This method should be employed on branches which are too low or too crowded or on those which would have to be removed later to improve the shape of the tree, as after a branch is pruned it makes little growth for several weeks or months or even for a year or more after the fruit ripens, and by this time it may be well overgrown by surrounding branches.

"As good results have been obtained from girdling as from other methods. A branch one to three inches in diameter was selected on each of a number of trees and a band of bark removed in September. These branches produced good crops the following spring, even when no fruits at all were borne on the remainder of the tree. Such favorable results, however, were obtained on varieties which are inclined to bear well and where the band of bark removed was wide enough to prevent the new bark from growing over the area too rapidly. Bands one-eighth and even one-quarter of an inch in diameter were overgrown so quickly that no effect was seen on the branch. Bands from one-half to three-quarters of an inch produce the best results, as they do not heal over until after the blossoming season, the callus growing downward over the wound at the rate of one inch a year. . . . As removing enough bark to induce fruiting is very injurious to the branch, this practice is most profitably employed on undesirable branches which are to be removed later.

"Root pruning has been recommended, although no definite results have been noted from the experiments with it. It is best accomplished by cutting into the soil with a sharp spade about two feet inside the tips of the branches. In extreme cases the cutting may encircle the tree to a depth of eight or ten inches in heavy soil and even deeper in light soil where the root system is considerably below the surface. Cutting at such intervals as to sever the roots for one-half to two-thirds of the distance around the tree will induce blossoming under normal conditions without seriously checking the growth or thrift of the tree."

Experience in Florida has shown that girdling, to be effective, must be done in late summer. No one yet has had sufficient experience to recommend it as an orchard practice. Like root-pruning, the use of salt, and several other unusual practices, it may prove of decided value when its proper method of use has been determined. Every grower should conduct a few carefully arranged experiments along such lines as these, even though on a limited scale.

In India, the only pruning usually given the mango consists in cutting out dead wood. Since the fruit is produced at the ends of the branchlets, general pruning of the top cannot be practiced as with northern fruits. In Florida, however, several growers have found it desirable to prune out a certain number of branches from the center of the tree, so as to keep the crown open and admit light and air.

Propagation

Like many other fruit-trees, the mango has been propagated in the tropics principally by seed. In some instances seedling trees produce good fruits; this is particularly true of certain races, such as the Manila or Philippine. But in order to insure early bearing, productiveness, and uniformity of fruit, it is necessary to use vegetative means of propagation. Inarching, budding, and grafting are the methods most successfully employed.

The seedling races of the tropics are, so far as has been observed, polyembryonic in character. Three to ten plants commonly grow from a single seed. Since these develop vegetatively from the seed tissues, they are not the product of sexual reproduction, but may be compared to buds or cions from the parent tree. Most of the grafted Indian varieties, on the other hand, have lost this characteristic. When their seeds are planted a single young tree develops, and this is found to differ from its parent much as does a seedling avocado or a seedling peach. Usually the fruit is inferior, and the tree may be quite different in its bearing habits.

Dr. Bonavia, a medical officer in British India who did much to stimulate interest in mango culture, at one time took up the question of seedling mangos and wrote several articles advocating their wholesale planting. He argued that not only would many new varieties, some of them superior in quality, be obtained in this way, but also earlier and later fruiting kinds, and perhaps some suited to colder climates.

Just what percentage of seedling mangos will produce good fruit depends largely on their parentage. Seedlings of the fibrous mangos of the West Indies are invariably poor, while those from budded trees of such varieties as Alphonse and Pairi, although in most instances inferior or rarely equal or superior to the parent, are practically never so poor as the West Indian seedlings. At the Saharanpur Botanic Gardens, in northern India, some experiments were conducted between 1881 and 1893 to determine the average character of seedlings from standard grafted varieties. The results led to the conclusion that seedlings of the Bombay mango were fairly certain to produce fruit of good quality. An experimenter in Queensland, at about the same time, reported having grown seedlings of Alphonse to the fourth generation, all of which came true to the parent type.

Experience in the United States has shown, however, that degeneration is common. A number of seedlings of Mulgoba have been grown in Florida, but very few have proved of good quality. There is a tendency for the fruits to be more fibrous than those of the parent. The whole question is probably one of embryogeny. When monoembryonic seeds are planted, the fruit is likely to be inferior to that of the parent, if the latter was a choice variety; with polyembryonic seeds, even though of fine sorts like the Manila, the trees produce fruit closely resembling that of the parent.

The embryogeny of the mango cannot be discussed at great length here. It is not yet thoroughly understood, although it has been studied by several investigators. The most recent account and the only one which has been undertaken with the horticultural problems in mind, is that of John Belling, published in the Report of the Florida Agricultural Experiment Station for 1908. Belling says:

"In the immature seed of the sweet orange E. Strasburger has shown by the microscope, and Webber and Swingle have proved by their hybridizing experiments that besides the ordinary embryo which is the product of fertilization, the other embryos present in the young or mature seeds arise by the outgrowth of nucellar cells into the apical part of the embryo-sac. The first-mentioned embryo, when present, is liable to any variation which is connected with sexual multiplication, — the vicinism of H. De Vries. The remaining embryos, on the other hand, presumably resemble buds from the tree which bears the orange in whose seed they grow, in that they inherit its qualities with only a minor degree of variation."

The behavior of the mango has suggested a similar state of affairs. Belling goes on to quote Strasburger's account of the embryogeny of the mango, and describes his own investigations:

"Even in the unopened flower bud the nucellar cells at the apex of the embryo sac which are separated from the sac only by a layer of flattened cells, are swollen with protoplasm. In older fruits it may be noticed that the cells around the apical region of the sac except on the side near the raphe are also swollen. The adventitious embryos arise from these swollen cells, which in fruits 7 mm. long with ovules 3 mm. long divide up, sometimes forming the rudiments of a dozen or more embryos, but often fewer. The nucleated protoplasm on the embryo-sac wall is undivided into cells, and is thick opposite the places where embryo formation is going on."

Belling worked with fruits of the No. 11 mango, seedling

race of Florida identical with the common mango of the West Indies, Mexico, and Central America. He was not able to determine whether the egg-cell develops into an embryo, or whether all of the embryos are adventitious, — the egg-cell being crowded out or destroyed in some other way. If the fertilized egg-cell develops and is represented in the mature seed, the plant arising from it should exhibit variation; but the seedling races are so constant that it seems probable that the egg-cell is lost at some stage in the development of the fruit, and that all of the embryos are normally adventitious. There is as yet no proof, however, that fruits will develop on this or other mangos unless the flowers are pollinated. The subject is an important one and will repay further investigation.

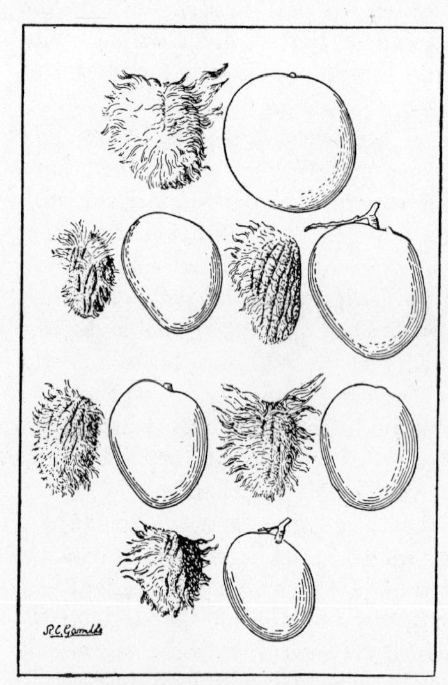

Fig. 10. Seedlings of grafted Indian mangos usually do not produce fruit exactly like the parent. Each of the fruits here shown represents a tree grown from a seed of the Mulgoba mango. The variations in size and shape of fruit, and in the amount of fiber around the seed, are noteworthy. (× $\frac{1}{6}$)

It has been observed in Florida that monoembryonic grafted varieties, such as Mulgoba, will, when grown from seed, sometimes revert to polyembryony in the first generation (Fig. 10).

PLATE VI. *Left*, the Sandersha mango; *right*, the ambarella.

G. L. Chauveaud[1] has advanced the theory that polyembryony is a more primitive state than monoembryony, which would seem to be borne out by this observation; for it must be true that the choice mangos of India which have been propagated by grafting for centuries are less primitive in character than the semi-wild seedling races.

Inarching is an ancient method of vegetative propagation. While several writers have attempted to show that it was not known in India previous to the arrival of Europeans, and that the Jesuits at Goa were the first to apply it to the mango, others have held the belief, based on researches in the literature of ancient India, that the Hindus propagated their choice mangos by inarching for centuries before any Europeans visited the country.

This method of propagation is still preferred to all others in India and a few other countries. In the United States it has been superseded by budding.

For the production of stock plants on which to bud or graft choice varieties, seeds of any of the common mangos are used. No preference for any particular race has yet been established. It is reasonable to believe, however, that there may be important differences among seedling races in vigor of growth and perhaps in their effect on the productiveness and other characteristics of the cion. The subject has never been investigated and deserves attention.

Seeds are planted, after having the husk removed, in five- or six-inch pots of light soil or in nursery rows in the open ground. They are covered with 1 inch or $1\frac{1}{2}$ inches of soil. In warm weather they will germinate within two weeks, and must be watched to prevent the development of more than one shoot. Polyembryonic mangos will send up several; all but the strongest one should be destroyed. If grown in pots and intended for budding, the young plants may be set out in the

[1] Compt. Rend. 114, 1892.

field in nursery rows when they are a foot high. If destined for inarching they must be kept in pots.

Inarching is more successful in the hands of the tyro than budding or crown-grafting. It can be recommended when only a few plants are desired, and when the tree to be propagated is in a convenient situation. G. Marshall Woodrow thus describes inarching as it is done in India. A slice is cut from the side of a small branch on the tree it is desired to propagate, and a slice of similar size — 2 to 4 inches long and deep enough to expose the cambium — is cut from the stem of a young seedling supported at a convenient height upon a light framework of poles. The two cut surfaces are bound together with a strip of fiber from the stem of the banana, or with some other soft bandage. Well-kneaded clay is then plastered over the graft to keep out air and water. The soil in the pot must be kept moist. After six to eight weeks the cut surfaces will have united.

Inarching may be done at any time in strictly tropical climates, but the best time in the hot parts of India is the cool season. Toward the northern limits of mango cultivation the middle of the rainy season is better.

The graft is sometimes allowed to remain attached to the parent tree for too long a time, with the result that swellings, due to the constriction of the bandages, occur at the point of union. It is better to remove the grafted plant fairly early and place it in the shade for a few weeks. It is detached from the parent tree by severing the branch which has been inarched to the seedling at a point just below the point of union with the latter. This leaves the young branch from the tree it was desired to propagate growing upon a seedling; the top of the latter is cut out, and the branch from the old tree takes its place, ultimately forming the crown of the mature tree.

The age of the stock is not important. Plants three weeks to three years old have been used with success. If kept in pots too long, however, the plants become pot-bound and lose their vigor; hence it is desirable to graft them when young and get them into the open ground as soon as possible. Seeds planted in June and July make strong plants ready for inarching by November. December and January are good months in which to inarch, and such plants should be ready to set out in the field by the following July.

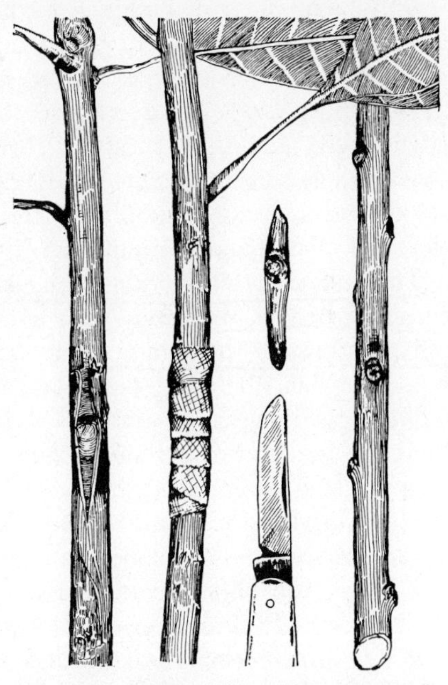

Fig. 11. Shield-budding the mango. On the left, a bud properly inserted; next, an inserted bud wrapped with a strip of waxed cloth; above the knife-point, a properly cut bud; and on the right, budwood of desirable character.

Inarching, as practiced in other countries, differs in no essentials from the Indian method above described.

Shield-budding is the method employed by nurserymen in Florida. In the hands of a skillful propagator who has made a careful study of this method, it gives excellent results. In inexperienced hands it usually proves altogether unsatisfactory. Particularly is experience required to enable the propagator to recognize the proper type of budwood, and to know when the stock plants are in the proper state of vegetative activity. By careful experimenting with stock plants and budwood of different conditions of growth throughout a season or two, a good propagator should be able to bud mangos successfully;

but comparatively few men have yet devoted the requisite time and study to the subject. Thus there are at present only a few propagators in the United States who can produce budded mango trees economically and in quantity.

Various methods of budding, beginning with the patch-bud, have been tried at different times, but shield-budding (Fig. 11) is the only one which has proved altogether satisfactory for nursery purposes. The method is the same as that used with citrus fruits and the avocado. Having been less extensively practiced, however, mango budding is less thoroughly understood, and it is not a simple matter to judge the condition of the stock plants and the budwood without experience.

The best season for budding the mango in Florida is generally considered to be May and June, but the work is done successfully all through the summer. It is necessary to bud in warm weather, when the stock plants are in active growth.

When seedlings have attained the diameter of a lead-pencil they can be budded, although they are commonly allowed to grow a little larger than this. The proper time for inserting the buds is when the plants are coming into flush, *i.e.*, commencing to push out wine-colored new growth. When they are in this stage, the bark separates readily from the wood; after the new growth has developed further and is beginning to lose its reddish color, the bark does not separate so easily and budding is less successful.

The budwood should be taken from the ends of young branches, but usually not from the ultimate or last growth; the two preceding growths are better. It is considered important that budwood and stock plant be closely similar, in so far as size and maturity of wood are concerned. If possible, branchlets from which the leaves have fallen should be chosen. In any event, the budwood should be fairly well ripened, and the end of the branchlet from which it is taken should not be in active growth.

The incision is made in the stock plant in the form of a T or an inverted T, exactly as in budding avocados or citrus trees. The bud should be rather large, preferably $1\frac{1}{2}$ inches in length. After it is inserted it should be wrapped with waxed tape or other suitable material. A formula for use in preparing waxed tape will be found under the head of avocado budding.

After three to four weeks the bud is examined, and if it is green and seems to have formed a union, the top of the stock plant is cut back several inches to force the bud into growth. A few weeks later the top can be cut back still farther, and eventually it may be trimmed off close above the bud, — this after the bud has made a growth of 8 or 10 inches.

J. E. Higgins [1] describes a method of shield-budding which has been successful in the Hawaiian Islands. So far as known, it has not been used on the mainland of the United States. Higgins says, "Budding by this method has been successfully performed on stocks from an inch to three inches in diameter. . . . Wood of this size, in seedling trees, may be from two to five years old. It is essential that the stocks be in thrifty condition, and still more essential that they should be in 'flush.' If not in this condition, the bark will not readily separate from the stock. It has been found that the best time is when the terminal buds are just opening. . . . The budwood which has been most successfully used is that which has lost most of its leaves and is turning brown or gray in color. Such wood is usually about an inch in diameter. It is not necessary in this method of budding that the budwood shall be in a flushing condition, although it may be of advantage to have it so. . . . The incision should be made in the stock about six inches in length. . . . The bud shield should be three to three and a half inches long, with the bud in the center." After-treatment of the buds is the same as with the Florida method which has been described; in fact the Hawaiian method seems

[1] Bull. 20, Hawaii Agr. Exp. Sta.

distinct only in the size of stock plant and budwood, and the consequent larger size of the bud.

Crown-grafting (Fig. 12) is not commonly practiced in Florida, but it has been successful in Porto Rico. It has also been employed with good results by H. A. Van Hermann of Santiago de las Vegas, Cuba, and it is said to have proved satisfactory in Hawaii and in India. W. E. Hess, formerly expert gardener of the Porto Rico Agricultural Experiment Station, who has had much experience with the method, says that it has proved more successful in Porto Rico than budding, and is at the same time superior to inarching because of the greater rapidity with which trees can be produced in large quantities. As in budding, success seems to depend mainly on the condition of stock and cion at the time the graft is made. Provided the stock is in flush, the work can be done at any season of the year. For cions, tip ends of branchlets are used. They should be of about the diameter of a lead-pencil; of grayish, fully matured,

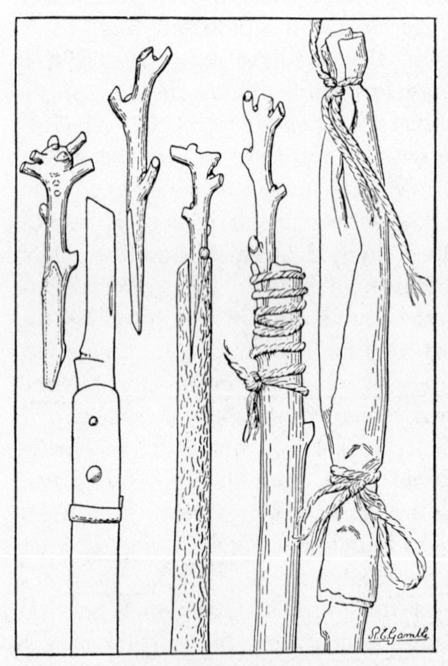

FIG. 12. Crown-grafting the mango. On the left, two cions of proper size and character; in the center, a cion inserted and another tied in place; and on the right, the covering of waxed paper which protects the cion while it is forming a union with the stock.

dormant wood; and from 3 to 5 inches in length. A slanting cut 1 to 2 inches long is made on one side, tapering to a point at the lower end of the cion. The stock may be of almost any size. When young plants are used they are cut back to 1 foot above the ground, and a slit about 1 inch long is made through the bark, extending downward from the top of the stump. The cion is then forced in, with its cut surface next to the wood, and is tied in place with soft cotton string. No wax is used. The graft is inclosed in three or four thicknesses of oiled paper which is wound around the stock and tied firmly above and below. This is left on for twelve to twenty days, when it is untied at the lower end to admit air. Fifteen or twenty days later the cions will have begun to grow and the paper can be removed entirely.

This method is applicable not only to nursery stock but also to old trees which it is desired to topwork. In this case about half of the main branches of the tree should be cut off at three or four feet from their union with the trunk. It is necessary to leave several branches to keep the tree in active growth; this also has a beneficial effect on the grafts by protecting them from the sun. When the cions are well established, these branches may be removed or they also may be grafted if more limbs are necessary to give the tree a good crown. The cions are inserted under the bark at the cut ends of the limbs, exactly as described for young stocks, but larger cions may be used.

In Florida many large trees have been topworked by cutting off several of the main branches, close to their union with the trunk, and allowing a number of sprouts to come out. When these have reached the proper size, they are budded in the same manner as seedlings.

Throughout the tropics there are many thousands of seedling mango trees which are producing fruit of inferior quality. By topworking, these trees could be made to yield mangos of the choicest Indian varieties. The work is not difficult and the value of the tree is increased enormously. Perhaps no other field in tropical horticulture offers such opportunities for immediate results as this.

The Mango Flower and Its Pollination

The scanty productiveness of many Indian mangos has been attributed by several writers to defective pollination. A. C. Hartless, superintendent of the Government Botanical Gardens at Saharanpur, India, discussed the matter at some length in the Agricultural Journal of India, April, 1914. The writer has personally investigated the subject in Florida, and the results have been published in Bulletin 542 of the United States Department of Agriculture. Burns and Prayag have written on the structure and development of the mango flower in the Agricultural College Magazine, Poona, India, March, 1911.

Fig. 13. A bisexual mango flower. (× 4)

The mango is polygamous and produces its flowers on terminal panicles varying in length from a few inches up to two feet. Each panicle carries from 200 or 300 up to more than 4000 flowers, of which only 2 or 3 per cent are perfect in some varieties, or as many as 60 to 75 per cent in others. The character of the panicle and the number of flowers produced upon it differs according to the variety.

The individual flower (Fig. 13) is subsessile, 6 to 8 millimeters in diameter when the corolla is outspread; the calyx composed of five ovate-lanceolate, finely pubescent, concave sepals; and corolla of five elliptic-lanceolate to obovate-lanceolate petals, 3 to 4 millimeters long, whitish, with three or four fleshy orange ridges toward the base, and inserted at the base of a fleshy, almost hemispherical disk, obscurely 5-lobed and usually about 2 millimeters in diameter. In the perfect flower the disk is surmounted by a globose-oblique ovary 1 millimeter broad, with a slender lateral style about 2 millimeters high. To one side and inserted upon the disk is the single fertile stamen, composed of a slender subulate filament about 1.5 millimeters long, surmounted by an oval purplish red anther 0.5 millimeter long, which dehisces longitudinally. Occasionally two such stamens are produced. The whorl is completed by staminodes of varying prominence, short and subulate in some varieties, larger and capitate in others, some even becoming fertile and producing a few pollen-grains. In the staminate flower the ovary is wanting.

Several writers have affirmed that the mango is largely if not solely wind-pollinated. It seems evident, however, that it has none of the characteristics of an anemophilous plant, but, on the other hand, presents well-developed adaptations to insect pollination. In anemophilous or wind-pollinated flowers, the pollen is usually abundant in order to compensate for the enormous loss in transport; the pollen-grains are dry and incoherent, so that they may easily be carried by the wind; and the stigmas are commonly bushy and freely exposed, so as to have every chance of catching the floating grains. The mango shows none of these adaptations. It produces comparatively few pollen-grains, often not more than 200 or 300 to an anther. These grains show a decided tendency to cling together, especially in damp weather; and even in dry sunny weather it is difficult to dislodge them with a strong draft of air. The stigma is small and not provided with projections of any sort to assist in catching pollen.

The production of nectar for the attraction of insects also indicates that the mango is entomophilous. Observations have shown that the flowers are visited by numerous insects of the orders Diptera, Hymenoptera, Lepidoptera, and Coleoptera, ranking in the order given as to the number of visits. Pollen-grains have been observed adhering to the bodies of many species belonging to these orders.

In spite of numerous insect visits, however, a large number of the stigmas are never pollinated, and it seems probable that very little pollen is transferred from one flower to another. Most of the stigmas receive their pollen from the anther (rarely is more than one fertile) of the same flower. Cross-pollination is in all probability uncommon. In damp cloudy weather the pollen-grains swell and are much more difficult to dislodge than when the weather is dry and sunny. After a heavy dew they will be found in this swollen condition, but when the sun comes out they return to their normal dry form. Protection of the flowers from dew and rain by means of a canvas shelter did not increase the production of fruit in the case of an experiment carried out in Florida.

Sometimes there is considerable differentiation in the size of the pollen-grains. In most varieties the larger number, however, are uniform in shape and size, plump and apparently perfect. They can be germinated in sugar solution of the proper density, and there is nothing to suggest that impotency is common.

From the fact that pollination ordinarily is scanty, it might be assumed that productiveness could be increased by making it more abundant. This has not, however, been found to be the case, except when the pollen was obtained from a tree of a different variety (cross-pollination); under these conditions there was a somewhat better yield. The total number of flowers produced is so enormous that it is of little importance whether all are pollinated or not. Seedling mangos, which are not pollinated more abundantly than budded varieties, nor furnished with a greater number of anthers, nor, so far as can be ascertained, with pollen of greater potency, often set many more fruits than they can carry to maturity. This has been noted also with several grafted kinds, such as Bennett and Cambodiana.

Sometimes the entire tree comes into bloom at one time,

covering itself with flowers; again, one side of the tree may flower, while the other shows no buds; or the flowering may be confined to a small section of the tree, probably the branchlets arising from one large limb. This behavior of the mango corresponds to the growth habit of the tree which is mentioned but not explained by A. F. W. Schimper.[1] When one side of the tree flowers independently, it might be expected that the remainder would flower at another time, but this is not always the case.

Some varieties develop all their flowers within ten days after the first buds open; others, such as Sandersha and Julie, push out flower-panicles during a period of several weeks, or even months; thus, in 1915 there was not a single day between the middle of January and the latter part of May on which flowers could not be found on the old Sandersha tree in the Plant Introduction Garden at Miami, Florida. This feature is of importance in that it gives the tree a greater opportunity to set fruit. Often the attacks of the anthracnose fungus are severe when the tree is in bloom, and the entire crop of flowers is destroyed. In some varieties this means a crop failure, since the tree will not produce any more flowers that season; but in the Sandersha (if early in the season) it need mean only the loss of the flowers which were present at that particular time. Those developed later might enjoy more favorable weather, with consequent freedom from the anthracnose peril, and a crop of fruit would result. Anthracnose, one of the greatest enemies of the mango, is discussed under the heading pests and diseases.

Some varieties which fruit heavily are characterized by a high percentage of perfect flowers. Others which are known to be unusually regular in fruiting, although they may not produce such heavy crops, have relatively few perfect flowers. The Philippine race of seedlings, which sometimes bears heavily,

[1] Plant Geography.

commonly has more perfect than staminate flowers. Most of the Indian varieties have fewer perfect flowers than the seedling races.

The experiments conducted in Florida indicate that the scanty fruiting of many varieties is not due to any morphological defect in the pollen or to defects in the mechanism of pollination. While such factors as lack of pollinating insects and loss of pollen through rains or moist weather probably lessen the production of fruit in some seasons, from a practical standpoint the question of pollination seems relatively unimportant. The problem is more probably a physiological one, connected with nutritional conditions as influenced by changes in soil-moisture and food-supply, principally the former. Suggestions are given under the heading culture for encouraging the formation of fruit-buds on soils or under climatic conditions which normally tend to produce vegetative growth to the detriment of reproduction.

The Crop

In the tropics seedling mangos usually come into bearing four to six years from the time of planting. More time than this may be required in some instances. Certain races are more precocious than others. In Florida, growth is less rapid than in the tropics and fruiting is delayed in consequence.

Budded trees should fruit at an earlier age than seedlings. As regards a given variety or race, they usually do so; but grafted or budded trees of some varieties do not fruit so early as seedlings of certain races. In Florida, dwarf kinds such as D'Or and Julie sometimes fruit the second year after planting. Haden has produced good crops four years from planting. Mulgoba should fruit at four to six years of age. Malda and several other sorts have been grown in Florida ten years or more without having fruited as yet. At Saharanpur, India,

A. C. Hartless has found that it commonly requires four to nine years for inarched trees to come into bearing.

The yield of many budded varieties is uncertain, while of many seedling races it is uniformly heavy. Seedling trees in Cuba and other parts of tropical America often carry as much fruit as the branches will support. Budded mangos sometimes bear heavily one season and nothing the next. The following table prepared by A. C. Hartless shows the behavior of the orchard of grafted trees in the Botanical Garden at Saharanpur, India, during a period of twenty-seven years. Numerous varieties are included; and it is probable that some bore more regularly than others; but the table takes account of the crop as a whole:

TABLE III. SHOWING THE BEARING OF MANGO TREES

YEAR	CHARACTER OF CROP	YEAR	CHARACTER OF CROP	YEAR	CHARACTER OF CROP
1886	Fair	1895	Extremely light	1904	Very heavy
1887	Almost a failure	1896	Very light	1905	Light
1888	Good	1897	Fair	1906	Good
1889	Complete failure	1898	Excellent	1907	Very light
1890	Light	1899	Fair	1908	Good
1891	Poorest on record	1900	Below average	1909	Very poor
1892	Heavy	1901	Very light	1910	Very poor
1893	Heavy	1902	Fair	1911	Poor
1894	Very light	1903	Very light	1912	Excellent

Records from Lucknow, India, show that during a period of thirty years there were nineteen in which the crop was poor, six in which it was fair, and five in which it was heavy. At Nagpur during a period of nine years there were six in which the crop was poor and three in which it was good.

In Florida Mulgoba has, up to the present, produced a good crop about once in four years.

These figures would be discouraging, were it not for the

certainty that much can be done to increase the likelihood of good crops by attending to cultural details and by planting varieties known to be productive. The extensive tests which have been made in Florida have brought to light a number of choice sorts which combine excellent quality of fruit with a degree of productiveness far above the average. Amini, for example, has borne much more regularly than Mulgoba. In Porto Rico also it has done remarkably well. Sandersha has produced a fair crop nearly every year. Cambodiana has also given a good account of itself. Pairi has fruited much more regularly than Mulgoba and is almost as good in quality. When reasonably productive kinds are planted, and their cultural requirements are thoroughly understood, such records as that of Saharanpur should no longer be encountered.

The varieties now grown in Florida supply the market with ripe fruit from July to October. The main season is August and September. Cambodiana is one of the earliest varieties. Sandersha is probably the latest. A few of its fruits ripen as late as the first half of October. In India a kind known as Baramassia (more likely a number of different mangos known under the same name) is said to mature fruits throughout most of the year, doing this by producing two or three light crops. It is probable, however, that many statements regarding this variety are exaggerated, for it seems to be known much better by reputation than by the personal experience of those who describe it. A variety in northern India, Bhaduria, ripens later than most others. In this part of India the mango season extends from May to October.

The Indian method of picking and ripening the mango, and the type of carrier employed in shipping the fruit, are described by G. Marshall Woodrow. He says:

"The mango is gathered as soon as the fruit comes away freely in the hand. . . . When gathered too early the sap exudes freely, does not agglutinate, and the fruit shrivels. The collection of the fruit should be by hand as far as practicable; a bag-net with the mouth distended by a circle of cane, and suspended by a strap from the shoulder, leaves both hands free to gather. None must be allowed to fall to the ground; all should be handled as gently as eggs because a slight bruise brings on decay quickly. To bring down the higher fruit a bag-net 15 inches in depth, the mouth distended by a circle of cane, traversed by and bound to a light bamboo and having a piece of hoop iron bound across the mouth of the bag at right angles to the bamboo forms an efficient apparatus for the purpose; the hoop iron breaks the stalk, and the fruit falls into the net and is gently lowered to the ready baskets. It is then carried to the fruit room and arranged in single layers, with soft dry grass above and below. The room must be well ventilated and cool, yet not subject to decided changes of temperature; a moist atmosphere hastens ripening and decay, coolness and fresh air retard destructive changes.

" For transport, small baskets fit to contain a dozen mangos should be provided, each with a lid and some hay for packing at top and bottom. Each basket should be filled so as to prevent motion of the fruit, choice specimens being separately wrapped in soft paper. Twelve small baskets may be packed firmly into one large one, and the load becomes sufficient for a man to carry when the basket has been raised on to his head. By this means bruised and damaged fruit is reduced to the lowest terms, and repacking for distribution is avoided."

A. C. Hartless of Saharanpur says: "It is a common practice here to ripen the fruit artificially. This is done to save the expense of watching and protecting from predatory animals and birds. When the fruits attain the desired size they are taken off and packed in straw in closed boxes where they will ripen. The taste may in this way differ slightly from those ripened on the tree, but it is not uncommon for the fruits on the same tree to differ materially in taste." C. Maries reports that the variety Mohur Thakur is ripened on the tree at Darbhanga, small bamboo baskets being placed around the fruits to keep flies and moths from eating them. When the basket falls to the ground the fruit is ripe and ready for eating.

Some varieties will keep much longer after picking than others. William Burns,[1] in his article on the Pairi mango,

[1] Agricultural Journal of India, p. 27, 1911.

says that Alphonse can be kept two months, if properly stored. Pairi, on the other hand, will only remain in good condition for eight days. C. F. Kinman points out that the Indian mangos have proved to be much better keepers in Porto Rico than the native seedlings. The flavor and keeping quality of a fruit depend, of course, largely on the degree of maturity at which it is picked. For local use the fruit, with the exception of Sandersha, should be allowed to color fully and to soften slightly on the tree, while for shipping to market it must be picked before it is fully colored. Some varieties, such as Amini, develop an objectionable flavor if left on the tree until fully ripe.

Fig. 14. Florida-grown mangos packed for shipment.

From Florida the Indian varieties have been shipped successfully to northern markets (Fig. 14). The fruit is picked when it has begun to acquire color, but before it has softened in the slightest degree. It is then wrapped in tissue-paper of the kind used in shipping citrus fruits, and is packed in tomato baskets. Mangos of moderate size, such as Mulgoba, will pack twelve to a basket. A small amount of excelsior is used above and below them. Six of these baskets are placed in a crate for shipment. Sometimes tomato baskets are dispensed with and the fruit is packed in a crate with a partition in the center, using an abundance of excelsior between each tier or layer.

Numerous storage tests have been made at the Porto Rico Agricultural Experiment Station (Bull. 24). Mangos of different varieties were placed in (*a*) warm storage at 80 to

83° F., and (*b*) cold storage at 40 to 47° F. Some of the results were as follows:

Amini. — Fruits which were ready for eating when taken from the tree remained in the warm room in good condition about four days. Fruits which were well colored but had not softened on the tree began to decay in seven to ten days. All of these fruits developed attractive color in storage. In the cool room fruits which were ready for eating when removed from the tree remained in good condition eleven to eighteen days. Those which were mature when taken from the tree, but which had not commenced to soften, were ready for eating twenty days after being put in storage, and did not show signs of decay until six days later.

Cambodiana. — Fruits which had fallen from the tree due to ripeness remained in the warm room five days in good condition. Those which were picked when soft on one side remained six to eight days without decaying perceptibly. Those picked when about half colored remained in good condition eight days only. Fruits ripened on the tree and placed in the cool room kept only five or six days. Those which had colored on the tree but had not begun to soften were ripe nineteen days after being placed in the cool room, and remained in good condition until the twenty-sixth day; they were not so good, however, as those ripened on the tree.

Sandersha. — Fruits picked just before they began to soften and placed in the warm room were ready for eating nine days later, and remained in good condition three days. Fruits picked similarly mature and placed in the cool room remained in good condition for nearly five weeks, at the end of which time the flavor was better than that of tree-ripened specimens.

"Fancy" mangos have been shipped successfully from India to London, from Jamaica to London, and from the French West Indies to Paris. When care is used in packing and picking the fruit, the loss in transit is not heavy. The selection of varieties having unusually good shipping qualities will do more than anything else to encourage export trade of this sort. When the fruit has only to be shipped from Florida to New York, keeping quality is not so important. Some mangos which have been placed on the market have made an unfavorable impression because they were improperly ripened. More attention must be given to methods of ripening in the future, so that the fruit may reach the consumer in full possession of its delightful flavor and aroma.

Pests and Diseases

The commonest and most troublesome enemy of the mango in tropical America is anthracnose. This is a parasitic fungus (*Colletotrichum gloeosporioides* Penz.) which attacks many different plants, and is particularly known as the cause of wither-tip in citrus fruits. It is a species of wide distribution which springs up with no evident center of infection whenever the weather is warm and moist. On the flowers and flower-stalks of the mango it appears in the form of small blackish spots. Often it causes many of the flowers to drop. On the leaves, spots and sometimes holes are produced; these begin as minute black dots and enlarge until they are about an eighth of an inch in diameter. Young fruits may be attacked and made to drop in large numbers, while older fruits become spotted with black or streaked, and their keeping qualities are impaired.

S. M. McMurran, who studied anthracnose control methods in Florida and reported his results in Bulletin 52 of the United States Department of Agriculture, says:

"Spraying before the buds begin to grow is of no value so far as protecting the inflorescence, and later the young fruit, is concerned. These must be kept covered with the fungicide (Bordeaux mixture) while growing, if fungous invasion is to be prevented. The difficulty of so protecting the inflorescence is at once apparent. Elongations of the panicle continue for a period ranging from 10 to 15 days. Those which were sprayed every third day were practically all disease-free when the flowers began to open. This, however, required four sprayings in one case and six in the other. Those sprayed every fourth day showed but little more disease than those sprayed every third day, but those on which the spray was applied at five and six day intervals had traces of disease, showing that they were less perfectly protected.

"The spraying of the inflorescence at least three times, beginning when the buds are just swelling and repeated every fourth day until the flowers open, will help to prevent the dropping of fruit caused by the disease on the peduncles and pedicels.

"The inflorescence may be kept in a clean condition up to the time of blooming; but, when this takes place, immediately there are hundreds of points which are not covered by the fungicide and are open to infection . . . spraying is of little or no value in controlling the blossom blight form of the disease, and profitable sets of fruit can be expected only during seasons which are dry at blooming time, unless varieties which are resistant to the disease are developed and cultivated."

This disease is a serious obstacle to the production of marketable mangos in the West Indies. J. B. Rorer,[1] who conducted spraying experiments in Trinidad, found, however, that "All of the sprayed trees set more fruit than the control trees, and the greater part of the fruit ripened without infection, while the fruit on unsprayed trees was for the greater part spotted or tear-streaked. The fruit from sprayed trees matured a little later than that from the unsprayed and was somewhat larger in size. The foliage of sprayed trees was much heavier than that of the unsprayed." If fruit is not sprayed to keep it clean while it is developing, it not only is less attractive when placed on the market, but is subject to decay.

Anthracnose does not appear to be mentioned by Indian writers on mango culture. It is known, however, to be serious in Hawaii as well as in tropical America. Bordeaux mixture used in its control can be made according either to the 4–6–50 or the 5–5–50 formula, using a small amount of whale-oil soap to make it adhere more tenaciously to the foliage.

Ethel M. Doidge, in the Annals of Applied Biology (1915) describes a disfiguring and rotting disease of mangos which occurs in South Africa. It is caused by *Bacillus mangiferæ*, an organism which is carried by water or may be transported from tree to tree by the wind. Woody tissues are not affected, but small angular water-soaked areas appear on leaves, longitudinal cracks are produced on petioles, and discolored spots on twigs and branches; while on the fruit the first sign of the disease is a small discolored spot. This spreads, becoming

[1] Trinidad and Tobago Bull. 5, 1915.

intersected with cracks, and may extend some distance into the flesh. No means of controlling this bacterial disease has yet been discovered.

Of the insects which attack the mango, the fruit-flies (Trypetidæ) rank first in importance. Belonging to this family are the Mediterranean fruit-fly (*Ceratitis capitata* Wiedemann), which has become a very serious pest in Hawaii and several other regions; the Queensland fruit-fly (*Batrocera tryoni* Froggatt), distributed throughout Malaysia and Australia; the mango fruit-fly (*Dacus ferrugineus* Fabricius), which occurs from India to the Philippines; the Mexican fruit-fly (*Anastrepha ludens* Loew); and *Anastrepha fraterculus* Wiedemann, another Mexican species, now distributed throughout Central and South America and the West Indies. Several other species have also been reported as attacking the mango. The females of these flies insert their eggs beneath the skin or in the flesh of the fruit, and the larvæ render it unfit for human consumption. Control is difficult; the sweetened arsenical sprays have met with varied success, and control by means of parasites is receiving attention.

In some parts of India the mango hopper (various species of Idiocerus) is troublesome. H. Maxwell-Lefroy [1] writes:

"These insects resemble the Cicadas superficially but are much smaller, being one-sixth of an inch in length. They are somewhat wedge-shaped with wings sloped at an angle over the back. Large numbers are found on the mango trees throughout the hot weather but especially at the flowering season when there is a flow of sap to the flowering shoots. These insects pass through their active life on the tree, sucking the juice of the soft shoots and causing them to wither. . . . There is only one effective treatment which must be adopted vigorously. This is spraying with strong contact poison such as crude oil emulsion or sanitary fluid."

[1] Indian Insect Pests.

Another serious pest in India is the mango weevil (*Sternochetus mangiferæ* Fabricius, better known as *Cryptorhynchus mangiferæ*). It is not limited to India, but is found also in the Straits Settlements, the Philippines, South Africa, and Hawaii. In the last-named country it has become formidable. "The insect is a short, thick-set weevil, dark brown in color, one-third of an inch in length. . . . The grubs bore in the kernels of the mango fruit when it is growing large; these grubs pupate inside the fruit and as the mango ripens become beetles, eating their way out through the pulp of the fruit, which they spoil." Maxwell-Lefroy recommends that all infested fruits be destroyed, and that weevils hiding in the bark of the tree be killed in August. Kerosene emulsion is useful in destroying those which are on the bark. It is also advisable to cultivate or flood the ground beneath the trees, in order to kill weevils which may be lurking there.

In Florida, red-spiders and thrips are responsible for extensive injury to foliage, leading to disturbances of the general health of the trees; but contact sprays, *e.g.*, lime-sulfur or nicotine, properly applied, will effect complete eradication.

The mango bark-borer (*Plocœderus ruficornis* Newman) is a formidable enemy of the mango in the Philippines. This is a large beetle. C. R. Jones [1] says of it:

"The mango bark borer, while a comparatively unknown pest outside the vicinity of Manila, is exceedingly dangerous, largely on account of its feeding habits, which make detection difficult. The beetle has, so far as we know, no natural enemies, being fully protected both in the larval and pupal stages. Physical remedies are, therefore, necessary, such as the removal of larvæ and pupæ from their burrows by hand."

The mango shoot psylla (*Psylla cistellata* Buckton) is reported only from India. "It injures the terminal shoots by producing imbricated pseudo-cones of a bright green or yellow

[1] Philippine Bur. Agr. Circ. No. 20.

color in which the larval and pupal stages are passed." *Dinoderus distinctus* is a beetle which attacks branches of the mango in India. *Sternochetus gravis* is the mango weevil of northern India, similar to the common mango weevil described above. These and many other insects reported as attacking the mango in various parts of the world are described in "A Manual of Dangerous Insects," published by the United States Department of Agriculture (1917). The scale insects are particularly numerous, and cannot be listed here. Several of them are common in the mango orchards of Florida. The genera Aspidiotus, Chionaspis, Coccus, Pulvinaria, and Saissetia are well represented in different parts of the world. Generally speaking, their control by spraying is relatively simple.

Races and Varieties

The classification of mangos must be considered from two distinct standpoints. First, there are numerous seedling races; and second, there are horticultural groups of varieties propagated by grafting or budding.

The seedling races have not been studied in all parts of the tropics. Most of those in America are now fairly well known, but they are probably few compared to those of the Asiatic tropics. The latter region has not been explored thoroughly.

So far as known, all the seedling races are polyembryonic. Individuals reproduce the racial characteristics with remarkable constancy. Numerous writers have said that these races (incorrectly termed varieties) come true from seed, and that there is no need of grafting or budding. There is enough variation among the seedlings, however, to make some of them more desirable than others. When one has been propagated by budding or grafting it becomes a true horticultural variety.

The classification of mangos has been discussed by Burns and Prayag in the Agricultural Journal of India (1915); by

P. H. Rolfs in Bulletin 127, Florida Agricultural Experiment Station; and by the author in the Proceedings of the American Pomological Society for 1915 and 1917.

The abundance of grafted mangos has led Indian investigators to neglect the seedling races. Doubtless some of the horticultural groups of grafted varieties represent seedling races. C. Maries, in the Dictionary of the Economic Products of India, grouped the named varieties with which he was familiar in five "cultivated raccs." Probably some of these represent seedling races. The antiquity of its culture in India and the extensive employment of vegetative means of propagation have placed the mango on a different footing from that which it occupies in regions where it has been grown relatively a short time and propagated principally by seed. In India, the horticultural varieties are most prominent; elsewhere, seedling races (see definition of a race in the discussion of avocado races) are more in evidence.

The mangos of the Malayan Archipelago have been less thoroughly studied, from a pomological standpoint, than those of any other region. The botanist Blume (Museum Botanicum Lugduno-Batavum) viewed them botanically, and described as botanical varieties a number of forms which are in all probability analogous to the seedling races of other regions. In addition to races, there are a number of distinct species of Mangifera in the Malayan region which bear fruits closely resembling true mangos. These must be studied in connection with any attempt to straighten out the classification of horticultural or pomological forms.

Cochin-China appears to be the home of a race of mangos which is unusual in character, and which is certainly one of the most valuable of all. This is the Cambodiana. By some botanists it is considered a distinct species of Mangifera. It seems to be identical with the race grown in the Philippine Islands. The latter has been carried to tropical America, where it is known as Manila (Mexico) and Filipino (Cuba). David Fairchild, who studied this race in Saigon, Cochin-China, and introduced it into the United States, describes it as a mango of medium size, yellow when ripe, furnished with a short beak, and having a faint but agreeable odor. The flesh varies from light to deep orange in color, and is never fibrous. The flavor is not so rich as that of the Alphonse, but is nevertheless delicious. One of the plants grown from the seed sent to the United States by Fairchild has given rise to the horticultural variety Cambodiana, now propagated vegetatively in Florida.

There appear to be several different forms of this race. Three forms are grown in the Philippines, where they are distinguished by separate names. P. J. Wester states:

"There are three very distinct types of mangos in the Philippines: the Carabao, the Pico (also known as Padero), and the Pahutan, in some districts called Supsupen and Chupadero. The Carabao is the mango most esteemed and most generally planted." He further says, "Although uniform as types, there is considerable variation in the form and size of the fruit and presence of fiber and size of seed in both the Carabao and Pico mangos, and careful selection will not only bring to light varieties much larger than the average fruit of these types, but also those having a much smaller percentage of fiber and seed than the average fruit."

The seedling mangos of the Hawaiian Islands have been given some attention by Higgins. In Bulletin 12 of the Hawaii Agricultural Experiment Station he describes a number of them. Judging from his illustration, the Hawaiian Sweet mango is the common seedling race of the West Indies.

The French island of Réunion is said to be the source of several seedling races which have been introduced into tropical America. Paul Hubert[1] says the mango has become thoroughly naturalized in this island. He mentions thirteen varieties

[1] Fruits des Pays Chauds.

which are the most common; the names of several are the same as those of well-known varieties in the French West Indies.

Little is known of the mangos cultivated on the African coast and in Madagascar.

The seedling races of Cuba and those of Florida are practically the same, seeds having carried from the former region to the latter. The principal race is the one known in Cuba as mango (in contradistinction to manga, the race second in importance), and in Florida as No. 11. This is the common race of Mexico and many other parts of tropical America. For convenience it may be termed the West Indian. The tree is erect, 60 to 70 feet in height, with an open crown. The panicle is 8 to 12 inches long, with the axis reddish maroon in color. The fruit is strongly compressed laterally, with curved and beaked apex. It is yellow in color, often blushed with crimson; the fiber is long and coarse, and the quality of the fruit poor, although the flavor is very sweet.

The manga race of Cuba is less widely grown in other regions, although it is well represented in Florida. The tree is spreading, 35 to 40 feet high, with a dense round-topped crown. The panicle is 6 to 10 inches long, stout, pale green in color, often tinged with red. The fruit is plump, not beaked, yellow in color, with long, fine fibers through the flesh. Two forms of this race are common, *manga amarilla* and *manga blanca*. The former, known in Florida as turpentine or peach mango, has an elongated fruit, deep orange yellow in color, with bright orange flesh. The latter, known in Florida as apple or Bombay mango, has a roundish oblique fruit, bright yellow in color with whitish yellow flesh.

The Filipino (Philippine) race probably reached Cuba from Mexico, and thence was carried to Florida. It is the most delicious and highly esteemed of seedling mangos in all of these regions. Indeed, it ranks in quality with many of the choice grafted varieties from India. The tree is erect, 30 to 35 feet high, with a dense oval crown. The panicle is 12 to 24 inches long, pale green, sometimes tinged with red. The fruit is strongly compressed laterally, sharply pointed rather than curved or beaked at the apex, lemon-yellow in color, with deep yellow flesh almost free from fiber. In Florida there are comparatively few trees of this race.

In addition to the above, there are several other races of limited distribution in Cuba. The biscochuelo mango of Santiago de Cuba is an excellent fruit, worthy of propagation in other regions. The mango Chino of the Quinta Aviles at Cienfuegos (a remarkable mango orchard established years ago) is a large fruit always in great demand in Habana markets. It is not, however, of rich flavor or fine quality. Manga mamey, also of the Quinta Aviles at Cienfuegos is of better quality than mango Chino, but is not so well known in Habana.

In Jamaica the No. 11 race is esteemed above most other seedlings. It had its origin in one of the grafted trees found on a captured French vessel and brought to the island in 1782, as related on a foregoing page.

The seedling races of Porto Rico have been treated in detail by G. N. Collins[1] and more recently by C. F. Kinman. The most prolific and popular race is known as *mango blanco*. The *mangotina* is found near Ponce; it is rather inferior in quality. The *redondo* is a seven-ounce fruit, lacking in richness. The *largo* has a small oval fruit with much fiber. The name *piña* is applied to several distinct forms, the commonest being a long fruit of inferior quality. None of these Porto Rican forms seems to merit propagation.

In Mexico the principal races are the common West Indian, and the Manila or Filipino. The latter is grown principally in the state of Vera Cruz. Its culture should be extended to other parts of the country, as well as to other tropical countries where it is not now grown.

[1] Bull. 28, U. S. Dept. Agr.

There is one race in Brazil which is of exceptional value. This is the *manga da rosa* (rose mango), grown commercially in the vicinity of Pernambuco and to a less extent at Bahia and Rio de Janeiro. While frequently propagated by grafting, it is polyembryonic and should come true to race when grown from seed. It is heart-shaped, slightly beaked, and of good size. Its coloring is unusually beautiful. The fiber is coarse and rather long, but not so troublesome as in many seedling races. The flavor is rich and pleasant. This mango is believed to have been brought to Brazil from Mauritius. The *espada* race of Brazil is of little value: its fruit is slender, curved at both ends, green in color, and of poor quality.

The horticultural varieties of the mango are numerous. C. Maries reported having collected nearly 500, of which 100 were good. Many of these were, however, of limited distribution and little importance. More recent Indian writers catalog from 100 to 200 varieties. The author has published in the Pomona College Journal of Economic Botany (December, 1911) a descriptive list of about 300, which includes the best-known from all parts of the world. Some of these, however, are probably seedling races, not horticultural varieties propagated by grafting or budding. Many writers have made no distinction between races, in which the seedlings reproduce the characteristics of the parent, and varieties, which can be propagated only by vegetative means.

The confusion which involves mango nomenclature in India is rather appalling. There can be no doubt that in numerous cases the same name is applied to several distinct varieties, and it is equally certain that one variety in some instances has several different names. In addition, some of the kinds catalogued by Indian nurserymen probably never existed outside of their own imaginations. There are only a few varieties which are well known and highly esteemed in India. Most of these have been introduced into the mango-growing regions of the

Western Hemisphere by the Office of Foreign Seed and Plant Introduction of the United States Department of Agriculture.

The varieties described in the following pages are the best which have been tested in Florida up to the present. Most of them are well-known Indian sorts. They are few in number, but it is not possible to include in such a work as this a fully complete list. The classification here made into groups based on natural resemblances throws related varieties together and should aid the prospective planter to gain an idea of the more salient characteristics of each. Only the most important varieties in each group are described.

Mulgoba group.

In this group the tree is usually erect, with a broad, dense crown. The leaves are slender, smaller (especially in the variety Mulgoba) than in some of the other groups, the primary transverse veins 22 to 24 pairs, moderately conspicuous. The panicle is usually slender, frequently drooping, 12 to 18 inches in length, the axis and laterals varying from pale green tinged pink to rose pink, the pubescence heavier than in most other groups. The flowers are usually very abundant on the panicle. The staminodes are strongly developed, often capitate, one or two sometimes fertile. In general, varieties of this group require the stimulus of dry weather to make them flower profusely, and they show a decided tendency to drop most of their fruits. Haden, however, holds its fruits well. The fruit is usually oval. It varies in color from dull green to yellow blushed red, and lacks a distinct beak. The flesh is deep yellow to orange-yellow, variable in quality. The seed is normally monoembryonic.

Mulgoba (Fig. 15). — Form oblong ovate to ovate, laterally compressed; size medium to above medium, weight 9½ to 14½ ounces, length 3½ to 4½ inches, breadth 3 to 3½ inches, base flattened, with the stem inserted obliquely in a very shallow cavity; apex rounded to broadly pointed, the nak a small point on the ventral surface about $\frac{1}{2}$ inch above the longitudinal apex; surface slightly undulating, deep to apricot-yellow in color, sometimes overspread with scarlet around base and on exposed side, dots few to numerous, small, lighter in color than surface; skin thick, tough, tenacious, flesh bright orange-yellow, smooth and fine in texture, with a pronounced and very agreeable aroma, very juicy, free from fiber, and of rich piquant flavor; quality excellent; seed oblong to oblong-reniform, plump, with sparse, stiff, short fibers $\frac{1}{8}$ inch long over the surface. Season in Florida July to September.

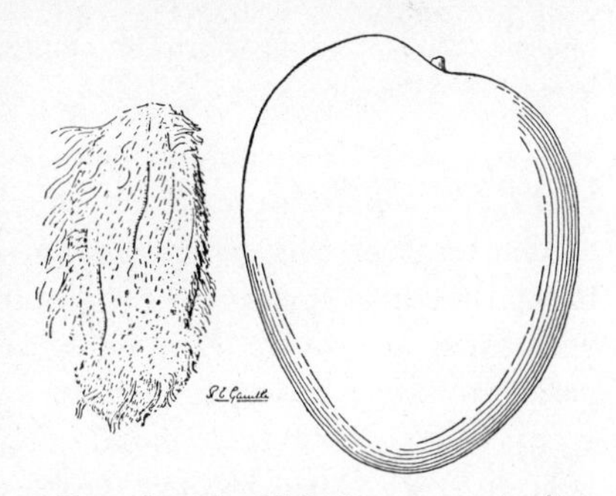

FIG. 15. The Mulgoba mango. ($\times \frac{2}{5}$)

Introduced into the United States in 1889 from Poona, India, by the United States Department of Agriculture. This was the first grafted Indian variety to fruit in the United States. In attractive coloring, delicate aromatic flavor, and freedom from fiber, Mulgoba is scarcely excelled, but it has proved irregular in its fruiting habits and for this reason cannot be recommended for commercial planting expect in regions with dry climates. The tree does not come into bearing at an early age. The name Mulgoba (properly Malghoba) is taken from that of a native Indian dish, and means "makes the mouth water."

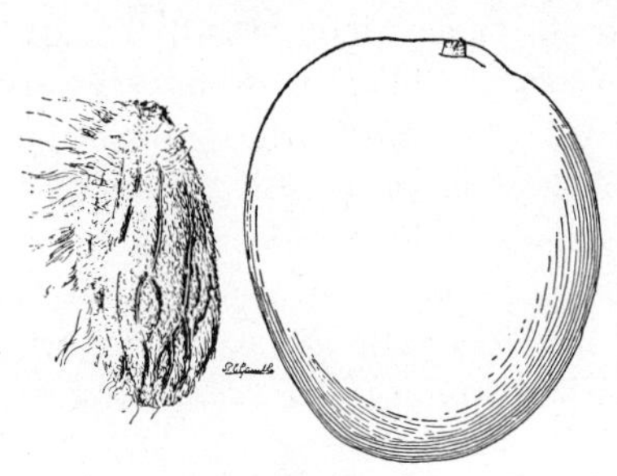

FIG. 16. The Haden mango. ($\times \frac{1}{3}$)

Haden (Fig. 16). — Form oval to ovate, plump; size large to very large, weight 15 to 20 ounces, sometimes up to 24 ounces, length 4 to $5\frac{1}{2}$ inches, breadth $3\frac{1}{2}$ to $4\frac{1}{2}$ inches, base rounded, the stem inserted almost squarely without depression; apex rounded to broadly pointed, the nak depressed, $\frac{3}{4}$ inch above the longitudinal apex; surface smooth, light to deep apricot-yellow in color, overspread with crimson-scarlet, dots numerous, large whitish yellow in color, skin very thick and tough; flesh yellowish orange in color, firm, very juicy, fibrous only close to the seed, and of sweet, rich, moderately piquant flavor; quality good; seed oblong, plump, with considerable fiber along the ventral edge and a few short stiff bristles elsewhere. Season in Florida July and August.

Originated at Coconut Grove, Florida, as a seedling of Mulgoba. First propagated in 1910. The fruit is not so fine as that of Mulgoba, but the tree is a stronger grower, comes into fruit at an early age, and bears more regularly.

Alphonse group.

The trees of this group are usually broad and spreading in habit, but in a few cases, *e.g.*, Amini, they may be rather tall, with an oval crown. The foliage is abundant, bright to deep green in color, the leaves medium to large in size, with primary transverse veins 20 to 24 pairs, fairly conspicuous. The panicle is large, very broad toward the base, stiff, sometimes stout, 10 to 18 inches long, the axis and laterals pale green to dull rose-pink in color, glabrate to very finely and sparsely pubescent. The flowers are not crowded on the panicle. The staminodes are poorly developed, rarely capitate. Most varieties of this group are not heavy bearers. Flowers are often produced sparingly, or on only one side of the tree, but a much higher percentage of flowers develops into fruits than in the Mulgoba group. Under average conditions, most of the varieties bear small to fair crops. The fruit is longer than broad, usually oblique at the base, and lacks a beak. The stigmatic point or nak often forms a prominence on the ventral surface above the apex. The color varies from yellowish green to bright yellow blushed scarlet. The flesh is orange colored, free from fiber, and is characterized by rich luscious flavor, in some varieties nearly as good as that of Mulgoba. On an average, the quality of fruit is better than in any other group. The seed contains but one embryo.

Amini (Fig. 17). — Form oval, laterally compressed; size small to below medium, weight 6 to 8 ounces, length 3 to $3\frac{1}{4}$ inches, breadth $2\frac{1}{2}$ to $2\frac{3}{4}$ inches, base obliquely flattened, cavity none; apex rounded, the nak conspicuous and $\frac{5}{16}$ inch above the end of the fruit; surface smooth, deep yellow in color overspread with dull scarlet particularly around the base, dots numerous, small, pale yellow; skin thick and firm; flesh bright orange-yellow in color, melting, very juicy, strongly aromatic, free from fiber, and of sweet unusually spicy flavor; quality excellent; seed oblong-oval, very thin, with only a few short fibers on the ventral edge. Season in Florida June and July.

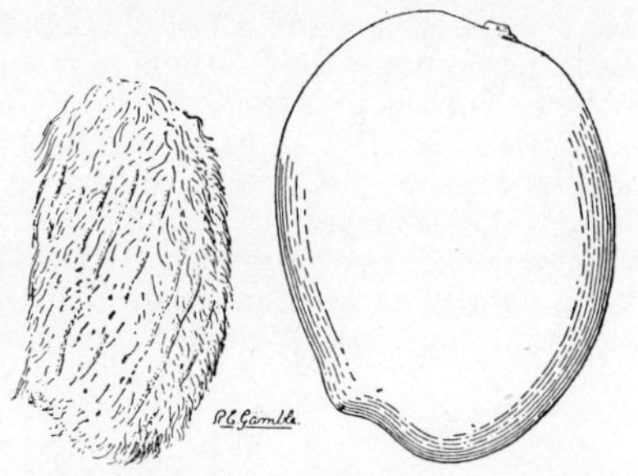

Fig. 17. Amini mango. (× about ½)

Introduced into the United States in 1901 by the United States Department of Agriculture (S. P. I. No. 7104) from Bangalore, India. One of the most satisfactory Indian varieties tested in Florida and the West Indies. It is more regular in bearing than many others, and the aroma and flavor of the fruit are excellent. Not to be confused with *Amiri*, which has sometimes been sold under the name *Long Amini*. *Amin* (Sanskrit) means a tall, pyramidal mango tree; *amin* (Arabic) means constant, faithful.

Bennett (Fig. 18). — Form ovate-oblique to ovate-cordate, very plump; size below medium to medium, weight 7 to 12 ounces, length 3 to 3¼ inches, breadth 2¾ to 3¾ inches, base obliquely flattened, cavity almost none; apex broadly pointed, the nak level or slightly depressed, about ¾ inch above end of fruit; surface smooth, yellow-green to yellow-orange, dots few, light yellow; skin thick and tough, not easily broken; flesh deep orange, free from fiber, firm and meaty, moderately juicy, of pleasant aroma and sweet, rich, piquant flavor; quality excellent; seed oblong-reniform, thick, with short stiff fibers over the entire surface. Season in south Florida late July and August.

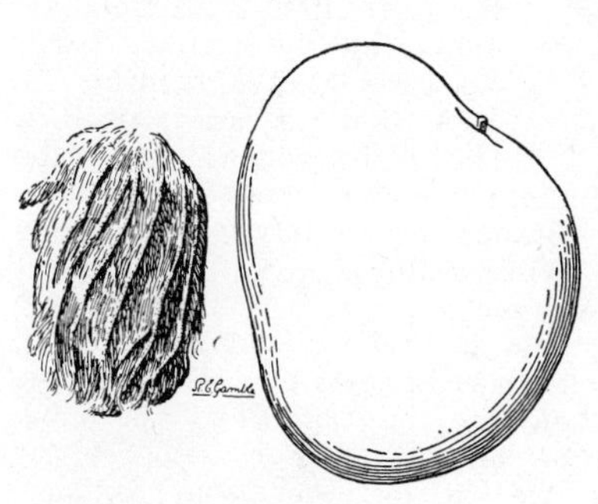

Fig. 18. The Bennett mango. (× ⅖)

Introduced into the United States in 1902 by the United States Department of Agriculture (S. P. I. 8419 and 8727) from Goregon, near Bombay, India. Syn. *Douglas Bennett's Alphonse*. This is one of the esteemed Alphonse mangos of western India. Some of the fruits produced in Florida have been characterized by hard sour lumps in the flesh, hence the variety has not made such a favorable impression as would otherwise have been the case. The tree is vigorous, and bears more regularly than Mulgoba. The Alphonse mangos are supposed to have been named for Affonso (Alphonse) d'Albuquerque, one of the early governors of the Portuguese possessions in India. The name has been corrupted to *Apoos*, *Afoos*, *Hafu*.

Pairi (Fig. 19). — Form ovate-reniform to ovate-oblique, prominently beaked; size below medium to medium, weight 7 to 10 ounces, length 3 to 3½ inches, breadth 2⅞ to 3¼ inches; base obliquely flattened, cavity none; apex rounded to broadly pointed, with a conspicuous beak slightly above it on the ventral side of the fruit; surface smooth to undulating, yellow-green in color, suffused scarlet around the base, the dots few, small, whitish yellow; skin moderately thick; flesh bright yellow-orange in color, firm but juicy, of fine texture, free from fiber, of pronounced and pleasant aroma and sweet, rich, spicy flavor; quality excellent; seed thick, with short bristly fibers over the entire surface. Season in south Florida July and August.

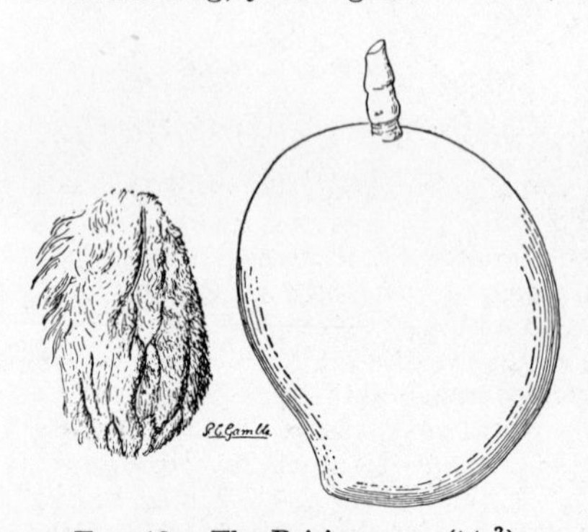

Fig. 19. The Pairi mango. (× ⅖)

Introduced into the United States in 1902 from Bombay, India, by the United States Department of Agriculture (S. P. I. 8730); a variety (S. P. I. 29510) introduced under the same name in 1911 from Poona, India, has proved to be slightly different. Syns. *Paheri*, *Pirie*, *Pyrie*. Ranks second only to Alphonse in the markets of Bombay, India. William Burns says, "Personally I prefer the slightly acid Pairi to the heavier and more luscious Alphonse." Two subvarieties are known in India, *Moti Pairi* and *Kagdi Pairi*. The tree is a good grower, and resembles Bennett in productiveness, although it sometimes fruits more heavily. The word *Pairi* is probably a corruption of the Portuguese proper name *Pereira*.

Rajpuri. — Form roundish ovate to ovate-reniform, beaked; size below medium to medium, weight 8 to 12 ounces, length 3¼ to 3¾ inches, breadth 3 to 3½ inches; base flattened, scarcely oblique, cavity none; apex bluntly pointed, with the prominent nak to one side; surface smooth, green-yellow to yellow in color, over-spread with scarlet on exposed side and around base; dots small, numerous, whitish; skin moderately thick; flesh deep yellow in color, free from fiber, juicy, with pronounced aroma and rich piquant flavor; quality excellent; seed oblong-elliptic, thick, with short stiff fibers over the surface. Season July and August in Florida.

Introduced into the United States in 1901 from Bangalore, India, by the United States Department of Agriculture (S. P. I. 7105). Syns. *Rajpury*, *Rajapuri*, *Rajabury*, and *Rajapurri*. A fruit of fine quality, with aroma and flavor distinct from that of other mangos. Its fruiting habits have proved fairly good. *Rajpur*, name of a town in India (perhaps *Rajapur?*).

Sandersha group.

The tree is erect, stiff, with the crown less broad than in the Mulgoba group and usually not so umbrageous. The foliage is fairly abundant, deep green in color, the leaves comparatively small but broad, with primary transverse veins 18 to 24 pairs, moderately conspicuous. The panicle is small to large, broad toward the base, 8 to 18 inches long, stiff, the axis and laterals deep magenta-pink to bright maroon, the pubescence very minute and inconspicuous. The flowers are abundant but not closely crowded on the panicle. The staminodes are weakly developed, rarely capitate or fertile. Varieties of this group often flower in unfavorable weather, and they remain in bloom during a long period. On the whole, the group is characterized by a higher degree of productiveness than any other class of Indian mangos yet grown in the United States. The fruit is long, usually tapering to both base and apex and terminating in a prominent beak at the apex, large in size, deep yellow in color, the flesh orange-yellow, and free from fiber. The somewhat acid flavor makes the mangos of this group more valuable as culinary than as dessert fruits. The seed is long, containing normally one embryo, the cotyledons often not filling the endocarp completely.

Sandersha (Fig. 20). — Form oblong, tapering toward stem and prominently beaked at the apex; size large to extremely large, weight 18 to 32 ounces, length $6\frac{1}{2}$ to 8 inches, breadth $3\frac{3}{4}$ to $4\frac{1}{4}$ inches; base slender, extended; apex broadly pointed, with the nak forming a prominent beak to the ventral side; surface smooth, yellow to golden yellow in color, sometimes blushed scarlet on exposed side, dots numerous, small, yellow-gray; flesh orange-yellow in color, meaty, moderately juicy, free from fiber, and of subacid, slightly aromatic flavor; dessert quality fair, culinary quality excellent; seed long, slender, slightly curved, with fiber only along the ventral edge. Season in south Florida August and September.

Introduced into the United States in 1901 from Bangalore, India, by the United States Department of Agriculture (S. P. I. 7108). Syns. *Soondershaw*, *Sandershaw*, *Sundersha*. A variety introduced from Saharanpur, India, under the name Sundershah (S. P. I. 10665) is probably distinct. The tree has remarkably good fruiting habits. Etymology of name unknown.

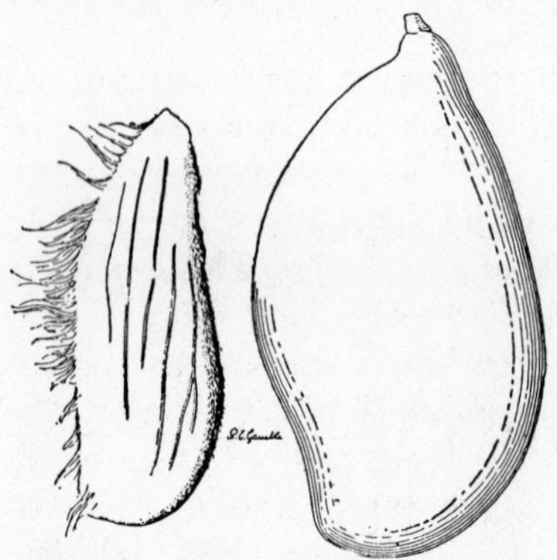

FIG. 20. The Sandersha mango. The fruit is not so richly flavored as that of Mulgoba or Pairi, but is excellent for cooking. (× $\frac{1}{3}$)

Totapari. — Form oval to oblong-reniform, beaked; size medium, weight 10 to 12 ounces, length $4\frac{1}{5}$ to 5 inches, breadth 3 to $3\frac{1}{2}$ inches; base rounded, the stem inserted squarely; apex broadly pointed, with the nak forming a prominent beak to the ventral side; surface smooth, greenish yellow in color, overspread with scarlet on exposed side; skin moderately thick and tough; flesh bright yellow in color, unusually juicy, free from fiber, moderately aromatic, and of subacid, moderately rich flavor; dessert quality fair, culinary quality good; seed oblong, rather thin, with small amount of fiber on edges. Season in south Florida August and September.

Introduced into the United States in 1902 from Bombay, India, by the United States Department of Agriculture (S. P. I. 8732). Syn. *Totafari.* The tree does not bear as well as Sandersha, nor is the fruit quite as good. The name means "parrot's beak."

Cambodiana group.

In this group the tree is erect, with the crown usually oval, never broadly spreading, and densely umbrageous. The foliage is abundant, deep green in color, the leaves medium sized to rather large, with primary transverse veins more numerous than in other groups, commonly 26 to 30 pairs, quite conspicuous. The odor of the crushed leaves is distinctive. The panicle is very large, loose, slender, 12 to 20 inches in length, and laterals pale green to dull magenta-pink, very finely pubescent. The staminodes are poorly developed, rarely capitate or fertile. The varieties of this group usually bloom profusely; those from Indo-China are productive, while the Philippine seedlings in Florida sometimes bear excellent crops and in other seasons drop all their flowers. Three to five fruits, or even more, may develop on one panicle. In form the fruits are always long, strongly compressed laterally, and usually sharply pointed at the apex, lemon-yellow to deep yellow in color, with bright yellow flesh almost free from fiber and of characteristic sprightly subacid flavor, lacking the richness of some of the Indian mangos. The seed is oblong, normally polyembryonic.

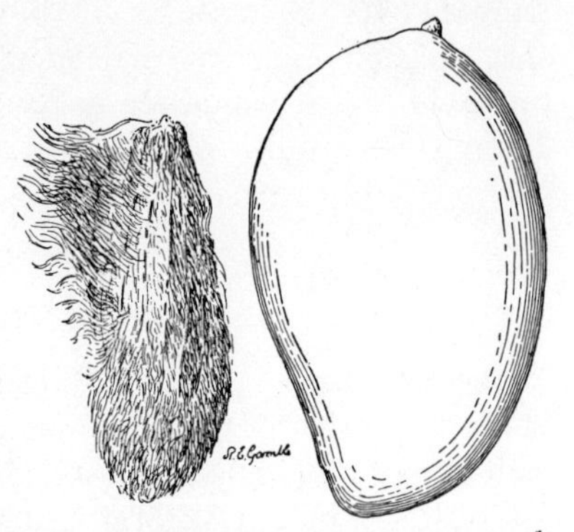

FIG. 21. The Cambodiana mango. (× $\frac{1}{3}$)

Cambodiana (Fig. 21). — Form oblong to oblong-ovate, compressed laterally; size below medium to medium, weight 8 to 10 ounces, length $3\frac{3}{4}$ to $4\frac{1}{2}$ inches, breadth $2\frac{1}{2}$ to $2\frac{3}{4}$ inches; base rounded, the stem inserted squarely or slightly to one side without depression; apex pointed, the nak a small point $\frac{1}{2}$ inch above the longitudinal apex; surface smooth, yellow-green to deep yellow in color, dots almost wanting; skin very thin and tender; flesh deep yellow in color, very juicy, free from fiber, and of mild, subacid, slightly aromatic flavor; quality good; seed elliptic-oblong, thick, with short fiber on ventral edge. Season in Florida late June to early August.

Originated at Miami, Florida, from a seed introduced in 1902 from Saigon, Cochin China, by the United States Department of Agriculture (S. P. I. 8701). A later importation of seeds from the same region (S. P. I. 11645) has given rise to another variety propagated by budding which differs slightly from the one here described. The tree bears more regularly than most of the Indian varieties. Named for Cambodia, a region of French Indo-China.

CHAPTER IV

RELATIVES OF THE MANGO

WHILE the mango is the leading fruit of the Anacardiaceæ or Cashew family, yet other species are more or less cultivated and should be briefly discussed here. The family comprises as a whole some 400 species in about 60 genera, widely distributed over the earth, mostly in warm countries. Some of the species (as poison ivy and sumac) are poisonous; but it is probable that it comprises many comestible products of value. The pistachio-nut is one of them.

THE CASHEW (Fig. 22)

(*Anacardium occidentale*, L.)

The Brazilians are the only people who fully appreciate the cashew. Father J. S. Tavares, whose studies of Brazilian fruits are probably the most exhaustive as well as the most interesting which have been published, says of this tree: "It furnishes food and household remedies to the poor, a refreshing beverage to the sick, a sweetmeat for tables richly served, and resin and good timber for industrial uses."

The readiness with which the cashew grows and fruits in a semi-wild state has kept it from receiving the horticultural attention which other and more delicate species have enjoyed. In nearly all regions where it is grown, it is more common as a naturalized plant than in the fruit garden. It does not object to such treatment, but multiplies rapidly, grows vigorously, and yields abundantly of its handsome fruit.

To see the cashew at its best, one must visit the markets of Bahia or some other city of the Brazilian coast. Here, during the short season in which they ripen, immense heaps of cashews are piled up on every side. Its brilliant shades of color, varying from yellow to scarlet, and its characteristic and penetrating aroma combine to make this one of the most enticing of all tropical fruits.

FIG. 22. Foliage, flowers, and fruit of the cashew (*Anacardium occidentale*). The kidney-shaped seed (properly speaking, the fruit) contains an edible kernel of delicious flavor, while the fleshy portion (fruit-stalk) above it is filled with aromatic juice, and may be used in many ways. (× about ¼)

The cashew is a spreading evergreen tree growing up to 40 feet in height. One of the early voyagers, Father Simam de Vasconcellos, speaks of it as "the most handsome of all the trees of America," for which extravagant statement Father Tavares takes him to task. The cashew cannot fairly be called handsome; indeed, it is oftentimes awkward or ungainly in habit, with crooked trunk and branches. The leaves, which are clustered toward the ends of the stiff branchlets, are oblong-oval or oblong-obovate in form, rounded or sometimes emarginate at the apex, and acute to cuneate at the base. They vary between 4 and 8 inches in length, and 2 and 3 inches in breadth.

The flowers are produced in terminal panicles 6 to 10 inches long. The cashew, like the mango, is polygamous; that is, some of the flowers are unisexual (staminate) and others bisexual, both types being produced on the same panicle. The calyx is five-partite, the corolla $\frac{1}{3}$ inch broad, with five linear-lanceolate, yellow-pink petals. The stamens are usually nine in number, all fertile. The ovary is obovoid, with the style placed to one side.

The fruit is peculiar. The part which would be taken for the fruit at first glance is in reality the swollen peduncle and disk, while the fruit proper is the kidney-shaped cashew-nut attached to its lower end. The fleshy portion may be termed the cashew-apple, in order to distinguish it from the true fruit, or cashew-nut. It differs in size, being sometimes as much as $3\frac{1}{2}$ inches in length, while it may be less than 2 inches. The surface is commonly brilliant yellow or flame-scarlet in color. The skin is a thin membrane, easily broken; the flesh light yellow in color and very juicy. The kidney-shaped nut which is attached to its lower end contains the single oblong seed.

The cashew was formerly thought, by some writers at least, to be indigenous both in America and Asia. It has been shown, however, that it was originally confined to America, whence it was carried to Asia and Africa by early Portuguese voyagers. Jacques Huber [1] considered it indigenous on the campos (plains) and dunes of the lower Amazon region and the north Brazilian coast in general. It spread very early to other parts of the tropical American seacoast, and probably was introduced into the West Indies by the Indians who reached those islands from the South American mainland before the arrival of Europeans. Gabriel Soares de Souza, one of the earliest chroniclers of Brazil, found the tree growing both wild and cultivated on the coast of Bahia in the sixteenth century. He mentions a "fragrant and delicious wine" which the Indians prepared from the fruit.

[1] Boletim do Museu Goeldi, 1904.

At the present time the cashew is common on the mainland of tropical America from Mexico to Peru and Brazil. It is abundant also in the West Indies. In Africa it is found on both the east and west coasts, and in Madagascar. In southern India it has become thoroughly naturalized in many of the coastal forests. It is grown in the Malay Archipelago, and is said to be abundant in Tahiti. In Hawaii it is not very common.

Regarding its occurrence in India, Dymock, Warden, and Hooper (Pharmacographia Indica) say:

"It was not known in Goa A.D. 1550; but Christopher a Costa saw it in Cochin shortly after this. . . . In 1653 only a few trees existed on the Malabar coast; since then it has become completely naturalized on the western coast, but is nowhere so abundant as in the Goa territory, where it yields a very considerable revenue. It is planted upon the low hilly ridges which intersect the country in every direction, and which are too dry and stony for other crops. The cultivation gives no trouble, the jungle being simply cut down to make room for the plants."

In the United States the culture of this tree is limited to the coast of Florida, south of Palm Beach and Punta Gorda, approximately. There are sturdy fruiting trees both at Palm Beach and Miami. In California all experiments up to the present time have indicated that the climate is not warm enough for it.

In Mexico and Central America the cashew is common on the seacoast but is rarely found at elevations higher than 3000 feet. At altitudes of 5000 or 6000 feet the climate appears to be too cool for the tree.

The English name cashew is an adaptation of the Portuguese *cajú*. The latter was taken by the earliest settlers in Brazil from the Tupi name *acajú*. In the Spanish-speaking countries of tropical America the usual name is *marañon*, presumably from the Brazilian state of Maranhão. The name *pajuil* is used in Porto Rico, while in Guatemala the similarity of the cashew to its relative the mombin (*Spondias Mombin*) is recognized in the common name *jocote marañon* (the mombin being called simply *jocote*). In India the form *kaju* (*gajus* in the Malayan region) has appeared, in addition to a number of names not derived from the American *caju*. In French the cashew-apple is called *pomme d'acajou*, and the nut *noix d'acajou*. The latter is termed *castanha* (chestnut) in Brazil.

In many regions the nut is more extensively used than the apple or fleshy portion. In Brazil this is not the case.

The cashew-apple is soft, juicy, acid, and highly astringent before maturity, retaining sufficient astringency when fully ripe to lend it zest. Owing to its remarkably penetrating, almost pungent aroma, the jam or sweetmeat made from it possesses a characteristic and highly pleasing quality. It is also used to supply both a wine and a refreshing beverage, similar to lemonade, which the Brazilians know as *cajuada*. The wine, which is manufactured commercially in northern Brazil, retains the characteristic aroma and flavor of the fresh fruit. The preserved fruit in various forms also is an article of commerce.

In several countries the cashew-nut is produced commercially and exported to Europe and North America. According to Consul Lucien Memminger, shipments to the United States from the Madras Presidency in India during the year 1915 totaled 2288 cwt., valued at $28,063. "About 15,000 cwt. of these nuts are now exported in an average season to England, France, and America, the principal port of shipment being Mangalore."

The cashew-nut is kidney-shaped, and about an inch in length. The soft, thick, cellular shell or pericarp incloses a slightly curved, white kernel of fine texture and delicate flavor. To prepare the nuts for eating, they are roasted over a charcoal fire. The shell contains cardol and anacardic acid substances which severely burn the mouth and lips of any one who attempts to bite into a fresh nut. Since these principles are decomposed

PLATE VIII. *Upper*, the cherimoya at its best; *lower*, the soursop and other fruits.

by heat, the roasted nut can be eaten without the slightest inconvenience or danger. The kernel is said to contain: fats 47.13 per cent, nitrogenous matter 9.7 per cent, and starch 5.9 per cent. An analysis made in Hawaii by Alice R. Thompson showed the presence of protein to the amount of 14.43 per cent, ash 2.58 per cent, fat 4.56 per cent, and fiber 1.27 per cent.

The cashew is not particular in regard to the soil on which it grows, but it is intolerant of frost and can only be cultivated successfully in regions where temperatures much below the freezing point are rarely experienced. An account of its culture in southwestern India is given in the Daily Consular and Trade Reports for November 3, 1914:

> "Cashew-nut trees can be grown successfully on any soil. They thrive in sandy places as well as on stone, and are not fastidious in point of soil, but are generally grown where no other crop can be produced. In this district there are many sand hills, especially below Ghats, which are utilized for this crop. Along seacoasts which are exposed to severe gusts of wind, the plants never attain the form of a tree, but keep along the ground, producing small branches.
>
> "Seeds . . . are usually planted in the month of June, at a distance of about 15 feet each way. In many cases this distance proves to be insufficient. The plants are watered the first year only. No other care is taken of them. The plantation is usually inclosed by walls.
>
> "The plants begin to bear from the third year and continue till the age of about fifteen, at which stage the trees exude a gummy substance in large quantities and then die."

In other regions the trees live to a greater age than fifteen years. Reports from many parts of the world indicate that they may come into bearing the second or third year. P. W. Reasoner recommended the cashew for cultivation in northern greenhouses, because of its habit of bearing at an early age.

In Brazil the cashew flowers in August and September and ripens its fruit from November to February. In southern India the flowering season is December and January, and the fruit ripens in March. An Indian writer estimates the yield of a mature tree at 115 to 150 pounds of fruit yearly. "To get one maund (28 pounds) of kernels about 1½ candies (115 pounds) of seed nuts are required."

Very few pests have been reported as affecting the cashew. Father Tavares [1] mentions a fungus parasite which attacks the branchlets, leaves, and flowers at Bahia, Brazil. The red-banded thrips (*Heliothrips rubrocinctus* Giard.) sometimes attacks the tree in the West Indies. H. Maxwell-Lefroy mentions two other species of thrips which have been found on the cashew in Mysore, India: these are *Idolothrips halidaji* Newm. and *Phloeothrips anacardii* Newm. (?).

Seedling cashew trees differ in the character and quantity of fruit they yield. In Brazil the trees which produce the largest and finest fruits are distinguished with varietal names. Some of these trees acquire local reputations.

Recently P. J. Wester has shown that the cashew can be shield-budded. By employing this method, it is easily possible to propagate choice varieties originating as chance seedlings. The reader is referred to Wester's publication "Plant Propagation in the Tropics," [2] one of the most valuable contributions which have been made to tropical pomology.

The method of budding the cashew is essentially the same as that described in the chapter on the avocado. Wester says in brief: "Use nonpetioled, mature budwood which is turning grayish; cut the bud 1½ to 1¾ inches long; insert the bud in the stock at a point of approximately the same age and appearance as the cion."

THE IMBU (Fig. 23)

(*Spondias tuberosa*, Arruda.)

Of the several fruits belonging to the genus Spondias which are grown in various parts of the tropics, the imbu, although

[1] In Broteria, xiv, January, 1916.
[2] Bull. 32, Philippine Bur. Agr., 1916.

relatively little known, is perhaps the best. It merits cultivation wherever climate and soil are suited to its growth.

The imbu grows spontaneously upon the catingas or dry plains of northeastern Brazil. Rarely is it cultivated, since the wild trees furnish more fruit than can be consumed. It has been planted, however, in a few localities where the wild trees are not found. It was introduced into the United States in 1914, but so far as is known, has not been planted in other countries. In view of its abundance in its native home, it is strange that a fruit of such good quality should have escaped the attention of horticulturists until very recently.

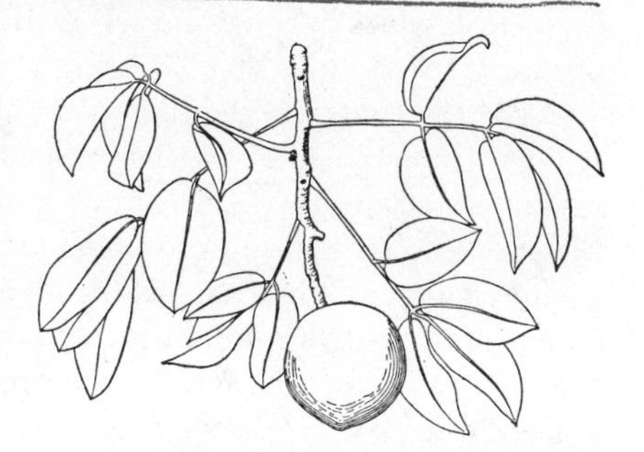

FIG. 23. Fruiting twig of the imbu (*Spondias tuberosa*). (× about $\frac{1}{3}$)

The imbu tree is distinguishable from other growths on the catinga by its low spreading crown, which is often 25 feet in diameter. The roots are swollen (whence the specific name *tuberosa*), and are said by M. Pio Correa to be used as food in times of scarcity. The leaves are 4 to 6 inches long, with five to nine oblong-ovate leaflets, equilateral or nearly so, subserrate or entire, and from 1 to $1\frac{3}{4}$ inches in length. The small white flowers are borne in panicles 4 to 6 inches long. Like those of other species of Spondias, the flowers are composed of a calyx having four or five segments and a corolla of four or five valvate petals. The stamens are eight to ten in number, the styles three to five.

The fruit is produced on slender stems, mainly toward the ends of the branches. Some trees are so productive that the fruit, when allowed to fall, forms a carpet of yellow upon the ground. In general appearance the imbu may be likened to a Green

Gage plum. It is oval, about 1½ inches in length, and greenish yellow in color. The skin is thicker than that of a plum, and quite tough. The flavor of the soft, melting, almost liquid flesh is suggestive of a sweet orange. If eaten before it is fully ripe, the fruit is slightly acid. The seed is oblong and about ¾ inch in length.

In its native home the imbu is eaten as a fresh fruit, and also furnishes a popular jelly. It is used besides to make *imbuzada*, a famous dessert of northern Brazil. This is prepared by adding the juice of the fruit to boiled sweet milk. The mixture is greenish white in color and when sweetened to taste is relished by nearly every one.

While the tree is susceptible to frost, it cannot be considered strictly tropical. In south Florida young plants have withstood temperatures of 28° above zero without serious injury. Little is known regarding its adaptability to various soils and alien climates. While the wild trees are found on very dry soil in a region of little rainfall, it is possible that other conditions will prove suitable. A few bearing trees were seen by the writer in the city of Bahia, Brazil, where the humidity is great and the annual rainfall about 60 inches. In south Florida it has been tried at Miami, but has not done well. Its failure there has been attributed to the large amount of lime contained in the soil, but it is not certainly known that this is the limiting factor. The soil on the Brazilian catingas is a gravelly loam, sometimes mixed with clay, sometimes sandy.

Fruit from the wild trees varies in size, color, and quality. It should be easy to propagate the best seedlings by cuttings; at least, other species of Spondias are propagated in this way. Mature wood is used. At Miami, Florida, the imbu has been inarched on the ambarella (see below). Seeds are easily induced to grow, and should be germinated in flats or boxes of light soil.

THE AMBARELLA (Plate VI)

(*Spondias cytherea*, Sonnerat.)

This is probably the most widely cultivated species of Spondias, although it is not so extensively distributed, in its wild state, as the yellow mombin. It is known in many tropical countries and can be cultivated successfully as far north as southern Florida. While not generally considered a fruit of excellent quality, an occasional tree is much superior to the average and is worth propagating.

The ambarella is an erect, stately, semi-deciduous tree, usually stiff in appearance. It reaches a maximum height of 60 feet. The leaves are large, commonly 8 to 12 inches long; the leaflets, 11 to 23 in number, are oval to oblong in outline, $2\frac{1}{2}$ to 3 inches in length, remotely serrate, and acuminate at the apex. Like those of the imbu, they are equilateral or nearly so. The small whitish flowers are produced in large loose panicles 8 to 12 inches in length.

The fruit is oval or slightly obovoid in form, 2 to 3 inches long, and orange-yellow in color. The skin is as thick as that of the mango, but tougher. The flesh is firm, very juicy, and of pale yellow color. Its subacid flavor suggests that of the apple; sometimes, however, it is resinous or pungent. The seed is large, oval, 1 inch in length, covered with stiff spines or bristles to which the surrounding flesh clings tenaciously.

Although larger than those of other species of Spondias, the fruits of the ambarella are not usually so pleasantly flavored as are choice imbus or the best red mombins (see below). They are produced in long pendent clusters of two to ten. In Florida they ripen during the winter: in Tahiti the season is said to be May to July, and in Hawaii November to April. The composition of the fruit, according to an analysis by Alice R. Thompson of Hawaii, is as follows: Total solids 14.53 per cent, ash 0.44 per cent, acids 0.47 per cent, protein 0.50 per cent, total sugars 10.54 per cent, fat 0.28 per cent, and fiber 0.85 per cent.

Ambarella is the Sinhalese name used in Ceylon, and is preferred as being more euphonious and attractive than the name Otaheite-apple. The latter term is current in some of the British colonies, but is sometimes applied also to a different fruit, the ohia. Jew-plum is another name for the ambarella, used in Jamaica. The French call the fruit *pomme Cythere.* In Polynesia its name is *vi* or *evi,* the former word (spelled wi) being used in Hawaii. In Brazil the Portuguese name is *cajá-manga.* *Spondias dulcis* Forst. is a botanical synonym of *S. cytherea* Sonnerat.

The tree is considered indigenous in Polynesia. It was brought to Jamaica in 1782, and again in 1792 (on this second occasion by Captain Bligh, who introduced the breadfruit into the West Indies from Tahiti). It has not become popular in Cuba, nor is it commonly grown on the mainland of South America, with the exception of certain parts of Brazil. In South Florida it is successful as far north as Palm Beach. No trees are known to have reached fruiting size in California. The winters there are probably too cool for it.

While the tree thrives best on deep rich soils, it has been successful in Florida (though not reaching large size) upon shallow sandy land. Thomas Firminger says that the seeds do not germinate readily, and that plants "are usually obtained by grafting upon seedlings of *S. mangifera.*" P. J. Wester has found that the species can be shield-budded in the same manner as the avocado; he says, "Use nonpetioled, slender, mature, but green and smooth budwood; cut large buds with ample wood-shield, $1\frac{1}{2}$ to $1\frac{3}{4}$ inches long; insert the buds in the stock at a point of approximately the same age and appearance as the cion."

Early travelers who visited Polynesia spoke of this fruit in high terms. More recently, however, it has been likened to a "very bad mango," and several writers have adjudged that it did not merit cultivation. Much depends on the variety; while the average may be poor, an occasional one is good. Only superior kinds propagated by some vegetative means should be planted. As yet no attempt has been made to find the best varieties and establish them as horticultural forms.

The Red Mombin (Plate VII)

(*Spondias Mombin*, L.)

No other species of Spondias is so extensively used in tropical America as this. In many parts of Mexico and Central America it is a fruit of the first importance. It occurs in a wide range of seedling races or forms, and is capable of great improvement by selection and vegetative propagation. While scarcely so good as the imbu, the better varieties are pleasantly flavored and attractive in appearance.

The red mombin is a small tree, often spreading in habit. The trunk is thick and the branches are stout and stiff. Its native home is tropical America, where it reaches a maximum height of about 25 feet. The leaves are 5 to 8 inches long, with 16 to 21 oblong-elliptic, oblique, subserrate leaflets 1 inch to $1\frac{1}{2}$ inches in length. The purplish maroon flowers are produced in small unbranched racemes about $\frac{1}{2}$ inch long.

The fruits, borne singly or in clusters of two or three, are quite variable in size and form. Commonly they are oval or roundish, but they may be oblong, obovoid, or somewhat pyriform. They range from 1 to 2 inches in length, and from yellow to deep red in color. The seed is oblong, $\frac{1}{2}$ to $\frac{3}{4}$ inch long, and rough on the surface. The season of ripening in most parts of tropical America is August to November.

In most Spanish-speaking countries this species is known as *ciruela* (plum), a name which has been corrupted in the Philippines to *siniguelas*. In parts of Mexico and in Guatemala it is known by the Aztec name *jocote* (xocotl). The common name in the French colonies is *prunier d'Espagne, prunier rouge,* and *mombin rouge,* and in the British colonies it is sometimes called Spanish-plum. *Spondias purpurea,* L. is a botanical synonym of *S. Mombin,* L.

J. N. Rose[1] describes a number of different forms observed in Mexico. These races (perhaps species in some instances) deserve further study.

The red mombin is abundant in Mexico and Central America from sea-level up to elevations of 5000 or 6000 feet. The value of the annual crop in Mexico is estimated at more than $70,000. The fruit may be eaten fresh or may be boiled and dried, in which latter condition it can be kept for several months. When fresh it has a subacid spicy flavor somewhat resembling that of the cashew, but less aromatic. Some varieties are sour, and others have very little flesh; the best are pleasantly flavored and have about the same amount of flesh and seed as a very large olive.

In Cuba several seedling races are grown. They are usually distinguished as *ciruela roja, ciruela amarilla,* and so on. In Brazil the species appears to be little known. It is successfully cultivated in south Florida, as far north as Palm Beach or perhaps farther. Varieties from high elevations in tropical America should prove slightly hardier than those from the seacoast. No trees have been grown to fruiting age in California, so far as is known. In favorable situations they might succeed there if given protection during the first few winters.

The tree is semi-deciduous. The leaves fall toward the end of the cool season and are soon replaced by new ones.

The character of the soil does not seem to be important. Good trees can be found growing on shallow sandy land, on gravel, and on heavy clay loam. A rich, moist, fairly heavy

[1] The Useful Plants of Mexico; contributions from the U. S. Nat. Herbarium, V, 4, 1899.

loam perhaps suits it best. Cuttings take root so readily that large limbs, cut and inserted in the ground as fenceposts, will often develop into flourishing trees. P. J. Wester recommends that cuttings 20 to 30 inches long, of the previous season's growth (or even older wood) should be set in the ground to a depth of about 12 inches, in the positions which the trees are to occupy permanently. The rainy season is the best time to do this. The trees should stand about 25 feet apart, unless the soil be very poor, in which case 20 feet will be sufficient. No horticultural varieties have as yet been established. By selecting from the existing seedlings in tropical America, many good ones could be obtained.

The Yellow Mombin

(*Spondias lutea*, L.)

This species is generally considered inferior in quality to the red mombin. Its cultivation is much less extensive, but it occurs abundantly as a wild tree in many tropical regions. The name hog-plum, which has been applied to it in the West Indies, has perhaps given it a lower reputation than it merits, but the term does not, as Cook and Collins point out, cast any reflection on the character of the fruit, inasmuch as it refers only to the fact that hogs are extremely fond of it, and fatten on the fruit which falls to the ground from wild trees in the forest.

The tree is tall and stately in appearance, and under favorable conditions it may reach 60 feet in height. The leaves are 8 to 12 inches long, composed of 7 to 17 ovate-lanceolate or lanceolate-serrulate leaflets, oblique at the base and 2½ to 4 inches in length. The yellowish white flowers are borne in loose panicles 6 to 12 inches long. The fruit is ovoid, commonly 1 inch in length, bright yellow, with thin skin, and an oblong seed of relatively large size. The flesh is yellow, very soft and juicy, and of subacid, rather pungent flavor. Many varieties are scarcely pleasant to the taste, others are sweet and agreeable. The fruit is usually eaten fresh. Its composition, according to an analysis by Alice R. Thompson of Hawaii, is as follows: Total solids 11.47 per cent, ash 0.65 per cent, acids 0.98 per cent, protein 1.37 per cent, total sugars 9.41 per cent, fat 0.56 per cent, and fiber 1.16 per cent.

The species is considered to be cosmopolitan in the tropics. In Spanish-speaking countries it is called *jobo,* while in Brazil it is known as *cajá.* In the French colonies the names *mombin jaune* and *prune Myrobalan* are current. *S. Mombin,* Jacq. (not L.) is a botanical synonym of *S. lutea,* L.

Occasional trees are seen in cultivation throughout tropical America. Cook and Collins report that it is planted extensively in Porto Rico. In south Florida it succeeds, but has never become common. In California no trees of fruiting age are known. The species is rather susceptible to frost; it is found in the tropics only at low elevations, and probably will not withstand temperatures much below freezing point, particularly when young.

The method of propagation is the same as that used for the red mombin (see above), *i.e.,* by cuttings of mature wood.

CHAPTER V

THE ANNONACEOUS FRUITS

THE annonas are tropical fruits composed of more or less coherent fleshy carpels or parts. More than 50 species are known, several of which are widely cultivated for their fruits. The family comprises 40 to 50 genera. One of them, Asimina, is native in temperate North America, and one species (*Asimina triloba,* known also as papaw but very different from the papaya) occurs as far north as New York and Michigan.

THE CHERIMOYA (Plate VIII)

(*Annona Cherimola*, Mill.)

"Deliciousness itself" is the phrase Mark Twain used to characterize the cherimoya. Sir Clements Markham quotes an even more flattering description:

"The pineapple, the mangosteen, and the cherimoya," says Dr. Seemann, "are considered the finest fruits in the world. I have tasted them in those localities in which they are supposed to attain their highest perfection, — the pineapple in Guayaquil, the mangosteen in the Indian Archipelago, and the cherimoya on the slopes of the Andes, — and if I were called upon to act the part of a Paris I would without hesitation assign the apple to the cherimoya. Its taste, indeed, surpasses that of every other fruit, and Haenke was quite right when he called it the masterpiece of Nature."

The cherimoya is essentially a dessert fruit, and as such it certainly has few equals. Although its native home is close to the equator, it is not strictly tropical as regards its requirements, being, in fact, a subtropical fruit, and attaining perfection only where the climate is cool and relatively dry. At home it grows on plateaux and in mountain valleys where proximity to the equator is offset by elevation, with the result that the climate is as cool as that of regions hundreds of miles to the north or south.

Commercial cultivation of the cherimoya has been undertaken in a few places. This fruit has not, however, achieved the commercial prominence which it merits, and which it seems destined some day to receive.

That it should be unknown in most northern markets, notwithstanding that it grows as readily in many parts of the tropics and subtropics as the avocado, can only be due to the inferiority of the varieties which have been disseminated, to tardiness in utilizing vegetative means of propagation, and to insufficient attention to the cultural requirements of the tree. The best seedling varieties must be brought to light, they must be propagated by budding or grafting, and a careful study made of pollination, pruning, fertilization of the soil, and other cultural details as yet imperfectly understood. There is no reason why, when this has been done, cherimoya culture should not become an important horticultural industry in many regions. Experience in exporting the fruit from Madeira to London, and from Mexico to the United States, has shown that it can be shipped without difficulty. The demand for it in northern markets, once a regular supply is available, is certain to be keen.

The cherimoya is a small, erect or somewhat spreading tree, rarely growing to more than 25 feet high; on poor soils it may not reach more than 15 feet. The young growth is grayish and softly pubescent. The size of the leaves varies in different varieties; in some they are 4 to 6 inches long, in others 10 inches. In California a variety (originally from Tenerife,

Canary Islands) with unusually large leaves has been listed by nurserymen under the name *Annona macrocarpa*. In form the leaves are ovate to ovate-lanceolate, sometimes obovate or elliptic; obtuse or obtusely acuminate at the apex, rounded at the base. The upper surface is sparsely hairy, the lower velvety tomentose. The fragrant flowers are about an inch long, solitary or sometimes two or three together, on short nodding peduncles set in the axils of the leaves. The three exterior petals are oblong-linear in form, greenish outside and pale yellow or whitish within; the inner three are minute and scale-like, and ovate or triangular in outline. As in other species of Annona, the stamens and pistils are numerous, crowded together on the fleshy receptacle.

The fruit is of the kind known technically as a syncarpium. It is formed of numerous carpels fused with the fleshy receptacle. It may be heart-shaped, conical, oval, or somewhat irregular in form. In weight it ranges from a few ounces to five pounds. Sixteen-pound cherimoyas have been reported, but it is doubtful whether they ever existed in reality. The surface of the fruit in some varieties is smooth; in others it is covered with small conical protuberances. It is light green in color. The skin is very thin and delicate, making it necessary to handle the ripe fruit with care to avoid bruising it. The flesh is white, melting in texture, and moderately juicy. Numerous brown seeds, the size and shape of a bean, are embedded in it. The flavor is subacid, delicate, suggestive of the pineapple and the banana.

The cherimoya is sometimes confused with other species of Annona. W. E. Safford,[1] who has studied the botany of this genus thoroughly, writes:

"For centuries the cherimoya has been cultivated and several distinct varieties have resulted. One of these has smooth fruit, devoid of protuberances, which has been confused with the inferior fruit of both *Annona glabra* and *A. reticulata*. The

[1] In Bailey, Standard Cyclopedia of Horticulture.

last two species, however, are easily distinguished by their leaves and flowers; *Annona glabra*, commonly known as the alligator apple or mangrove annona, having glossy laurel-like leaves and globose flowers with six ovate petals, and *A. reticulata* having long narrow glabrate leaves devoid of the velvety lining which characterizes those of the cherimoya."

Annona Cherimola, Mill. is the *Annona tripetala* of Aiton; the plant which has been offered in California under the name *A. suavissima* is a horticultural form of *A. Cherimola*. (The orthography *Anona Cherimolia* was used until Safford showed that it is incorrect.)

The country of origin of the cherimoya remains somewhat in doubt. Alphonse DeCandolle, after weighing all the available evidence, said, "I consider it most probable that the species is indigenous in Ecuador, and perhaps in the neighboring part of Peru." The presence of the fruit in Mexico and Central America since an early day has led other botanists to assume that it might also be indigenous in the latter countries. Recently Safford has re-sifted the evidence and has reached the conclusion that "De-Candolle is in all probability correct in attributing it to the mountains of Ecuador and Peru. The common name which it bears, even in Mexico, is of Quichua origin . . . and terra-cotta vases modeled from cherimoya fruits have been dug up repeatedly from prehistoric graves in Peru."

The name by which this fruit is known in Spanish-speaking countries, *cherimoya* or *chirimoya*, is derived (as mentioned above, quoting Safford) from the Peruvian name *chirimuya*, signifying cold seeds. The English frequently spell the word cherimoyer. The name custard-apple is often used in the British colonies; its application is not confined, however, to this one species, but extends to other annonas. The French use the name *cherimolier*, or more frequently *anone*. The name cherimoya or one of its variants is sometimes applied to other species of Annona.

From its habitat in South America, the cherimoya early spread northward into Mexico; much later it passed into the West Indies, the southern part of South America, and across the seas to the islands near the African coast, to the Mediterranean region, and to India, Polynesia, and Africa.

At present it is naturalized in many parts of Mexico and Central America. Throughout this region it occurs most abundantly at elevations of 3000 to 6000 feet, occasionally ascending (in Guatemala) to 8000 feet. On the seacoast it is not successful as a fruit-tree, and is rarely grown. The regions which produce the finest cherimoyas in Mexico lie at elevations of 5000 to 6000 feet and are characterized by comparatively dry cool climates. Excellent cherimoyas are grown at Querétaro and in the vicinity of Guadalajara. The fruit is highly esteemed in the markets of Mexico City, where it sells at high prices. While not grown commercially on a scale comparable with the avocado, its culture in certain regions is important, and regular shipments are made to the principal markets of the country.

In Jamaica, where the cherimoya was introduced by Hinton East in 1785, there are now many trees in the mountainous parts of the island. The fruit is highly esteemed in the markets of Kingston. In Cuba it is almost unknown. There are a few trees in Oriente Province and perhaps elsewhere, but the markets of Habana are not familiar with it. It may be mentioned that *Annona reticulata* is often called cherimoya in Cuba, which has led some writers to assume wrongly that the true cherimoya is commonly cultivated in the island.

In Argentina, cherimoya culture is conducted commercially in several places, notably the Campo Santo district in the province of Salta. The fruit is shipped to Buenos Aires, where it is marketed at very profitable prices. In Brazil it is not commonly grown; in fact it is not known in most parts of the Republic.

In 1897 M. Grabham wrote a short article in the Journal of the Jamaica Agricultural Society on the cultivation of the cherimoya in Madeira. He asserted that "many of the estates on the warm southern slopes of the island, formerly covered with vineyards, have now been systematically planted with the cherimoya" and went on to state that "the fruits vary in weight between three and eight pounds, exceptionally large ones may reach 16 pounds and over." This article, which has been widely quoted, has been responsible for the current belief that cherimoya culture in Madeira is more extensive than in any other part of the world, and that exceptionally fine varieties have been developed.

Charles H. Gable, an American entomologist and horticulturist who worked in the island during 1913 and 1914, has dispelled these illusions. Gable writes:

"I found the cherimoya industry in Madeira very primitive indeed. No effort has been made to commercialize the growing of this fruit. Most of the trees are volunteers which have sprung up from dropped seeds, or else they have been planted for shade, with perhaps a vague notion that they might some day produce fruit. . . . I do not know any one in Madeira (and I have been over the entire island) who has more than a dozen trees in bearing, and only a few have that many. Most of the important islanders have at least one tree. . . . At least 95 per cent of all those on the island are seedlings. Occasionally old trees are top-worked by a method of cleft-grafting, but this is not highly successful. . . . There is no uniformity in the quality of the fruits. Every gradation is found between smooth-surfaced and very rough fruits. In those which resemble each other externally there may be great differences in quality, acidity, number of seeds, and other characteristics. I never got so I felt competent to pick out a good fruit in the market. . . . The rough type attains the greatest size. The largest specimen I was able to find weighed three and a half pounds. . . . I hesitate to make an estimate, but I do not believe more than a thousand dozen fruits are exported from the island in a year. . . . The trees receive no intentional cultivation. Vegetables are often planted beneath them. A species of scale insect and the mealy bug infest many of them. . . . The trees do not seem to do well above 800 feet elevation. The ripening season is from the last of November until the first of February."

In the Canary Islands the cherimoya is not cultivated commercially, but it is grown on a limited scale. Georges V. Perez writes: "Ever since I can remember it has been cultivated in the gardens of Orotava as a delicious and perhaps unequalled tropical fruit."

In the Mediterranean region there are several localities in which it can be grown successfully. A. Robertson-Proschowsky, who has experimented with many tropical and subtropical plants at Nice, France, finds that the fruits, if caught by cold weather before they mature, do not ripen perfectly. If, however, the winter is mild and warm they may mature satisfactorily, even if very late. Robertson-Proschowsky believes that the cherimoya is well suited for cultivation in sheltered spots along the Côte d'Azur (French Riviera), and he recommends it as a fruit worthy of serious attention in that region.

It is cultivated on a limited scale in southern Spain and in Sicily. L. Trabut[1] of Algiers writes: "Lovers of the anona will find in the markets of Algiers, during November and December of each year, a few good fruits which are sold at 30 centimes to 1 franc each. These fruits come from gardens along the western coast, where there are some magnificent trees." He further says: "It seems evident that the moment has come to extend cherimoya culture. It is not more difficult than orange culture, and at present promises to be more remunerative." Trabut recommends that the tree be planted in Algeria on the coast only, since the climate of the interior is too cold.

The cherimoya has been planted in several parts of India but has not become a common fruit in that country. H. F. Macmillan says that it is "now cultivated in many up-country gardens in Ceylon." It was introduced into the latter island as late as 1880. In parts of Queensland, Australia, it is successfully grown.

In Hawaii it has become well established. Vaughan Mac-

[1] Bull. 24, Service Botanique, Algeria.

Caughey[1] says: "It was introduced into the Hawaiian Islands in very early times, and is now naturalized, particularly in certain parts of the Kona and Ka-u districts on the island of Hawaii." He adds that cherimoyas are rarely seen in the markets of Honolulu, but that trees are found in gardens throughout the city.

Nowhere in Florida is the cherimoya a common fruit. Trees in limited numbers have been planted in several parts of the state, notably in the Miami region. While they grow vigorously they do not fruit so freely, nor is the fruit of such good quality, as in many other countries. It is probable that the climate of south Florida is too tropical for this species.

As regards California, it is believed that the first cherimoyas planted in the state were brought from Mexico by R. B. Ord of Santa Barbara in 1871. A few years later Jacob Miller planted a small grove on his place at Hollywood, near Los Angeles. In the relatively short time since these first plantings were made, the cherimoya has become scattered throughout southern California, from Santa Barbara to San Diego. The climate and soil of the foothill regions seem to be peculiarly suited to it. A few commercial plantings have been made, notably at Hollywood, but since they are composed entirely of seedlings they have not proved remunerative. Had budded trees of desirable varieties been planted, the results would have been different. In the largest commercial planting, that of A. Z. Taft at Hollywood, one seedling, more productive than the remainder, produced one year about one-fourth the entire crop of the grove. Out of eighty trees comprised in the planting, only five produced more than a few fruits. By top-working the unproductive trees to a productive and otherwise desirable variety, they could have been made valuable.

For sheltered situations throughout the foothill tracts of southern California, cherimoya culture holds great promise.

[1] Torreya, May, 1917.

As soon as budded or grafted trees of good varieties are available, many small orchards should be established quickly.

The cherimoya is commonly eaten fresh: rarely is it used in any way except as a dessert fruit. Alice R. Thompson, who has analyzed the fresh fruit in Hawaii, finds that it contains: Total solids, 33.81 per cent, ash 0.66 per cent, acids 0.06 per cent, protein 1.83 per cent, total sugars 18.41 per cent, fat 0.14 per cent, and fiber 4.29. It will be noted that the sugar-content is high, while that of acids is low. The percentage of protein is higher than in many other fruits.

Cultivation.

The climatic requirements of the cherimoya have been indicated in the discussion of the regions in which it is cultivated. It is essentially a subtropical fruit, and in the tropics succeeds only at elevations sufficiently great to temper the heat. It thrives best in regions where the climate is relatively dry. In the southern part of Guatemala, where the annual rainfall is about 50 inches but where there is a long dry season, it is extensively grown and the fruit is of excellent quality; but in the northern part of the same country, where the rainfall is nearly 100 inches, distributed throughout the year, the tree cannot be grown successfully. In the highlands of Mexico it is best suited where the climate is dry, free from extremes both of heat and cold, and where abundant water is available for irrigating. The climate of southern California, except in sections subject to severe frosts, seems almost ideal for it. In many places frost is the limiting factor, for the cherimoya, while the hardiest of its genus, does not endure temperatures lower than 26° or 27° above zero without serious injury. Young plants will, of course, be hurt by mild frosts which mature trees would ignore; in fact, temperatures lower than 29° or 30° are likely to injure them.

Like other annonas, the cherimoya prefers a rich loamy soil. It can be grown, however, on soils of many different types.

In California it has done well on heavy clay (almost adobe), while in Florida it makes satisfactory growth on shallow sandy soils. H. F. Schultz considers the ideal soil to be a fairly rich, loose sandy loam, underlaid with gravel at a depth of two to three feet. He says: "Some of the best Campo Santo and Betania (Argentina) groves are located on such land, which is furthermore characterized by a liberal outcropping of scattered rocks." Carlos Wercklé states that the tree does well in Costa Rica on "stony cliffs." He reports that it is more productive under these conditions than when grown on richer soil, and himself considers it partial to mountain slopes on which there is much limestone rock.

Experience in California has shown that the cherimoya requires cultural treatment similar to that given the citrus fruits. Budded trees should be planted in orchard form about 20 to 24 feet apart; seedlings about 30 feet apart, since they grow to larger size. Irrigations, followed by thorough cultivation of the soil, are given at intervals of two weeks to one month. While the trees are young, more frequent irrigations are necessary. In Argentina, according to H. F. Schultz, it is the custom to irrigate the trees at intervals of six to twelve days. In Mexico two weeks is considered the proper interval.

In California, stable manure has been used for young trees with excellent results, and occasionally for bearing groves. Little attention has been devoted to the subject; hence it is not possible to give specific directions for the use of fertilizers. A writer in the Queensland Agricultural Journal recommends that each tree be given annually 1 to 3 pounds of superphosphate, 2 to 6 pounds of meat-works manure with blood, and 1 to 2 pounds of sulfate of potash.

The pruning of cherimoyas has received little attention as yet in the United States. In Argentina it is considered that trees which are kept low and compact are both more precocious and longer lived than those which are tall and open in habit.

In Guatemala the most productive trees are usually those which have been cut back heavily. It is possible that fruitfulness can be increased by severe pruning. The matter deserves careful investigation. The tree being semi-deciduous, pruning should be done after the leaves have dropped and before the new foliage makes its appearance.

Propagation.

In many regions seed-propagation is the only method which has been used with this plant. In the United States, in Madeira, in Algeria, and in the Philippines, cherimoyas have been grafted and budded successfully; one or the other of these methods should be employed to perpetuate choice varieties.

If kept dry the seeds will retain their viability several years. Given warm weather or planted under glass, they will germinate in a few weeks. Under glass they may be sown at any time of the year; if in open ground, they should be planted only in the warm season. Seeds should be sown in flats of light porous soil containing an abundance of humus, and should be covered to a depth of not more than $\frac{3}{4}$ inch. When the young plants are three or four inches high, they may be transferred into three-inch pots. Good drainage must be provided, and they should not be watered too copiously. When eight inches high they may be shifted into larger pots, or set out in the open ground. In the latter case, they must have careful attention, and, preferably, shade, until they have become well established.

For stock-plants on which to bud or graft the cherimoya, several species of Annona have been employed. *A. reticulata, A. glabra,* and *A. squamosa* are all recommended by P. J. Wester. In Florida *A. squamosa* has proved to be a good stock when a dwarf tree is desired; *A. glabra* tends to outgrow the cion. In California, seedling cherimoyas as stock-plants have given the best results.

Shield-budding has worked very satisfactorily in the United

States. In several other regions horticulturists have found grafting more successful. Budding is best done at the beginning of the growing season, when the sap is flowing freely. Stock-plants should be $\frac{3}{8}$ to $\frac{1}{2}$ inch in diameter. Well-matured budwood from which the leaves have dropped is preferable, and it should be gray, not green, in color. The buds should be cut $1\frac{1}{2}$ inches in length, and should be inserted exactly as in budding avocados or mangos. Waxed tape, raffia, and soft cotton string have proved satisfactory for tying. Three or four weeks after insertion of bud, the wrapping should be loosened and the stock lopped at a point 5 or 6 inches above the bud. Wrapping should not be removed entirely until the bud has made a growth of several inches.

For grafting, two-year-old seedlings are to be preferred (for budding they may be somewhat younger). The cleft-graft is the method usually employed. The cion should be well-matured wood from which the leaves have dropped. C. H. Gable wrote from Madeira in 1914: "I have been surprised to find how easily the annona is grafted. My first few efforts were not very successful, but later I grafted them in all sizes from seedlings smaller than a lead pencil to old trees, and more than 90% have grown beautifully." Gable found it advisable after making the graft to paint the cion and the top of the stock (around the cleft) with melted wax, to prevent evaporation.

Old seedling trees can be top-worked without difficulty. For this purpose cleft-grafting is used more commonly than any other method.

The pollination of the cherimoya has been investigated in Florida by P. J. Wester, and in Madeira by C. H. Gable. It has been thought that the scanty productiveness of many trees might be due to insufficient pollination, and the investigations tend to confirm this belief. Gable reports that normally in Madeira not more than 5 per cent of the flowers produced develop into fruits. By hand-pollinating them, however, he was able to obtain thirty-six fruits from forty-five flowers.

After carrying on pollination experiments in Florida during several years, P. J. Wester[1] wrote: "The investigations indicate that the flowers of the cherimoya, the sugar-apple, the custard-apple and the pond-apple are proterogynous and entomophilous, though the pollinating agent of the last-named species has not been detected." A proterogynous plant, it may be remarked, is one in which the pistils are receptive before the anthers have developed ripe pollen, cross-pollination being therefore necessary, and some outside agency being required to effect it. In the case of the annonas the work is done by insects; hence the plants are termed entomophilous.

The pollination of the closely allied *Asimina triloba* is thus described by Delpino:[2] "The stamens project in the center of the pendulous protogynous (proterogynous) flower as a hemispherical mass, from the middle of which a few styles with their stigmas project. In the first (female) stage of anthesis the three inner petals lie so close to the stamens that insect visitors (flies) cannot suck the nectar secreted at the bases of the former without touching the already mature stigmas. In the second (male) stage the stigmas have dried up and the inner petals have raised themselves, so that the anthers, — now covered with pollen, — are touched by insects on their way to the nectar. Cross-pollination of the younger flowers is therefore effected by transference from the older ones."

Wester concluded that one cause of the unproductiveness of the cherimoya in Florida was the scarcity of pollinating insects. Even under the same conditions of environment, however, there are marked differences in productiveness among seedling trees. The subject deserves further investigation. Productive varieties especially should be studied, to determine whether

[1] Bull. of the Torrey Bot. Club, 37, 1910.
[2] Paul Knuth, Handbook of Flower Pollination.

or not they differ in any way from the typical less fecund form in manner of pollination.

The crop.

Seedling cherimoyas, when grown under favorable cultural conditions, begin to bear the third or fourth year after planting. Most of them, even at fifteen or twenty years of age, do not produce annually more than a dozen good fruits. Occasional trees are more satisfactory in this respect, and it is such trees which should be propagated by budding. The writer has observed one small tree in Guatemala which bore eighty-five fruits in a single season, and C. H. Gable found a tree in Madeira which bore three hundred.

In California the main season for cherimoyas is spring, usually March and April; but sometimes a few fruits mature in late autumn. In Argentina the season is February to July. Felix Foex states that there are ripe cherimoyas in Mexico throughout the year, owing to the presence of trees at different elevations. From personal observation the writer ventures to doubt whether this all-year season is a fact; in any event, they are not abundant during the entire year. In Madeira the fruit begins to ripen about the end of November and continues in season until early in February.

When fully mature or "tree-ripe," the fruits are picked and laid away to soften. If, however, they are to be shipped to distant markets they are packed as soon as removed from the tree, and dispatched at once so that they will reach their destination before they have become soft. When fully mature and ready to pick, they usually have a yellowish tinge. In Mexico they are packed for shipment in baskets, using hay or straw as a cushion. According to H. F. Schultz, the same method is used in Argentina, where twelve to fifteen dozen fruits are packed in a basket. Good ventilation should be insured, and the fruits should not be wrapped in paper. Cherimoyas exported from Madeira to London net the growers $1.00 to $1.20 a dozen. In Argentina the average price to growers is $2.20 a dozen.

Pests and diseases.

Although the cherimoya has up to the present suffered little from the attacks of insect and other pests in California and Florida, it is far from being exempt from them in regions where it has been grown extensively for a long period. In Hawaii, *Pseudococcus filamentosus* Cockerell is a serious enemy. Several other coccids have also been reported on the cherimoya, *Aulacaspis miranda* Cockerell and *Ceropute yuccæ* Coquillet are two which are mentioned from Mexico. Certain of the fruit-flies (Trypetidæ) are known to attack the fruits of the cherimoya. Throughout the warmer parts of America there are small chalcid flies, related to the wheat-joint worm and the grape-seed chalcid, which infest the seeds of annonaceous fruits. *Bephrata cubensis* Ashm. has been reported as attacking the cherimoya in Cuba. These insects are serious pests. In Argentina the attacks of borers are said to reduce the life of the average tree by half, making it thirty in place of sixty years.

Varieties (Fig. 24).

While there are important differences among seedling cherimoyas, affecting not only the productiveness and foliage of the tree but also the size, form, character of surface, color, quality, and number of seeds of the fruit, few named varieties have as yet been propagated. In the Pomona College Journal of Economic Botany (May, 1912) the author has described two, viz., Mammillaris and Golden Russet, which have been propagated in California on a limited scale. Neither of these, however, merits extensive cultivation; hence the descriptions will not be included in this work. It seems desirable, however, to repeat the botanical classification of seedling cherimoyas published by W. E. Safford in the Standard Cyclopedia of Horticulture. This comprises the following five forms:

Finger-printed (botanically known as forma *impressa*). — Called in Costa Rica *anona de dedos pintados*. The fruit is conoid or subglobose in shape, and has a smooth surface covered with U-shaped areoles resembling finger-prints in wax. Many seedlings of this type are of good quality, and contain few seeds.

Smooth (forma *lævis*). — Called *chirimoya lisa* in South America and *anon* in Mexico City. This form is often mistaken for *Annona glabra* and *A. reticulata* because of the general appearance of the fruit and on account of the name *anon*, which is also applied to *A. reticulata*. One of the finest types of cherimoya.

FIG. 24. Seedling cherimoyas, showing some of the common types. (× $\frac{1}{5}$)

Tuberculate (forma *tuberculata*). — One of the commonest forms. The fruit is heart-shaped and has wart-like tubercles near the apex of each areole. The Golden Russet variety belongs to this group.

Mammillate (forma *mamillata*). — Called in South America *chirimoya de tetillas*. Said to be common in the Nilgiri hills in southern India, and to be one of the best forms grown in Madeira.

Umbonate (forma *umbonata*). — Called *chirimoya de puas* and *anona picuda* in Latin America. The skin is thick, the pulp more acid than in other forms, and the seeds more numerous. The fruit is oblong-conical, with the base somewhat umbilicate and the surface studded with protuberances, each of which corresponds to a component carpel.

Hybrids between the cherimoya and the sugar-apple (*Annona squamosa*) have been produced in Florida by P. J. Wester and

PLATE IX. *Upper*, the home of the Fardh date; *lower*, in the date gardens of Basrah.

Edward Simmonds. The aim has been to develop a fruit having the delicious flavor of the cherimoya, yet adapted to strictly tropical conditions. Some of the hybrids have proved to be very good fruits, and further work along this line is greatly to be desired. Wester calls this new fruit *atemoya*. Hybrids between it and the sugar-apple, the bullock's-heart, and the pond-apple (all of which see below) have been obtained by him in the Philippines.

The Sugar-Apple (Fig. 25)

(*Annona squamosa,* L.)

With the exception of the little-known ilama (described later), the sugar-apple is the best of the tropical annonas. In its climatic requirements it resembles the bullock's-heart and the soursop, rather than the subtropical cherimoya. In precocity and productiveness it excels all of these species.

Fig. 25. The sugar-apple (*Annona squamosa*), a favorite fruit in India and many parts of tropical America. The tree succeeds particularly well in dry situations. ($\times \frac{1}{3}$)

The sugar-apple is more widely disseminated throughout the tropics than any other species of Annona, and in many regions is an important fruit. Particularly is it esteemed in India, where it is extensively grown. P. Vincenzo Maria wrote of it in 1672: "The pulp is very white, tender, delicate, and so delicious that it unites to agreeable sweetness a most delightful fragrance like rose water . . . and if presented to one unacquainted with it he would certainly take it for a blanc-mange."

The tree is smaller than that of most other species of the genus, its maximum height being 15 to 20 feet. Like the cherimoya, it is semi-deciduous. The leaves resemble those of *A. reticulata* except in their smaller size; they are lanceolate or oblong-lanceolate in form, acute or shortly acuminate at the apex and acute at the base, 2½ to 4 inches long, pale green on both surfaces, and glabrate or nearly so, except for the sparsely pubescent petiole. The flowers, which are produced singly or in clusters of two to four, resemble those of *A. reticulata*. They are greenish yellow in color, about an inch long, the three outer petals oblong, thick, rounded at the tips; the inner petals minute, ovate. The fruit is round, heart-shaped, ovate or conical, 2 to 3 inches in diameter, yellowish green in color. The surface is tuberculate and covered with a whitish bloom. The pulp is white, custard-like, sweet and slightly acidulous in flavor. The carpels, each of which normally contains a brown seed the size of a small bean, cohere loosely or not at all, the sugar-apple differing in this respect from the cherimoya, in which it is difficult to distinguish carpellary divisions in the flesh.

The sugar-apple is indigenous in tropical America. Its abundance in India at a very early period has led several botanists to assume that it was common to tropical America and tropical Asia. More recently, however, the belief has found acceptance that it was originally limited in its distribution to the New World. Alphonse DeCandolle, who discusses this subject at length, concludes: "It can hardly be doubted, in my opinion, that its original home is America, and in especial the West India islands."

The arguments advanced in favor of an Asiatic origin for the species were the occurrence of common names for it in Sanskrit; the fact of the tree growing wild in several parts of India; and the presence of carvings and wall-paintings, believed to represent the fruit, in the ruins of ancient Muttra and Ajanta. Yule and Burnell (Hobson-Jobson) suggest that it may have reached India from both of two directions: from Mexico *via* the Philippines and from Hispaniola (Santo Domingo, in the West Indies) *via* the Cape of Good Hope; in the former instance bringing with it the common name *ata*, or *ate*, which is still used in parts of Mexico (*e.g.*, the Huasteca region, near Tampico), and in the latter coming under the name *annona*. Safford is not certain that the name *ata* is of American origin; he suspects it may be derived from the Malayan word *atis*, meaning heart, and that it was carried to Mexico from the Philippines in early days.

In tropical America the sugar-apple is widely distributed. In the lowlands of Mexico it is a popular fruit, often cultivated and not infrequently found in a naturalized or wild state. It is grown from Central America southward to northern South America, extending there on the east into Central Brazil, where it is one of the important cultivated fruits. At Bahia, Brazil, it is said to have been introduced first in 1626 by the Conde de Miranda, after whom it is called *fructa do conde* (Count's fruit). In Cuba it ranks with the mango as one of the favorite fruits, and it is common in other islands of the West Indies.

In the Orient its cultivation is not limited to India, although it appears to be most extensive there. It is grown in the Philippines, in south China (where it is known as *fan-li-chi*, or foreign litchi), and in Cochin-China. In many islands of Polynesia it is abundant. Vaughan MacCaughey says: "It is common in many of the older Hawaiian gardens, not only in Honolulu, but also on the other islands of the group." In the French colonies near the African coast it is well known, and it is also reported from the mainland of Africa. Albert H. Benson[1] writes: "It is grown throughout a considerable part of coastal Queensland. . . . It is usually a heavy bearer, and is the variety (of annona) most commonly met with in our fruit stores." It is not known to have succeeded in the Medi-

[1] Fruits of Queensland, Dept. Agr. Brisbane, 1911.

terranean region, although it has been planted in several districts there.

So far as is known, the sugar-apple tree has never been grown to fruiting size in California: the climate appears to be too cool for it. In Florida, on the contrary, it is quite successful. P. W. Reasoner records that it has fruited as far north as Putnam County. On the east coast it occurs as far north as Cape Canaveral, and on the west it is found on the south side of the Manatee River. The zone in which it can safely be grown, however, lies farther south, viz., from Punta Gorda on the west coast and Palm Beach on the east to Key West. Throughout this part of Florida it succeeds admirably, and deserves greater popularity than it enjoys at present.

In addition to sugar-apple, a name probably of West Indian origin, the term sweet-sop is used in the British West Indies. In India it is called custard-apple by English-speaking people. Its commonest name in Hindustani is *sharifa* (meaning *noble*): but it is also called *sitaphal* (the fruit of Sita). The name *ata* is given it in parts of India. In the French colonies the names are *pomme-cannelle* (cinnamon apple) and *atte*. In the interior of Brazil the Portuguese name is *pinha;* on the coast *atta* and *fructa do conde* are also heard. In Mexico the Spanish terms are *anona, anona blanca,* and (erroneously) *saramuya* and *chirimoya.* In Cuba *anon* is the form generally used; this also appears in Costa Rica. The Aztec name used in ancient Mexico was *texaltzapotl,* meaning "zapote which grows on stony ground." The botanical synonyms of *A. squamosa,* L., are several; Safford lists *A. cinerea,* Dunal, *A. Forskahlii,* DC., and *A. biflora,* Moç. & Sessé.

The sugar-apple is preëminently a dessert fruit. Unlike the soursop, it is never made into preserves nor is it commonly used for sherbets. In composition it is similar to the cherimoya. Alice R. Thompson, who has analyzed the fruit in Hawaii, has found it to contain: Total solids 24.82 per cent, ash 0.67

per cent, acids 0.12 per cent, protein 1.53 per cent, total sugars 18.15 per cent, fat 0.54 per cent, and fiber 1.22 per cent. In spite of its similarity in most chemical constituents, the sugar-apple is not equal to the cherimoya in flavor. It has less piquancy, less character than the latter.

The climatic requirements of the tree are somewhat different from those of its congeners. It delights in a hot and relatively dry climate, such as that of the low-lying interior plains of many tropical countries. In Central America it is rarely seen at elevations greater than 2500 feet. In hardiness it ranks between the soursop and the cherimoya. Mature plants are not seriously injured by temperatures of 28° or 29° above zero; young ones may be killed at 30°.

G. Marshall Woodrow[1] says: "A deep, very stony soil with perfect drainage, enriched with decayed town sweepings, are the conditions enjoyed by this hardy fruit tree." In other regions it has been noted that it does well on rocky land, although it is probable that it prefers a loose sandy loam. Since it is rarely given systematic cultivation, little can be said regarding cultural methods. F. S. Earle has found in Cuba that it needs to be fertilized generously for the best results in fruit production, and he recommends a commercial fertilizer containing 3 per cent nitrogen, 10 per cent phosphoric acid, and 10 per cent potash. The sugar-apple withstands drought better than many other fruit-trees.

The methods of propagation employed are the same as with the cherimoya. Shield-budding has given the most satisfactory results in Florida. P. J. Wester has found that *A. reticulata* and *A. glabra* are congenial stock-plants; seedling sugar-apples are also used for the purpose, and are perhaps better than those of a different species.

Compared with other species of Annona, the sugar-apple bears heavily. This does not mean, however, that the trees

[1] Gardening in India.

habitually load themselves with fruit, for they rarely do so. A mature tree, fifteen feet in height, may produce several dozen fruits in a season. Usually all of them do not ripen at one time; thus the season is much longer than that of the cherimoya. In Florida it is common to pick ripe fruits during as many as six months out of the year. When the fruits are fully ripe, they burst open on the tree. They should be picked before reaching this stage and placed in the house, where they will soften in one to three days. After they have softened and are ready for eating, they must be handled with care. The fruit of the sugar-apple is not so well adapted to shipping long distances as that of the cherimoya.

Seedlings usually come into bearing when three or four years old. Some are much more productive than others, and there is much variation in the size and quality of fruit produced by different trees. When a tree has proved to be unusually good, it should be propagated by budding.

The Soursop (Plate VIII)

(*Annona muricata*, L.)

For the preparation of sherbets and other refreshing drinks, the soursop is unrivaled. Those who have visited Habana and there sipped the delectable *champola de guanábana* will agree with Cubans that it is one of the finest beverages in the world. Soursop sherbet is equal to that prepared from the best of the temperate zone fruits, if not superior to all other ices.

The tree is more strictly tropical in its requirements than the cherimoya or the sugar-apple. It withstands very little frost, and succeeds best in the tropical lowlands. Though widely disseminated, it is nowhere grown on an extensive scale. This is due, most probably, to the scanty productiveness which characterizes the species in general. There is an opportunity here for an excellent piece of work; by obtaining a productive variety and propagating it by budding, or by increasing the productiveness of the species through improved cultural methods, the soursop could be made profitable and of considerable commercial importance. In the large cities of tropical America there is a good demand for the fruits at all times of the year, a demand which is not adequately met at present.

The soursop is a small tree, usually slender in habit and rarely more than 20 feet high. The leaves are obovate to elliptic in form, commonly 3 to 6 inches long, acute, leathery in texture, glossy above and glabrous beneath. The flowers are large, the three exterior petals ovate-acute, valvate, and fleshy, the interior ones smaller and thinner, rounded, with the edges overlapping. The fruit is the largest of the annonas; specimens 5 pounds in weight are not uncommon and much larger ones have been reported. It is ovoid, heart-shaped, or oblong-conical in form, deep green in color, with numerous short fleshy spines on the surface. The skin has a rank, bitter flavor. The flesh is white, somewhat cottony in texture, juicy, and highly aromatic. Numerous brown seeds, much like those of the cherimoya, are embedded in it. The flavor suggests that of the pineapple and the mango.

Alphonse DeCandolle says that the soursop "is wild in the West Indies; at least its existence has been proved in the islands of Cuba, Santo Domingo, Jamaica, and several of the smaller islands." Safford states that it is of tropical American origin. The historian Gonzalo Hernandez de Oviedo, in his "Natural History of the Indies," written in 1526, describes the soursop at some length, and he mentions having seen it growing abundantly in the West Indies as well as on the mainland of South America. At the present day it is perhaps more popular in Cuba than in any other part of the tropics. In Mexico it occurs in many places, and the fruit is often seen in the markets. It is also grown in the tropical portions of South America. H. F.

Macmillan says that it thrives in Ceylon up to elevations of 2000 feet. It is cultivated in India, in Cochin-China, and in many parts of Polynesia. Vaughan MacCaughey states that it is the commonest species of Annona in the markets of Honolulu. Paul Hubert notes that it is cultivated in Réunion and on the west coast of Africa.

It will be observed that its distribution is limited to tropical regions. In the United States it can only be grown in southern Florida, where with slight protection it succeeds at Miami and even as far north as Palm Beach. Exceptionally cold winters, however, may kill the trees to the ground. In California it is not successful.

The name soursop is of West Indian origin, and is the one commonly used in English-speaking countries. In Mexico the fruit is known as *zapote agrio*, and more commonly as *guanábana* (sometimes abbreviated to *guanaba*), which is the name most extensively used in Spanish-speaking countries. *Guanábana* is considered to have come originally from the island of Santo Domingo. In the French colonies the common name is *corossol* or *cachiman épineux*. Yule and Burnell say: "Grainger identifies the soursop with the suirsack of the Dutch. But in this, at least as regards use in the East Indies, there is some mistake. The latter term, in old Dutch writers on the East, seems always to apply to the common jackfruit, the 'sourjack,' in fact, as distinguished from the superior kinds, especially the champada of the Malay Archipelago." In Mexican publications the soursop is sometimes confused with the soncoya (*A. purpurea*), though it actually differs widely from the latter both in foliage and fruit.

The soursop is more tolerant of moisture than the sugar-apple, and can be grown in moist tropical regions with greater success. Temperatures below the freezing point are likely to injure it, although mature trees may withstand 29° or 30° above zero without serious harm.

The soil best suited to this species is probably a loose, fairly rich, deep loam. It has done well, however, on shallow sandy soils in south Florida. F. S. Earle has found in Cuba that liberal applications of fertilizer will increase greatly the amount of fruit produced. The formula used is the same as that recommended for the sugar-apple. Little attention has yet been given to the cultural requirements of the plant.

The soursop, grown from seed, comes into bearing when three to five years old. The season of ripening in Mexico and the West Indies is June to September; in Florida it is about the same.

Mature trees rarely bear more than a dozen good fruits in a season. Oftentimes there are produced numerous small, malformed, abortive fruits which are of no value. These are due to insufficient pollination, only a few of the carpels developing normally, the remainder being unable to do so because they are not pollinated. The same phenomenon often occurs in the cherimoya, and, less commonly, in the sugar-apple and bullock's-heart.

Seedling trees differ in the amount of fruit they yield. Only the most productive should be selected for propagation. It may be possible still further to increase their productiveness by attention to pollination, and it has been shown that proper manuring is a great aid. Since the fruits are commonly of large size, it cannot be expected that so small a tree will produce many; still, the average seedling does not bear more than a small proportion of the crop it could safely carry to maturity, and the object of future investigations should be to obtain varieties which will be more productive.

In various parts of the world the tree is attacked by several scale insects, and the fruits by some of the fruit-flies, notably the Mediterranean fruit-fly.

Propagation of the soursop is usually effected in the tropics by seed. Choice varieties which originate as chance seedlings, however, can only be perpetuated by some vegetative means.

P. J. Wester has found that the species can be budded in the same manner as the cherimoya. He recommends as stock-plants the bullock's-heart and the pond-apple, both described below. Seeds are germinated in the same manner as those of the cherimoya.

The Bullock's-Heart (Fig. 26)

(*Annona reticulata*, L.)

The bullock's-heart, although widely grown, is a fruit of little value. Compared with the sugar-apple and the cherimoya it lacks flavor. An occasional seedling produces fruit of fair quality, but there is no reason why this species should be cultivated when the sugar-apple and the ilama can be produced on the same ground.

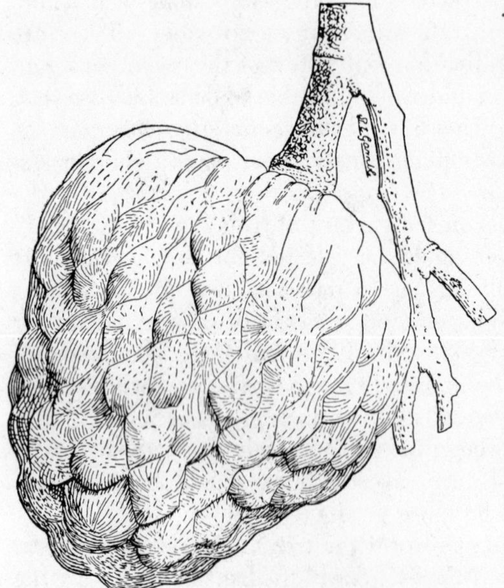

Fig. 26. The bullock's-heart (*Annona reticulata*), a fruit widely cultivated in the tropics. ($\times \frac{1}{2}$)

The tree is commonly 20 to 25 feet high. It is semi-deciduous, sometimes remaining devoid of foliage for several weeks. The leaves are oblong-lanceolate or lanceolate in form, commonly 4 to 6 inches in length, acute, and glabrate. The flowers are borne in small clusters upon the new branchlets. The three outer petals are oblong linear, about an inch long; the inner ones small, scale-like, and ovate in form. The fruit is usually heart-shaped (whence its common name), but it may be conical or oval. It weighs from a few ounces to 2 pounds, and requires a long time to reach maturity. The smooth surface, usually reddish-yellow or reddish-brown in the ripe fruit, is divided by impressed lines into rhomboidal or hexagonal areoles. The flesh, which contains numerous brown seeds the size of a small bean, is milk-white in color, granular near the thin skin, and sweet, even mawkish in flavor.

Safford says of this species: "Its fruit is inferior in flavor to both the cherimoya and the sugar-apple (*A. squamosa*), from the first of which it may be distinguished by its long, narrow, glabrate leaves, and from the second by its solid, compact fruit as well as its larger leaves. From *A. glabra,* with which it is also confused, it may be distinguished readily by its elongate narrow outer petals and its small, dark brown seeds."

The bullock's-heart is indigenous in tropical America. It is more abundant in the gardens of seacoast and lowland towns than its value warrants. From America it has been carried to the Asiatic tropics, and it is now cultivated in India, Ceylon, the Malay Archipelago, Polynesia, Australia, and Africa. Vaughan MacCaughey says that it is not very common in Hawaii, but may be found in a few gardens. In the Philippines and in Guam it has become spontaneous.

One West Indian common name of this fruit, custard-apple, is applied in India to *A. squamosa,* and sometimes in America to *A. Cherimola* and other species. In India *A. reticulata* is often termed *ramphal* (fruit of Rama). In Mexico the Spanish names are *anona* and *anona colorada;* the Aztec name, which appears in the early work of Francisco Hernandez, was *quauhtzapotl,* or tree zapote. In the French colonies the name *cachiman* or *cachiman cœur-de-bœuf* is generally used. In Brazil it is called in Portuguese *coração de boi.*

So far as is known the tree has never fruited in California, the climate of that state being probably too cold for it. It has been planted in protected situations there but no specimens have reached large size. In southern Florida it grows and fruits well. P. W. Reasoner,[1] who apparently confused this species with the cherimoya, says that it is confined to the same territory in Florida as the sugar-apple. Its requirements seem to be about the same as those of *A. squamosa*. It does not appear to be so partial, however, to a dry climate. The mature tree will withstand several degrees of frost without serious harm; a temperature of 27° or 28° usually does not injure it severely. In Ceylon, according to H. F. Macmillan, it does not grow at elevations above 3000 feet. In tropical America it ascends to the same altitude, or occasionally to 3500 feet.

The bullock's-heart prefers a deep rich soil with plenty of moisture. It is propagated by budding in the same manner as the cherimoya. P. J. Wester has found that it can be budded on the soursop, the pond-apple, and the sugar-apple, as well as on seedlings of its own species. As a rule, the trees bear more freely than those of the soursop and cherimoya, but not more so than the sugar-apple. There are as yet no named varieties in cultivation.

The Ilama (Fig. 27)

(*Annona diversifolia*, Safford)

The ilama is probably the finest annonaceous fruit which can be grown in the tropical lowlands; yet it has not, until very recently, been planted outside the region in which it is indigenous. Now that it has been called to the attention of horticulturists, its range should be extended rapidly to all parts of the tropics.

[1] Bull. 1, Div. Pomology.

The identity of the ilama, first mentioned by Francisco Hernandez toward the end of the sixteenth century, remained in doubt until W. E. Safford showed, in 1911, that it was a species which had not been described botanically. Safford named it *Annona diversifolia,* and brought together much information concerning its habits and the character of its fruit. These data were published in the Journal of the Washington Academy of Sciences, March 4, 1912. More recently the writer has been able to study the species in Mexico and Guatemala, and the United States Department of Agriculture has distributed several thousand plants in the warmest regions of the United States and in tropical America.

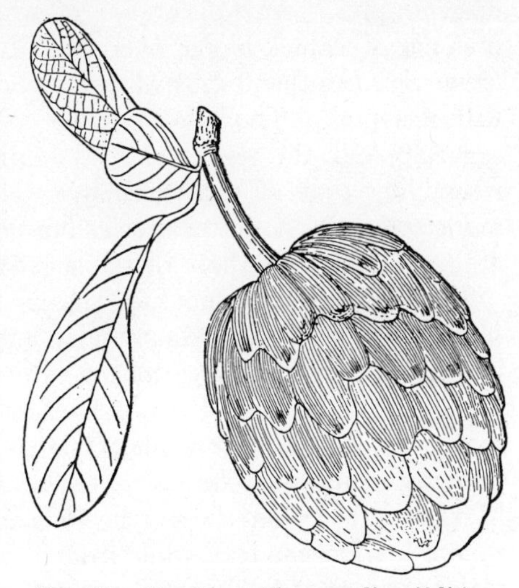

Fig. 27. The ilama (*Annona diversifolia*), an excellent fruit from southern Mexico and Central America. (× about $\frac{1}{3}$)

The tree grows to an ultimate height of 25 feet. It is slender in habit, the trunk not more than 10 inches thick, often branching from the ground to form three to six main stems. Some trees are erect, others spreading in habit. The foliage somewhat resembles that of *A. squamosa,* but the leaves are larger and of distinct form, being broadly elliptic to oblanceolate, rounded at the apex, and 4 to $5\frac{1}{2}$ inches in length. A distinguishing characteristic of this species is the presence of orbicular leaf-like bracts at the bases of the smaller branchlets. The flowers are maroon-colored, 1 inch long, with the three outer petals linear-oblong in form, the inner petals minute. The fruit is conical, oval, or round in form, the largest specimens weighing about 1½ pounds. The surface is rough, with the carpellary areas indicated by deeply incised lines; from each of the areoles thus formed rises a short thick protuberance. Sometimes these protuberances are suppressed, the fruit then being almost smooth. The color varies from pale green to magenta-pink. An appearance of whiteness is given by the presence of a thick bloom over the entire surface. In the pale green varieties the flesh is white; in the pink kinds it is tinged with rose-pink. The flavor is sweet, very similar to that of the sugar-apple in the green varieties; in the pink it is more acid, resembling that of the cherimoya. The seeds are about as numerous as in the latter species but larger in size. The fruits are used fresh, like those of the sugar-apple.

The ilama is indigenous in the mountains and foothills of southwestern Mexico, Guatemala, and Salvador, but is not known to occur at elevations greater than 2000 feet. It is found in the gardens of many Mexican and Central American towns, notably in Tapachula, Chiapas, where it is one of the principal cultivated fruit-trees. In Colima and Acapulco, Mexico, it is called *ilama* (the *ilamatzapotl* or "old women's zapote" of Hernandez), while from Tehuantepec to the Guatemalan border it is known as *papauce*. In Guatemala and Salvador it is named *anona blanca*.

The climatic requirements of the ilama are similar to those of the sugar-apple and the custard-apple. The species is found only at relatively low elevations, indicating that it prefers a hot climate. The amount of cold it will withstand has not yet been determined. The regions where it occurs most abundantly are dry during several consecutive months and subject to abundant rainfall the remainder of the year. In Guatemala it sometimes appears in places where there is little rainfall. The same is true as regards Tehuantepec, but in this region the trees

PLATE X. *Upper*, a date palm in full production; *lower*, the purple granadilla.

are irrigated. The best soil seems to be a deep, rich, rather loose loam.

Although propagated in Mexico and Central America by seed only, the ilama can probably be budded in the same manner as other annonas. By using this method of propagation, it will be possible to perpetuate the best varieties which originate as seedlings.

The trees come into bearing when three or four years old, and sometimes produce good crops. Productive trees often bear 100 fruits in a single season. There is, however, the same variation in this regard as with other annonas, though less as to the form and size of the fruit. The ripening season is short; July and August are the principal months. When the fruits are fully mature they crack open. They are commonly left on the tree until they reach this condition but it would be better to pick them a few days earlier. So handled, they require to be kept one to three days after being taken from the tree before they soften and are ready for eating.

The ilama may be termed the cherimoya of the lowlands. The cherimoya does not succeed in the tropics unless grown at elevations of 4000 to 6000 feet, where the climate is cool. The ilama, on the other hand, belongs to the lowlands, but is strikingly similar in character to a good cherimoya. It is a valuable recruit and one which cannot be too strongly recommended for cultivation throughout the tropics.

Minor Annonaceous Fruits

Pond-apple (*Annona glabra*, L.). — This species is of no value as a fruit, but has been used as a stock for other annonas. It grows wild in south Florida around the shores of Lake Okechobee and along the Indian and Caloosahatchee rivers; occurring also in the West Indies, on the mainland of tropical America, on the west coast of Africa, and in the Galapagos Islands. In

Florida it is often called custard-apple; in the West Indies alligator-apple and cork-wood. While tropical in nature, it withstands a few degrees of frost. It is swamp-loving, as the name here used indicates, and a vigorous grower. The tree is usually small, but sometimes reaches a height of 40 feet. The leaves are smooth, ovate to oblong or elliptic in form, acute to bluntish, glossy green above and paler beneath. The flowers are large, with the outer petals cream-colored, the inner smaller and narrower, whitish outside and blood-red within. The fruit is ovoid or heart-shaped, 2 to 4 inches long, smooth, yellowish when ripe, with soft yellowish flesh. Mexican writers have asserted that the tree is cultivated and the fruit sold in the markets. These statements are due to the confusion of *A. glabra* with other species of Annona, most probably the smooth-fruited forms of *A. Cherimola* and *A. reticulata*. *Annona palustris*, L. and *A. laurifolia*, Dunal are synonyms of *A. glabra*, L.

Wild cherimoya (*Annona longiflora*, Wats.). — This species comes from the state of Jalisco, Mexico. Horticulturally it is not yet well known, but it is said to have been introduced into California. Safford describes it as a shrub or small tree, with leaves resembling those of the true cherimoya but distinguished when mature by being glabrate or glabrescent between the lateral nerves. The flowers are often 2 inches long. The fruit is conical or ovate in form, the surface smooth to rough as in the cherimoya, which in flavor it resembles.

Mountain soursop (*Annona montana*, Macf.). — This species is native to the West Indies, where it is also known as *guanábana cimarrona* (Spanish, wild guanábana) and *corossolier bâtard* (French). It is a small forest tree with leaves resembling those of the soursop; the flowers also resemble those of that species. P. J. Wester,[1] who tested the fruit in the Botanic Garden at

[1] Philippine Agrl. Review, 2, 1916.

Buitenzorg, Java, was "surprised to find it of remarkably good quality considering that it is entirely unimproved and that it has never been recorded as edible. The fruit is about the size of a small custard-apple, with sparse, short prickles; greenish, and with yellowish, rather cottony but juicy and subacid, refreshing pulp, somewhat recalling the flavor of the soursop though inferior to that fruit." The tree is larger and more robust than *A. muricata.*

Soncoya (*Annona purpurea,* Moç. & Sessé). — This tree is little known outside of southern Mexico and Central America, where it is native. In Mexico it has been confused with the soursop, although neither foliage nor fruit resembles that of *A. muricata.* It is confined to the lowlands; a moist, hot climate suits it best. In Mexico it is sometimes called *cabeza de negro* (negro-head) and *ilama.* The leaves are large, oblong-elliptic to oblong-obovate in form, acuminate at the apex. The young branchlets are reddish pubescent. The flowers resemble those of the soursop. The fruit is round, sometimes as much as 6 inches in diameter, brownish gray in color and covered with pyramidal protuberances which terminate in short hooks curved toward the stem. The carpels, which separate readily, each contain an obovate brown seed about an inch long. The flesh is bright orange in color, soft, of pleasant flavor suggesting that of the northern papaw (*Asimina triloba*). The fruit is not highly esteemed, but is common in the markets of the regions where it is native. The tree is cultivated in Mexican and Central American dooryards. Because of its large size, its thick skin, the attractive color of its flesh, and its aromatic flavor, the soncoya is of interest in connection with the possibilities of annona breeding.

Posh-tê (*Annona scleroderma,* Safford). — This species, which grows wild in southern Mexico and Guatemala, is scarcely known in cultivation. It is remarkable for its thick, relatively hard shell, which makes it of possible value with regard to the production of annonas suitable for shipping to distant markets. This is a vigorous tree with large, thick, glabrous, oblong leaves and small cinnamon-brown flowers. The fruit is roundish oblate in form, about 3 inches in diameter, with dull green surface divided into areoles by small ridges, the shell being nearly $\frac{1}{4}$ inch thick. The seeds, which are embedded in the white melting pulp, are about the same size as those of the cherimoya. O. F. Cook[1] says: "The texture of the pulp is perfect, the flavor aromatic and delicious with no unpleasant aftertaste. It is much richer than the soursop, with a suggestion of the flavor of the matasano (*Casimiroa edulis*). . . . The most fragrant pulp is close to the rind. The seeds separate from the surrounding pulp more readily than in most annona fruits." The posh-té appears to be adapted to moist tropical regions most probably at elevations of less than 4000 feet.

Annona testudinea, Safford, the *anona del monte* of Honduras and Guatemala, is closely related to *A. scleroderma*. The fruit has soft, juicy pulp similar to that of the cherimoya but not quite so highly flavored. When fully ripe the surface takes on a brownish color. The external appearance of the fruit resembles that of the posh-té, although the ridges are not so pronounced. Both of these species merit horticultural attention.

Biribá (*Rollinia deliciosa*, Safford). — Jacques Huber[2] describes this as a medium-sized tree common in the orchards of Pará, Brazil. Its growth is rapid and it prospers equally well in sun and shade. "Of all the annonaceous fruits cultivated in Pará this seems best adapted to our (*i.e.*, the north Brazilian) climate, springing up almost spontaneously wherever seeds fall." The biribá has been referred incorrectly to *R. orthopetala*,

[1] Journal Wash. Acad. Sci., Feb. 19, 1913.
[2] Boletim Museu Goeldi, 1904.

A. DC., from which it can be distinguished by the decurved wings of its flowers. The leaves are obovate-oblong or elliptic in form, acuminate, 8 to 11 inches long, and nearly glabrous. The fruit is roundish oblate in shape, 3 to 5 inches in diameter, cream-yellow in color, with the areoles distinctly outlined. The flesh is white or cream-colored, juicy, sweet, and of pleasant flavor. In Pará it has been characterized as the finest annonaceous fruit of tropical America, but Florida-grown fruits do not entitle the species to this distinction: neither do specimens purchased in the markets of Rio de Janeiro, where they are sold under the name *fructa da condessa* (Countess' fruit). The tree is adapted only to tropical lowlands and to regions in the subtropics which are practically free from frost. At Miami, Florida, the mature tree has been killed by a temperature of 26.5° above zero.

CHAPTER VI

THE DATE

Plates IX–X

"HONOR your maternal aunt, the palm," said the prophet Muhammad to the Muslims; "for it was created from the clay left over after the creation of Adam (on whom be peace and the blessings of God!)." And again, "There is among the trees one which is preëminently blessed, as is the Muslim among men; it is the palm."

It is in this reverential aspect that the Semitic world has always regarded the date palm; and with sound reason, for its economic importance to the desert dweller as the source of both food and shelter is even greater than that of the coconut palm to the Polynesian.

Only in recent years, however, have oriental methods of date-culture been scientifically examined and tested by horticulturists. By far the greater part of this work must be credited to investigators in the United States. The first modern importation to this country was of palms rooted in tubs, shipped from Egypt to California in 1890. Better methods of shipping offshoots were gradually worked out, and introductions from all parts of the world have been made in ever-increasing numbers in the last quarter of a century.

Meanwhile, continued study has been given to methods of culture, with the result that the problems of the rooting of offshoots and the ripening of the fruit, which were at first serious sources of loss, have been brilliantly solved, and many others adequately dealt with. This work has been done by the United States Department of Agriculture, the experiment stations of California and Arizona, and many private growers; and any history of the progress of scientific date-culture will certainly record the names of such pioneers as Bruce Drummond, David Fairchild, R. H. Forbes, George E. Freeman, Bernard Johnston, Fred N. Johnson, Thomas H. Kearney, Silas C. Mason, James H. Northrop, F. O. Popenoe, Paul Popenoe, Walter T. Swingle, and A. E. Vinson.

As a result of the work not only of the Americans but of French horticulturists in North Africa and English in Egypt and India, the culture of the date palm is to-day perhaps better understood than that of any other fruit of which this volume treats. There is room, however, for immense improvement in method in practically all of the older date-growing regions, and the introduction of more scientific culture will add greatly to the national wealth in many parts of the Orient.

Such an important date-growing country as Egypt does not now produce enough dates for its own consumption; for although it is a moderate exporter it is still more of an importer of low-grade dates from the Persian Gulf. The markets of North America and Europe have scarcely been touched. Before the Great War the annual importation into New York was thirty to forty million pounds, — only five or six ounces a head of the country's population. This is a ridiculously low rate of consumption for a fruit possessing the food-value of the date, and which can be produced so cheaply. There would seem to be no reason why it should not become an integral part of the diet of American families, being eaten not as a dessert or luxury only, but as a source of nourishment. So regarded the market is almost unlimited, and considering how few are the areas available for growing first-class dates, over-production seems hardly possible.

The date palm characteristically consists of a single stem with a cluster of offshoots at the base and a stiff crown of pinnate leaves at the top. It reaches a maximum height of about 100 feet. If the offshoots are allowed to grow, the palm eventually becomes a large clump with a single base.

The plant is diœcious in character, *i.e.*, staminate and pistillate, or male and female, flowers are produced by separate individuals. The inflorescence is of the same general character in both sexes, — a long stout spathe which bursts and discloses many thickly crowded branchlets. Upon these are the small, waxy-white, pollen-bearing male flowers, or the greenish female blossoms in clusters of three. After pollination, two out of each three of the latter usually drop, leaving only one to proceed to maturity. Chance development of a blossom that has not been pollinated occasionally gives rise to unfounded rumors of the discovery of seedless dates; genuine seedless varieties have, however, been credibly reported.

The fruit varies in shape from round to long and slender, and in length from 1 to 3 inches. While immature it is hard and green; as it ripens it turns yellow, or, in some varieties, red. The flesh of the ripe fruit is soft and sirupy in some varieties, dry and hard in others. In many kinds, including most of those that ripen early, the sugar-content never attains sufficient concentration to prevent fermentation; the fruit of such varieties must, therefore, be eaten while fresh.

In cultivation about 90 per cent of the male palms are usually destroyed, since they can bear no fruit.

The presence of offshoots around the base is one of the simplest ways to distinguish the date palm, botanically known as *Phœnix dactylifera*, L., from the wild palm of India (*Phœnix sylvestris*, Roxb.) and the Canary Island palm (*P. canariensis*, Hort.); from the latter, which is often grown in the United States for ornamental purposes, it may also be distinguished by its more slender trunk, and by its leaves being glaucous instead of bright green.

Phœnix dactylifera is commonly supposed, following the

study of O. Beccari,[1] to be a native of western India or the Persian Gulf region. Evidently, long before the dawn of history, it was at home in Arabia, where the Semites seem to have accorded it religious honors because of its important place in their food supply, its diœcious character, and the intoxicating drink which was manufactured from its sap, and which in the cuneiform inscriptions is called "the drink of life."

Traditions indicate that when the Semites invaded Babylonia they found in that country their old friend the date palm, particularly at Eridu, the Ur of the Chaldees (Mughayr of modern maps) whence Abram set out on his migration to Palestine. It is even suggested that the Semitic immigrants settled at Eridu, which was then a seaport, on account of the presence of the date palms, one of which was for many centuries a famous oracle-tree. Several competent orientalists see in the date palm of Eridu the origin of the Biblical legend of the Garden of Eden.

In very early times the palm had become naturalized in northern India, northern Africa, and southern Spain. From Spain it was brought to America a few centuries ago.

In the last quarter of a century, United States governmental and private investigators have visited most of the date-growing regions of the Old World in search of varieties for introduction into this country, where, in California and Arizona, may now be found assembled all the finest ones that cultivation, ancient and modern, has yet produced.

Orthodox Muslims consider that the dates of al-Madinah, in Arabia, are the best in the world, partly for the reason that this was the home of the prophet Muhammad, who was himself a connoisseur of the fruit. Unbiased judgment, however, commonly yields the palm to the district of Hasa, in eastern Arabia, where the delicious variety Khalaseh grows, watered by hot springs. The district of greatest commercial importance is

[1] Malesia, iii.

that centering at Basrah, on the conjoined Tigris and Euphrates rivers, a region which contains not less than 8,000,000 palms and supplies most of the American market.

The region around Baghdad, while less important commercially, contains a larger number of good varieties than any other locality known. Date cultivation by Arabs is most scientifically carried on in the Samail Valley of Oman (eastern Arabia), where alone the Fardh dates of commerce are produced.

Serious attempts to put the date industry of northwestern India on a sound basis are being made, and with good prospects of success. Western Persia and Baluchistan produce some poor dates and incidentally a few good ones.

In Egypt there are nearly 10,000,000 palms, of which seven-tenths are widely scattered over Upper Egypt. Most of them are seedlings and practically all are of the "dry" varieties. On the whole, the Egyptian sorts are inferior.

The Saharan oases of Tripoli, Tunisia, and Algeria contain many varieties, of which one (Deglet Nur) is as good as any in the world, and is largely exported not only to Europe but to the United States, where it is marketed under the name of "Dattes Muscades du Sahara." Morocco grows good dates in the Tafilalet oases only, whence the huge fruits of one variety (Majhul) are shipped to Spain, England, and other countries. The date palms of southern Spain are seedlings and bear inferior fruit. Elsewhere about the Mediterranean the palm is grown mainly as an ornamental plant.

Intelligent culture of the date palm is now being attempted in some of the dry parts of Brazil, where it promises to attain commercial importance. It is doubtful whether the date will succeed commercially in any moist tropical region, although in isolated instances successful ripening of fruit has been reported in southern India, Dominica (British West Indies), Zanzibar, and southern Florida.

A large area in northern Mexico, not yet developed, is undoubtedly adapted to this culture; but experimental attempts with it on the Rio Grande in Texas have been abandoned. Arizona and California offer the best fields for date-growing in the United States, and in the Coachella Valley of California (a part of the Colorado River basin) conditions are particularly favorable. Residents of this valley are not exceeding the truth in asserting it to be the center of scientific date-growing at the present time.

Dates consist mainly of sugar, cellulose, and water. An average sample of fruits on the American market will show in percentages:[1] carbohydrates 70.6 per cent, protein 1.9 per cent, fat 2.5 per cent, water 13.8 per cent, ash (mineral salts) 1.2 per cent, and refuse (fiber) 10.0 per cent. Cane-sugar is found in dates; in a few varieties this is partly or wholly inverted by the time the fruit is fully ripe.

A diet of dates is obviously rich in carbohydrates but lacking in fats and proteins. It is, therefore, by no accident that the Arabs have come to eat them habitually with some form of milk. This combination makes an almost ideal diet, and some tribes of Arabs subsist on nothing but dates and milk for months at a time.

By Arabs, as well as by Europeans, the date is commonly eaten uncooked. Unsalted butter, clotted cream, or sour milk is thought to "bring out the flavor" and render the sugar less cloying. The commonest way of cooking dates is by frying them, chopped, in butter.

For native consumption around the Persian Gulf and in India, immature dates are boiled and then fried in oil. Jellies and jams are made from dates, and the fruit is also preserved whole. Again, they may be pounded into a paste with locusts (grasshoppers) and various other foodstuffs. The soft kinds are tightly packed into skins or tins, when they are easily transported and will keep indefinitely.

[1] U. S. Dept. Agr., Bull. 28.

Various beverages are made by pouring milk or water over macerated dates and letting slight fermentation take place. The sap of the plant provides a mild drink resembling coconut milk, which when fermented becomes intoxicating. From cull dates a strongly alcoholic liquor is distilled, which, flavored with licorice or other aromatics, becomes the famous (or rather perhaps, infamous) arrak, of which many subsequent travelers have confirmed the verdict of the sixteenth-century voyager Pedro Teixeira, himself probably no strict water-drinker, who said of it, "This is the strongest and most dreadful drink that was ever invented, for all of which it finds some notable drinkers."

Cultivation

While the date palm grows luxuriantly in a wide range of warm climates, it is, for commercial cultivation, adapted only to regions marked by high temperature combined with low humidity. Properly speaking, it belongs to the arid subtropical zone. A heavy freeze will kill back the leaves, but the plant may nevertheless be as healthy as ever in a year or two. Thus, date palms have withstood a temperature of only 5° above zero and have borne satisfactory crops in subsequent years. Ellsworth Huntington speaks of seeing the date palm in Persia where twenty inches of snow lay on the ground; many generations of natural selection in such an environment would doubtless produce a hardy race, but such a region would scarcely be thought adapted to commercial date-growing.

At the other climatic extreme, the date palm apparently finds no limit, being at its best where the summer temperature stays about 100° for days and nights together. The combination of warm days with cool nights is unsatisfactory; unless there is a prolonged season during which high temperatures prevail night and day, the best varieties of dates will not ripen successfully.

Humidity is an important factor with many varieties. Dates coming from the Sahara usually demand a dry climate; yet the Coachella Valley in California has sometimes proved too dry, and the fruit has shriveled on the tree unless irrigation was given while it was ripening. Persian Gulf and Egyptian varieties will endure more humidity, since they come from the seacoast or near it. Dew at night or rain coming late in the season when the dates are softening is almost ruinous to the crop, for which reason dates cannot be produced satisfactorily in some parts of Arizona. In regions of India where the summer rains begin in July, it has been possible to bring dates to maturity before the rains arrive.

In general, the best varieties require: (1) a long summer, hot at night as well as in the daytime; (2) a mild winter, with no more than an occasional frost; (3) absence of rain in spring when the fruit is setting; and (4) absence of rain or dew in the fall when the fruit is ripening. In regions lacking any of these characteristics, date-growing will be profitable commercially only if special care is taken to secure suitable varieties and to develop, by experiment, proper methods of handling them.

Date palms grow well in the stiff clays of the Tigris-Euphrates delta, in the adobe soils of Egypt, in the sand of Algeria, and in the sandy loam of Oman and of California. No one type of soil can be asserted to be necessary. Thorough drainage and aëration of the soil are desirable, but even in these regards the palm will stand considerable abuse, and is found to grow fairly well in places where the ground-water level is comparatively near the surface. Naturally, however, the palm responds to good treatment as do other plants. On the whole, it is probably best suited on a well-drained sandy loam.

The palm's tolerance of alkali has been noted from very early times, and has led Arab writers to believe that it throve best in alkaline soil. This is unlikely. Dates can indeed be grown successfully in ground the surface of which is white with alkaline efflorescence, provided the lower soil reached by the roots is less salty; but it is probable that the limit of tolerance is somewhere about 3 per cent of alkalinity, and the grower who looks for the best results should not plant on soil whose total alkaline-content exceeds one-half of 1 per cent. Naturally, old date palms will stand more alkali than young ones. It should be noted that the so-called black alkali, consisting of carbonates of sodium and potassium, is more harmful than the more or less neutral chlorids, sulfates, and nitrates of sodium, potassium, and magnesium which go by the name of white alkali.

If the irrigating water is free from alkalinity, it will, of course, help to counteract any alkali present in the soil; whereas the grower who needs to irrigate with brackish water must plant his palms in fairly alkali-free soil. Desert landowners sometimes calculate that soil which is too salty for anything else is good enough for a date plantation. This is short-sighted reasoning. Date-growing is, when rightly conducted, so profitable that it is worth giving the best conditions available, and the wise grower will plant his palms in his best soil. The ground should be tested to a depth of six or eight feet to determine its alkali-content, particularly if there is salt evident on the surface. Unless at least one stratum of alkali-free soil is found not far from the surface, the ground should not be used for date palms.

It is the custom in the United States to plant date palms 50 to the acre. The grower with plenty of land may find that 40 to the acre (33 feet apart each way) is more convenient. Arabs plant them much closer but do not cultivate their plantations frequently. The question of spacing is affected both by the nature of the soil and by the variety planted; according to Bruce Drummond, such kinds as Saidi and Thuri give the best results if spaced 35 or 38 feet apart.

Drummond gives the following advice about planting:

"The rooted offshoot when ready for transplanting should be pruned from three to five days before removing from the frame. The new growth should be cut back to one-half the original height, leaving from three to five leaf stubs to support the expanded crown of leaves. The holes in the field should be 3 ft. in diameter and 3 ft. deep, with from 12 to 16 in. of stable manure placed in the bottom of each, with 6 in. of soil on top, then irrigate thoroughly. The rooted palm when removed from the nursery should carry a ball of earth large enough to protect the small fibrous roots from exposure to the sun or dry winds. The average depth for planting should be 16 in., but this may be varied somewhat with the size of the shoot. In any case, the depth should be as great as can be without danger of covering the bud.

"It is not advisable to transplant rooted offshoots later than June. April and May are considered the best months of the entire year for the transplanting of either young or old date palms.

"In southern California, where the dry winds occur from March to June, the transplanted palms should be irrigated thoroughly every week; in sandy soil two irrigations a week should be given until new strong growth is established."

Arabs usually follow the basin method of irrigation, and it has been satisfactory in many other parts of the world. The most skillful American growers who irrigate in basins make them 15 feet square and a foot deep, filling them with a loose mulch of straw or stable manure.

Most American growers, however, prefer to irrigate in furrows, and use no mulch. The function of the mulch in reducing evaporation is covered by giving a thorough cultivation with a surface cultivator or spring-toothed harrow as soon as the ground has dried out enough to be workable. This involves cultivation of the ground every week or two.

Adequate fertilization of the soil is absolutely necessary in order to make date palms produce fruit as heavily as commercial growers desire and at the same time yield well in offshoots. Nitrogen-gathering cover-crops are much in favor, sesbania or alfalfa being preferred in California. The long roots of the latter are useful to break up any hardpan or layer of hard silt which may be present. Many growers plant garden-truck between the rows of palms, especially while the latter are young and making no financial return.

The soil in which date palms are usually grown is of a kind that benefits by the incorporation of rough material, and stable manure is, therefore, the fertilizer of first choice. Wheat-straw or similar loose stuff is frequently added with advantage. An annual application of fertilizer is required in most localities, and if the soil is sandy the grower must be more liberal. For palms producing offshoots, half a cubic yard a year is advised; for older palms a full yard is desirable: both in addition to such cover-crop as the grower may select.

In regard to irrigation, it is to be borne in mind that the soil must be kept moist during the entire year, and that the roots of the palm go deep. The character of the soil must be carefully and experimentally studied before the grower can be certain that he has arrived at the correct method for irrigation. The amount of water that the palm can stand in well-drained land is strikingly illustrated in the great plantings around Basrah, where fresh water is backed into the gardens by tidal flow, so that there are two automatic irrigations each day throughout the year.

In the Coachella Valley, with furrow irrigation, a twenty-four-hour flow each twelve days from April to November has generally been satisfactory, although in many soils weekly irrigation is required. During the winter the rainfall usually suffices. Each application of fertilizer must be followed promptly by several irrigations.

Pruning is not so important with date palms as with many fruit-trees. Dead leaves should be removed from young palms, and if the top growth is heavy the two lower rows of leaves may be removed when the palm is four years old. Regular pruning should begin about the sixth year, after which one row of leaves is usually removed at each midwinter. Drummond advises that "the leaves should not be pruned higher than the fruit stems of the former crop, which will leave about four rows of leaves below the new fruit stems, or approximately 30 to 36 expanded leaves."

Propagation

The date palm can be propagated in only two ways: by seed, and by the offshoots or suckers which spring up around the base or sometimes on the stem of the palm until it attains an age of ten to twenty years.

Seedlings are easily grown, but offer little promise to the commercial grower. Half of the plants will be males, and among the females there will be such a wide variation that no uniformity of pack or quality can be secured. In regions with a large proportion of seedling palms, such as Spain and parts of Egypt, there is practically no commercial date-culture. Most growers in California plant a few seedlings for windbreak or ornamental purposes. These yield a supply of males, but males can be secured better by growing offshoots from male palms of known value.

The multiplication of the date palm, therefore, is reduced in practice to the propagation of offshoots, and skill or lack thereof in this regard will determine largely the grower's success or failure at the outset.

In California at the present time the yield of offshoots is almost as valuable as that of fruit, and growers, therefore, desire to secure as many offshoots of their best varieties as possible. For this purpose ample fertilization and irrigation must be supplied. After the fourth or fifth year of a palm's life, the owner can usually take at least two offshoots a year from it for a period of ten years. The best size for offshoots at removal is when they weigh from ten to fifteen pounds (say 5 to 6 inches, is greatest diameter). The best season for the purpose is during February, March, or April.

Four or five days before the offshoots are to be removed from the mother-palm, their inner leaves should be cut back one-half and the outer leaves two-thirds of their length. It will be well worth while to have a special chisel made for removing offshoots. It should have a cutting bit of the best tool steel, 5 inches wide by 7 inches long, one side flat, the reverse beveled for 2 inches on the sides as well as on the cutting edge. The chisel should have a handle of soft iron 3 feet long and $1\frac{1}{4}$ inches in diameter, such as can be hammered with a sledge-hammer. The delicate operation of cutting is described by Bruce Drummond, who is the best American authority on the culture of the palm, as follows:

"To cut the offshoots from the tree the flat side of the chisel should always be facing the offshoot to be cut. Set the chisel well to the side of the base of the offshoot close to the main trunk. Drive it in with a sledge until below the point of union with the parent trunk; then by manipulating the handle the chisel is easily loosened and cuts its way out. Next reverse and cut from the opposite side of the shoot until the two cuts come together. This operation will in most cases sever the offshoot from the trunk. No attempt to pry the offshoot from the tree should be made, as the tissues are so brittle that the terminal bud may be ruined by checking or cracking. In cutting offshoots directly at the base of the palm the soil should be dug away until the base of the offshoot is located and enough exposed to show the point of union with the mother plant. Then the chisel can be set without danger of cutting the roots of the parent tree so much as to injure or retard its growth. The connection of the offshoot on such varieties as Deglet Nur is very small, and there is no necessity of cutting deeply into the trunk to sever the offshoot from the tree."

Once separated from its parent, the moist offshoot requires a period of seasoning before it is dry enough to be planted without danger of fermentation. Offshoots from the base of a palm are usually softer and sappier than those growing some distance above ground. The evaporation should amount to 12 or 15 per cent of the total weight, which will require at least ten to fifteen days to effect. Offshoots are usually left where cut, on the ground beneath the palm, to season.

PLATE XI. *Left,* a tropical substitute for the cantaloupe, the papaya; *right,* a papaya in bearing.

The Arabs plant offshoots at once in their permanent locations in the orchard, but the best results will be obtained by first rooting the young plants in a shed or frame where the two necessary conditions of high temperature and high humidity can be maintained. In California this is often done coöperatively.

A common type of shed for an individual grower is 12 by 20 feet in size with side walls 6 and 7 feet high respectively, presenting a roof-slope to the sun. The sides are usually of boards covered with tarred paper and the roof of 8- or 10-ounce canvas. In such a shed on an ordinary California summer day, the temperature will be about 115° and the humidity should be about 75.

The soil inside the shed should be a light sandy loam, well drained. Ten inches of the top soil should be removed and replaced with fresh stable manure, well packed, on which 2 inches of soil should be replaced. After a thorough flooding, the bed should be allowed to steam for a week, and then be flooded again, whereupon it is ready for the offshoots. These should be planted about 8 inches deep; in any case the bud must be above danger of flooding. During the summer the bed must be flooded at least twice a week, to keep the humidity at as high a point as possible. The offshoots must be kept in it until they are thoroughly rooted and have half a dozen new leaves. This may require one year or may need several years.

The causes that may lead to failure with offshoots are summarized by Drummond as: "(1) improper selection of the location for the nursery bed; (2) failure to construct the frame so nearly air-tight as to insure the necessary humidity and high temperature; (3) improper methods of cutting and pruning, and the neglect of seasoning before planting in the nursery-bed; and (4) the neglect of irrigation when necessary and failure to apply water properly. The points above mentioned are all essential to success, and to neglect one and observe the others may lead to as great a failure as to neglect them all." On the other hand, by using the proper care growers frequently succeed in making 90 to 95 per cent of their offshoots take root.

After they are removed to the open field, the young palms should be protected by wrapping during the following winter from the possibility of freezing, as they are tender at first. Newspaper is as good as anything for the purpose; canvas, burlap, and palm-leaves are also used.

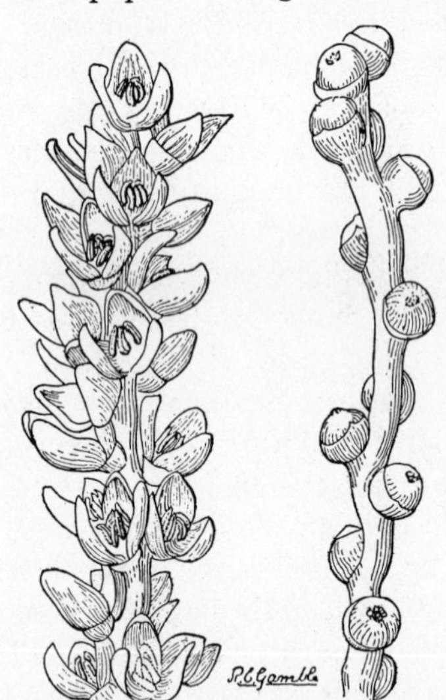

FIG. 28. On the left, a sprig of staminate or pollen-bearing flowers of the date palm; on the right, pistillate flowers which will, if properly pollinated, develop into fruits.

For security, the orchardist should allow one or two male date palms for each acre of fruit-bearing trees. Care should be taken to secure males that flower early in the season and yield abundant fertile pollen; sterility is common.

The female palm ordinarily blossoms between February and June (in California usually during March and April). Flowers appearing later than May 1 are not worth pollinating, so far as commercial production is concerned. Artificial pollination has been practiced since the dawn of history, and offers no difficulties.

The flowers of the two sexes can be distinguished readily (Fig. 28). The branchlets of the male inflorescence are only about 6 inches long, and are densely clustered at the end of the axis, while those of the female are several times as long and less densely clustered. The male blossoms are waxy white in color, the female more yellowish; while also the latter are much the less closely crowded together on the branchlets.

The presence of pollen in the male flower is in most cases easily to be detected by shaking a cluster of the blossoms.

As soon as the spathe containing the pollen-bearing flowers opens, it should be cut and put into a large paper bag to dry, the bag being stored, open, in a dry room. Thoroughly dry pollen will retain its vitality for many years, and a small quantity should be kept in a bottle from year to year, as a precaution. In case of need it can be used with a wad of cotton.

The pistillate flowers should be pollinated as soon as the spathes crack open, the plantation being inspected every day or two with this in view. The operation is preferably carried out about midday. The split female spathe is held open, and a sprig from the male flower gently shaken over it and then tied, open flowers downward, at the top of the female cluster. A single pollination with one sprig is enough for each cluster unless rain follows within twenty-four hours, in which case the operation should be repeated. The grower should keep the situation well in hand.

The grower must not let his young palms bear too many dates, particularly if he wants them to produce offshoots at the same time. Part of the female spadices (flower-stalks) should, therefore, be cut off. In most cases a palm may be allowed to bear its first two bunches of fruit in its fourth year, and three or four bunches in each of the next two years. If even a full-grown palm is allowed to bear to its limit in any year, it is likely to bear less the following season.

In case the grower should find himself absolutely without date pollen at a time when his pistillate trees are flowering, he may have recourse to the pollen of some other Phœnix, or even of a different genus of palms, Chamærops, Washingtonia, or whatever it may be. This will often enable him to save part, if not all, of the crop.

Yield and Season

Most varieties of date palm, if properly cared for, will begin to bear in the fourth year, and should yield a considerable return in the fifth and succeeding years. Under Arab treatment they usually take longer. References in the Code of Hammurabi (about 2000 B.C.) indicate that the Babylonians at that time could secure a paying crop in the fourth year; if so, they were better cultivators than their modern descendants.

Beginning with two small bunches, the grower may allow his palms to bear an increasing amount each year until maximum is reached. After the fifth, sixth, or seventh year, 100 pounds or thereabouts to a tree can be maintained steadily without difficulty by most varieties, and one or two offshoots a year will still be produced, given proper fertilization and irrigation. In many cases even larger yields can be obtained. If, however, the growing palm is not given proper culture, for instance is allowed to carry a full load of offshoots, and, simultaneously, to bear all the fruit that it can, it tends to become an intermittent bearer, bringing in a large crop one year and little or nothing the next. This should be avoided by eliminating the conditions named.

The season of ripening is from May to December, depending on variety and location. Fresh dates as early as May can be secured in favored locations in Arabia, where certain early kinds are grown. They have not yet been produced so early in the United States, where the first dates do not ripen until July. In many regions very late varieties will carry fruit into midwinter. In California and at Basrah the height of the season is September; in Egypt, August; in western Arabia, July; in Algeria, September or early October. As a general rule, the dates of best quality are late in ripening and the early dates are soft varieties which must be consumed fresh as they lack the necessary amount of sugar to keep without fermenting.

American growers will find an advantage in fairly early varieties (other considerations agreeing), as the crop can thus be disposed of without competition, say before November 1, at about which time dates from Persian Gulf or North African sources can be put on the market, possibly at lower prices.

Picking and Packing

The picking process offers no particular problems, although the methods are not the same with all varieties. Usually two persons can pick together conveniently, one holding the basket and the other gathering the dates and placing them in it. Under favorable conditions, some varieties will mature a whole bunch so evenly that it can be removed entire without loss, but in many cases it is necessary to pick out the different "threads" carrying dates, and cut them separately, leaving those whose fruit is not yet mature for another day. It is advisable, with kinds that permit of it, to leave the calyx on the fruit, since if this is pulled off it opens an avenue for the entrance of insects and dirt. Bunches left to ripen on the tree frequently need to be protected by a bag of cheese-cloth or similar material, to keep off birds and insects.

Dates grown for home use need no treatment after picking unless it be a washing to remove the dust. If they are to be kept for some time, they may well be pasteurized to free them of insect eggs and the bacteria of fermentation and decay. Small quantities of fruit can be treated successfully in the oven of a cookstove, pains being taken by regulating the aperture of the door, to keep the temperature between 180° and 190° for three hours. This may slightly alter the taste; sterilization by exposure overnight to the fumes of carbon bisulfide is easy and causes no change of flavor.

There are many advantages in ripening dates artificially rather than leaving them to mature on the tree; hence some method of artificial ripening has been practiced in most date-growing countries since the time of the earliest written records. Much careful experimentation has been done in this country, first by the Arizona Experiment Station and later by the United States Department of Agriculture. As a result, such simple, satisfactory, and inexpensive methods of maturing dates have been worked out that the commercial grower will do well to rely on them. The exact process differs with the variety and with the conditions under which the dates have to ripen; for the precise technique advisable in his case the grower must either refer to those who have had the experience he needs, or experiment on a few dates for himself, after he has grasped the general principles.

As W. T. Swingle points out, a date is botanically mature, or "tree ripe" as horticulturists say, as soon as it reaches full size and the seed is fully developed. At this stage, however, the date is still astringent and not eatable. Following this comes a process that may be called "ripening for eating," consisting of complex chemical transformations by which the sugars are altered and the tannin deposited in insoluble form in "giant cells." This final ripening is brought about by the combination of heat and a certain degree of humidity.

The principle underlying modern methods of artificial ripening is, therefore, to expose the dates to a constant high temperature, while holding them in the humid atmosphere which is created by the moisture they naturally give off as they dry and wrinkle.

For this purpose the dates are picked when they first begin to soften. Most varieties at this stage show translucent spots while the remainder of the berry is still hard and remains bright red or yellow in color. Dates taken from the tree in this condition will ripen successfully in three or four days if they are packed loosely, stems and all, into a tightly closed box and left at ordinary room temperature, the room being closed at night to keep out cold air. Commercial growers provide a special house, or a room built in the packing-shed for this purpose. This is so constructed as to be air-tight when closed, so that the temperature can be maintained at an even figure, without variation of more than a degree or two, by means of an electric light or a lamp with thermostat attachment such as is used in the incubators of poultrymen. Under such conditions, dates will be brought to a beautiful even maturity and practically without loss by keeping them from twenty-four to seventy-two hours at a temperature of 110° to 120°.

The skillful grower will control further the ripening of his dates by irrigation. In some climates, like that of Upper Egypt and of the Coachella Valley in some seasons, a typically "soft" date like Deglet Nur will mummify on the palm, as it matures, until it becomes a "dry" date. This can be avoided by keeping the palms well irrigated while the dates are ripening. On the other hand, "soft" varieties sometimes "go to pieces" and ferment on the tree, because of too much moisture; in this case the soil must be kept dry during the ripening season.

The packing of dates is a matter for the grower's own taste, or for standardization by the coöperative association to which he may belong. Good dates of standard varieties are usually packed in layers in one-pound cardboard boxes, like sweetmeats. In California, where home-grown dates bring fancy prices, great pains are taken with this finest quality of fruit, which is easily retailed at $1 a pound.

Most dates worth marketing in the United States are worth packing in cartons. In Arizona, berry-boxes have been used. The American standard for bulk shipment is the lug-box of 30 to 40 pounds' capacity. It is important, in any case, that the pack be uniform, both in size and variety; otherwise the grower can expect to receive only "cull" prices.

Many varieties, such as Zahidi, ripen well in the bunch and adhere indefinitely. It is probable that a profitable trade can be developed in marketing entire bunches of these, which the retail dealer can display in his store as he does a bunch of bananas. Dates of inferior quality can be worked up into various by-products, such as "date butter," or sweetmeats, or may be sold to bakers and confectioners. Culls are used in the Orient for the distillation of arrak, or as feed for live-stock. Soft early dates, which in many cases are of a beautiful color as well as delicious flavor but which lack keeping quality, probably could be sold in crates as are berries and be similarly handled as perishable fruits. Marketing should be carried on through a growers' coöperative association, which can guard the interests of all by insisting on proper standards.

For a bearing plantation with fifty palms to the acre, 100 pounds of fruit to a tree each year is a conservative estimate of the yield. This means 5000 pounds of fruit an acre each year, the retail value ranging from 2 cents a pound in the Orient to $1 a pound in the United States. Growers in the Coachella Valley have been able for some years to sell practically all the good dates they produce at 25 cents to 75 cents a pound at the plantation. Such a price is not likely to be maintained, since dates of many varieties can be grown, picked, and packed at a total cost of not more than 5 cents a pound; but there are no present indications of an early decrease in price. If it should fall to an average of 20 cents a pound, this would still allow the satisfactory gross income of $1000 an acre from fruit alone, while the offshoots of good varieties at present prices ($5 to $15 each) are a valuable factor and may be worth almost as much to the orchardist as the fruit. Offshoots, in fact, should more than pay the whole cost of running a young plantation, leaving the entire proceeds from the fruit as clear profit.

Pests and Diseases

There are two scale insects, found wherever dates grow, that are troublesome to the orchardist. The Parlatoria scale

(*Parlatoria blanchardii* Targ. Tozz.) remains dormant during the winter but is active in summer, sucking the plant juices from the leaves at the time when growth is most vigorous. The following description of the insect is condensed from T. D. A. Cockerell: To the naked eye the scales appear as small dark gray or black specks, edged with white. If the scale is lifted by means of a pin or the point of a knife, the soft, plump and juicy female, of a rose-pink color, is found underneath. The male scales, which are rarely seen, are much smaller and narrower than those of the female. About the middle of March the female lays eggs; the larvæ hatch a fortnight later, crawl about restlessly for a time, and then settle down for the remainder of their lives.

The treatment is by dipping the offshoots in a solution of 1 gallon of Cresolin, 4 gallons of distillate, and 95 gallons of water. Mature palms may be sprayed with the same mixture. By these methods this scale is eventually eliminated.

The more dangerous Marlatt scale (*Phoenicococcus marlatti* Ckll.) is wine-colored, and secretes a white waxy substance. It usually lives at the base of the leaves, "inside" the palm, where it is almost inaccessible, coming out at intervals to molt. It can be destroyed by dipping the offshoots and following this by periodic spraying.

Date palms in moist regions are often attacked by parasitic fungi, which, however, yield to bordeaux mixture or other standard fungicides.

In some regions the palm is attacked by a borer (Rhyncophorus) which, if not destroyed, is fatal to the tree. The only successful treatment seems to be to watch for the intruder and kill it before it has penetrated too far. Locusts, grasshoppers, rats, gophers, ants, bees, wasps, birds, and the like give trouble in various localities. The treatment resorted to against these pests in connection with other cultures will also serve for the date palm orchard.

Stored dates are likely to become infested with such common enemies of stored foods as the fig-moth (*Ephestia cautella* Walker) and the Indian meal-moth (*Plodia interpunctella* Hübner). The best protection against these is a packing-house that is reasonably insect-proof and is fumigated at the beginning of each season. The modern methods of preparing dates for the market usually include some system of disinfection which kills insect eggs. It is reported that in Egypt dates for export are dipped in dilute alcohol, or in alcohol and glycerine. "Dry" dates can be scalded; "soft" dates are, in America, frequently pasteurized by dry heat or by fumigation.

Varieties and Classification

Several thousand varieties of dates have been recognized, but those which have any commercial importance are limited to a few score, while those that are of real merit number only a few dozen, since many kinds owe their reputation not to excellence of flavor but, as do the Elberta peach and the Ben Davis apple, to good shipping and keeping qualities.

Varieties are usually classified as "soft" (or "wet") and "dry." Orientals classify them by color (yellow or red, before they are cured); by keeping quality; and as "hot" and "cold," according to whether a long-continued diet of them "burns" the stomach or not.

The classification of "soft" and "dry" (which sometimes has been complicated and confused by the insertion of an intermediate class of "semi-dry") is commercially convenient, but not absolute; for practically any soft date may become a dry date under certain atmospheric conditions, and most dry dates can be made soft by proper management and artificial maturation.

The dry dates predominate in most parts of North Africa, including Egypt, being preferred by the nomads because they are easily packed and not likely to spoil. On the other hand, practically all of the dates which the world recognizes as valuable are soft varieties. In the following list, which includes the most important kinds from throughout the world, there is only one unmistakably dry date (Thuri), which, though recognized as good in its Algerian home, is given a place in this list mainly because it has succeeded particularly well in California. There are three others (Asharasi, Kasbeh, and Zahidi) that would probably be considered dry, but cannot be unequivocably placed in that class. Asharasi and Kasbeh are much softer than the typical dry date, while Zahidi at one stage of its maturity is typically soft, and is widely sold in that condition, although if left long enough on the palm it becomes actually a dry date. All the other varieties in the list are typically soft, but most, if not all, of them will be converted into dry dates if left to ripen on the trees in a sufficiently hot and dry climate.

The American and European markets are accustomed only to soft dates, and as most of the good varieties are soft, growers will naturally give attention to soft kinds by preference. A market for dry dates, in America at least, will have to be created before any large quantity can be sold. Nevertheless, Americans who have eaten good dry dates usually like them, and frequently consider them preferable to those soft dates, such as Halawi and Khadhrawi, which (often under the trade name of Golden Dates) have until recently been almost the only varieties on the American market.

Amri. — Form oblong, broadest slightly above the center and bluntly pointed at the apex; size very large, length 2 to 2½ inches, breadth 1 to 1¼ inches; surface deep reddish brown in color, coarsely wrinkled; skin thick, not adhering to the flesh throughout; flesh about ½ inch thick, coarse, fibrous, somewhat sticky, and with much rag close to the seed; flavor sweet, but not delicate; seed oblong, 1¼ to 1½ inches long, rough, with the ventral channel broad and shallow, and the germ-pore nearer base than apex. Season late.

More extensively exported from Egypt than any other variety. It is not, however, a first-class date. It is large and attractive in appearance, but inferior in flavor. The keeping and shipping qualities are unusually good. Named probably from Amr, a common personal name.

Asharasi. — Form ovate to oblong-ovate, broadest near the base and pointed at the apex; size medium, length $1\frac{1}{8}$ to $1\frac{3}{8}$ inches, breadth $\frac{7}{8}$ to $1\frac{1}{4}$ inches; surface hard, rough, straw-colored around the base, translucent brownish amber toward the apex; skin dry, thin, coarsely wrinkled; flesh $\frac{1}{4}$ inch thick, at basal end of fruit hard, opaque, creamy white in color, toward tip becoming translucent amber, firm; flavor rich, sweet, and nutty; seed oblong-elliptic, pointed at apex, $\frac{5}{8}$ to $\frac{3}{4}$ inch long, smooth, the ventral channel almost closed, and the germ-pore nearer base than apex. Ripens midseason.

Syn. *Ascherasi.* The best dry date of Mesopotamia, if not of the world. It can be used as a soft date; having always some translucent flesh at the apical end of the fruit, it has by some writers been classed as semi-dry. Grown principally in the vicinity of Baghdad; now also in the United States, where it succeeds well. The name means Tall-growing.

Deglet Nur. — Form slender oblong to oblong-elliptic, widest near the center and rounded at the apex; size large, length $1\frac{1}{2}$ to $1\frac{3}{4}$ inches, breadth $\frac{3}{4}$ to $\frac{7}{8}$ inch; surface smooth or slightly wrinkled, maroon in color; skin thin, often separating from the flesh in loose folds; flesh $\frac{1}{4}$ inch thick, deep golden-brown in color, soft and melting, conspicuously translucent; flavor delicate, mild, very sweet; seed oblong-elliptic, pointed at both ends, about 1 inch long, with the ventral channel shallow and partly closed, the germ-pore at center. Season late.

Syns. *Deglet Noor, Deglet en-Nour.* This variety is considered the finest grown in Algeria and Tunisia, where its commercial cultivation is extensive, and it is highly esteemed in California, where it holds at present first rank among dates planted commercially. Its defects are a tendency to ferment if kept for several months, and the immense amount of heat required to mature it properly. The name is properly transliterated *Daqlet al-Nur*, meaning Date of the Light, an allusion to its translucency.

Fardh. — Form oblong, widest near the middle and rounded at the apex; size small to medium, length about $1\frac{1}{4}$ inches, breadth about $\frac{3}{4}$ inch; surface shining, deep dark brown in color, almost smooth; skin rather thin, tender; flesh $\frac{1}{8}$ to $\frac{1}{4}$ inch thick, firm, russet brown; flavor sweet with a rather strong after-taste; seed small, length $\frac{5}{8}$ inch. Ripens midseason.

Syn. *Fard.* This is the great commercial date of Oman, in eastern Arabia. It has recently been planted in California; American markets are thoroughly familiar with the fruit through the large importations which are annually made from Oman. While inferior in quality to many other varieties, Fardh holds its shape well when packed and keeps well. For these reasons it is a valuable commercial variety.

According to modern Omani etymologists, the name means The Separated, because of the way the dates are arranged in the bunch; but the ancients, who are entitled to more credit, spell it in a way that means The Apportioned.

Ghars. — Form oblong to obovate, narrowest near the rounded apex; size large to very large, length 1½ to 2 inches, breadth about ⅞ inch; surface somewhat shining, bay colored; skin soft and tender; flesh ⅜ inch thick, soft, sirupy, slightly translucent; flavor sweet and rich; seed oblong, ¾ to 1 inch long, with the ventral channel deep and sometimes closed near the middle, and the germ-pore at center. Season early.

Syns. *Rhars, R'ars.* One of the commonest soft dates in North Africa, esteemed for its earliness in ripening, its productiveness, and the ability of the plant to resist large amounts of alkali and much neglect. In California it has proved to be a strong grower, but the fruit is not so good as that of several other varieties, and also ferments easily. The name means Vigorous Grower.

Halawi. — Form slender-oblong to oblong-ovate, broadly pointed or blunt at the apex; size large, length 1¼ to 1¾ inches, breadth about ¾ inch; surface slightly rough, translucent bright golden-brown in color; skin thin but rather tough; flesh ⅛ to 3/16 inch thick, firm, golden-amber in color, tender; flavor sweet and honey-like, but not rich; seed slender oblong, ⅞ inch long, with the ventral channel broadly open. Ripens midseason.

This is the great commercial date of Mesopotamia, and probably the most important variety in the world, as regards quantity sold. It is grown chiefly around Basrah, at the head of the Persian Gulf. It has good keeping and shipping qualities, but is not esteemed by the Arabs for eating; in American markets, however, it is preferred to several other varieties because of its attractive color. Both in California and in Arizona Halawi has succeeded remarkably well. The name means The Sweet.

Hayani. — Form oblong-elliptic, broadest slightly below the center and rounded at the apex; size very large, length 2 to 2½ inches, breadth 1 to 1¼ inches; surface dark brown in color, smooth; skin thick, separating readily from the flesh; flesh about ¼ inch thick, light brown in color, soft; flavor sweet, lacking richness; seed oblong, sometimes narrowed toward the apex, 1¼ to 1⅜ inches long, with the ventral channel broad and deep, and the germ-pore usually ⅜ inch from the base. Ripens midseason.

Syns. *Hayany, Birket al Hajji, Birket el Haggi, Birket el Hadji,* and *Birkawi.* One of the most satisfactory Egyptian dates in California and Arizona. It is precocious and prolific, and has proved to be more frost-resistant than many other varieties. The plant is unusually ornamental in appearance. The variety is named after the village of Hayân.

Kasbeh. — Form oblong-ovate, widest near the base and broadly pointed at the apex; size large, about $1\frac{3}{4}$ inches long, $\frac{3}{4}$ inch broad; surface golden-brown to chestnut in color; skin thin but fairly tough; flesh $\frac{3}{16}$ inch thick, firm, but never hard, tender; flavor sweet, slightly heavy but not cloying; seed oblong-elliptic, almost an inch long, the ventral channel open and deep, the germ-pore nearer base than apex. Season late.

Syns. *Kesba, Kessebi, El Kseba.* A variety of ancient origin, extensively cultivated in Algeria and Tunisia. Before Deglet Nur came into the field it was considered the finest date in North Africa. It is valued in California, where it has been found to have excellent keeping and shipping qualities as well as good flavor. The name means The Profitable.

Khadhrawi. — Form oblong to oblong-elliptic, widest near the center and broadly pointed at the apex; size medium to large, length $1\frac{1}{4}$ to $1\frac{3}{4}$ inches, breadth $\frac{3}{4}$ to $\frac{7}{8}$ inch; surface translucent orange-brown in color, overspread with a thin blue-gray bloom; skin firm, rather tough; flesh, $\frac{3}{16}$ to $\frac{1}{4}$ inch thick, firm, translucent, amber-brown in color; flavor rich, never cloying; seed oblong-obovate to oblong-elliptic, $\frac{7}{8}$ inch long, the ventral channel narrow or almost closed. Ripens midseason.

Syns. *Khadrawi, Khudrawee.* One of the most important commercial varieties of Mesopotamia, ranking second only to Halawi. It is a better date than the latter, but not so highly esteemed on the American market because of its slightly darker color. In California it has been grown with great success. The name means The Verdant.

Khalaseh. — Form oblong to oblong-ovate, broadest near the center and rounded to broadly pointed at the apex; size medium, length $1\frac{3}{8}$ to $1\frac{5}{8}$ inches, breadth $\frac{3}{4}$ to $\frac{7}{8}$ inch; surface smooth, orange-brown to reddish amber in color, with a satiny sheen; skin firm, but tender; flesh $\frac{1}{4}$ inch thick, firm, tender, reddish amber in color, free from fiber; flavor delicate, with the characteristic date taste in a desirable degree; seed oblong-elliptic, pointed at both ends, $\frac{3}{4}$ to $\frac{7}{8}$ inch long, the ventral channel almost closed. Ripens midseason.

Syns. *Khalasa, Khalasi, Khalas.* The most famous date of the Persian Gulf region, and unquestionably one of the finest in the world. It is grown principally at Hofhuf in the district of Hasa; a few palms have been planted in the United States, and have produced fruit of superior quality. Khalaseh likes a dry situation and sandy soil. It is not a heavy bearer, but is precocious. The name means Quintessence.

Khustawi. — Form oblong-oval, broadest near center and rounded at apex; size small to medium, length 1 to $1\frac{1}{2}$ inches, breadth $\frac{3}{4}$ to $\frac{7}{8}$ inch; surface smooth, glossy, translucent orange-brown in color; skin thin and delicate; flesh $\frac{1}{4}$ inch thick, soft and delicate in texture, translucent golden-brown in color; flavor unusually rich yet not cloying, with the characteristic date taste in a desirable degree; seed oblong-obovate, $\frac{3}{4}$ inch long, pointed at both ends, with the ventral channel open. Ripens midseason.

Syns. *Khastawi*, *Kustawi*, originally *Khastawani* (Persian). A delicious dessert date from Baghdad. It has proved well adapted to conditions in the date-growing regions of America. It is not a heavy bearer, but the fruit possesses good keeping qualities. The name means The Date of the Grandees.

Majhul. — Form broadly oblong to oblong-ovate, broadest at center to slightly nearer base and broadly pointed at apex; size very large, length 2 inches, breadth $1\frac{1}{4}$ inches; surface wrinkled, deep reddish brown in color; skin thin and tender; flesh $\frac{3}{8}$ inch thick, firm, meaty, brownish amber in color, translucent, with no fiber around seed; flavor rich and delicious; seed elliptic, $1\frac{1}{4}$ inches long, with the germ-pore nearest the base and the ventral channel almost closed. Season late.

Syns. *Medjool*, *Medjeheul*. A variety of large size and good keeping qualities, from the Tafilalet oases in the Moroccan Sahara, whence the fruit is exported to Europe. Probably suited only to the hottest and driest regions in the United States. The name means Unknown.

Maktum. — Form broadly oblong to oblong-obovate, usually broadest near center and rounded at the apex; size medium, length $1\frac{1}{4}$ to $1\frac{1}{2}$ inches, breadth $\frac{7}{8}$ to 1 inch; surface somewhat glossy, translucent golden-brown in color; skin firm, wrinkled, rather thin; flesh $\frac{3}{8}$ to $\frac{5}{8}$ inch thick, soft, almost melting, light golden-brown in color; flavor mild, sweet, similar to that of Deglet Nur. Season late.

Syn. *Maktoom*, originally *Makdum*. A rare variety from Mesopotamia which has proved admirably adapted to conditions in California, although not resistant to frost. It is large and of fine quality. The palm is a vigorous grower. The name means The Bitten.

Manakhir. — Form oblong, rounded at the apex; size very large, length 2 to $2\frac{1}{4}$ inches, breadth slightly more than 1 inch; surface smooth, brownish maroon in color, with a purplish bloom; skin thin and tender; flesh $\frac{1}{4}$ inch thick, soft and melting, with fiber around the seed; flavor delicate, resembling that of Deglet Nur; seed oblong, 1 inch long, with the germ-pore nearer the base and the ventral channel frequently closed. Season late.

Syns. *Menakher*, *Monakhir*. A rare and large-fruited variety from Tunis, of which only a few palms exist in the United States. In this country it is not a date of the best quality. The name means The Nose Date.

Saidi. — Form oblong-ovate, broadest near the base and blunt at the apex; size large, length $1\frac{1}{2}$ inches, breadth about 1 inch; surface almost smooth, brownish maroon in color, overspread with a bluish bloom; skin thin, tender; flesh $\frac{3}{16}$ inch thick, red-brown in color, firm; flavor very sweet, almost cloying; seed oblong-elliptic, $\frac{7}{8}$ inch long, the germ-pore slightly nearer the base and the ventral channel almost closed. Ripens in midseason.

Syns. *Saidy*, *Wahi*. One of the most important varieties of Upper Egypt. It is not considered so good in quality as some of the Algerian and Mesopotamian varieties, but it is a heavy bearer, though it requires a hot climate to ripen perfectly. The name indicates that it comes from Said or Upper Egypt.

Tabirzal. — Form broadly oblong-obovate, broadest below center and broadly pointed at the apex; size medium, length $1\frac{1}{8}$ to $1\frac{1}{2}$ inches, breadth $\frac{7}{8}$ to $1\frac{1}{8}$ inches; surface translucent deep orange-brown in color, with a blue-gray bloom; skin thin and tender, coarsely wrinkled; flesh $\frac{1}{4}$ inch thick, soft and tender, translucent orange-brown in color; flavor distinctive, mild and pleasant, sweet but not cloying; seed broadly oblong, $\frac{5}{8}$ to $\frac{3}{4}$ inch long, with the ventral channel narrow. Season late.

One of the best dates grown at Baghdad. In the United States it is little known as yet. Originally *Tabirzad* (Persian) meaning Sugar Candy.

Thuri. — Form oblong, broadest near center and bluntly pointed at apex; size large, length $1\frac{3}{4}$ inches, breadth $\frac{3}{4}$ inch; surface reddish chestnut color, overspread with a bluish bloom; skin thin; flesh $\frac{3}{16}$ inch thick, firm and nearly dry but not hard or brittle, golden-brown in color; flavor sweet, nutty and delicate; seed oblong, 1 inch long, the ventral channel deep and partly closed, the germ-pore nearer the base. A midseason date.

Syns. *Thoory*, *Tsuri*. One of the best Algerian dry dates. It is large, not too hard, and of excellent flavor; the palm bears heavily and the clusters are of exceptional size. In California it has proved very satisfactory. The name means The Bull's Date.

Zahidi. — Form oblong-obovate, broadest near the rounded apex; size medium, length $1\frac{1}{4}$ inches, breadth $\frac{7}{8}$ inch; surface smooth, glossy, translucent golden-yellow in color, sometimes golden-brown; skin rather thick and tough; flesh $\frac{1}{4}$ inch thick, translucent golden-yellow close to the skin, whitish near the seed, soft, meaty, and full of sirup; flavor sweet, sugary, and not at all cloying; seed oblong, $\frac{3}{4}$ inch long, the ventral channel open. Season early.

Syns. *Zehedi*, *Zadie*, originally *Azadi* (Persian). A remarkable date, the principal commercial variety of Baghdad. It can be used as a soft date (as described above) or as a dry date, depending on the length of time it is allowed to remain on the palm. The tree is vigorous, hardy, resistant to drought, and prolific in fruiting. The name means Nobility.

CHAPTER VII

THE PAPAYA AND ITS RELATIVES

THE papaya (sometimes called papaw) and the passion-flowers are closely related, and the fruit-bearing kinds are treated together in this chapter. Some botanists place them all in one family even though the papaya is an erect plant and the passion-flowers are tendril-bearing vines; but recent botanists separate them into the Caricaceæ (or Papayaceæ) and Passifloraceæ. In botanical structure, the fruits are very similar, and they are related not distantly to the Cucurbitaceæ (pumpkins and melons).

THE PAPAYA (Plate XI)

(*Carica Papaya*, L.)

"There is also a fruite," wrote the Dutch traveler Linschoten in 1598, "that came out of the Spanish Indies, brought from beyond ye Philipinas or Lusons to Malacca, and fro thence to India, it is called Papaios, and is very like a Mellon. . . and will not grow, but alwaies two together, that is male and female . . . and when they are diuided and set apart one from the other, then they yield no fruite at all."

The facility with which the papaya is propagated by means of its seeds made possible its rapid dissemination throughout the tropics, when once the Discovery had opened up routes of travel between its native home in the Western Hemisphere and the regions in Asia, Africa, and Polynesia favorable to its growth. In many places it early attained the position of importance among cultivated fruits which it holds at the present day. Higgins and Holt say of it: "Excepting the banana, there is no fruit grown in the Hawaiian Islands that means more to the people of this territory than the papaya, if measured in terms of the comfort and enjoyment furnished to the people as a whole."

It may fairly be said, perhaps, that the northern cantaloupe is replaced in Hawaii and other tropical regions by the papaya, a fruit which, in its better varieties, is a worthy rival of the melon. It is adapted to a wide range of territory; it comes into bearing when a few months old; and it yields most abundantly of its handsome fruits. The presence of inferior varieties in many regions has detracted from the prestige of the papaya, but its intrinsic merit is beyond dispute. It is the duty of tropical horticulture to encourage the dissemination of the better forms and further to improve them by means of breeding. Considerable attention has already been devoted to this subject, but much remains to be done. The rapidity with which seedlings can be brought to fruiting stage makes papaya-breeding a much less tedious process than is the case with the hard-wooded tree-fruits.

It has always been a source of wonder to those unfamiliar with the species that a plant so large as a mature papaya could be produced in so short a time. The poet Waller [1] wrote in 1635 with but slight exaggeration of the literal fact:

> "The Palma Christi and the fair Papaw
> Now but a seed (preventing Nature's Law)
> In half the circle of the hasty year,
> Project a shade, and lovely fruits do wear."

The papaya, a giant herbaceous plant rather than a tree, grows to a height of 25 feet, and is often likened to a palm in general appearance, although there is no botanical relationship.

[1] Battle of the Summer Islands.

The trunk bears no lateral branches, but sometimes divides to form several erect stems, which produce at their tops large deeply-lobed leaves sometimes 2 feet across, upon hollow petioles 2 feet or more in length. The wood is fleshy, the bark smooth, grayish brown, marked by conspicuous leaf-scars.

The papaya is normally diœcious (Fig. 29) and produces its flowers in the uppermost leaf-axils, the staminate blossoms sessile on pendent racemes 3 feet or more in length, the pistillate ones subsessile and usually solitary or in few-flowered corymbs. The staminate flowers are funnel-shaped, about an inch long, whitish, the corolla five-lobed, with ten stamens in the throat; the pistillate flowers are considerably larger, with five fleshy petals connate toward the base, a large, cylindrical or globose, superior ovary, and five sessile fan-shaped stigmas.

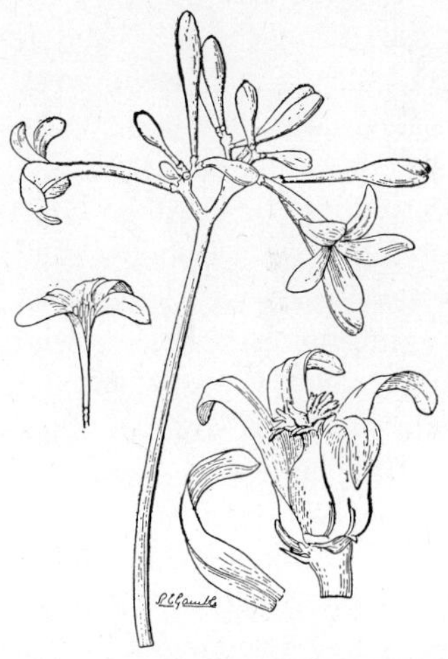

FIG. 29. Flowers of the papaya: the cluster and the single flower to its left are staminate (male), and the larger flower to the right is pistillate (female). Sometimes the organs of both sexes are found in the same flower, but this condition cannot be considered normal. (× about $\frac{1}{3}$)

The fruit is commonly spherical or cylindrical in form, round or obscurely five-angled in transverse section, from 3 up to 20 or more inches in length, and sometimes weighing as much as 20 pounds. In general character it strongly resembles a melon; the skin is thin, smooth on the exterior, orange-yellow to deep orange in color; the flesh, which is deep yellow to salmon-colored, being from 1 to 2 inches thick and inclosing a large, sometimes five-angled, cavity, to the walls of which are attached the numerous round, wrinkled, blackish seeds, the size of small peas, inclosed by a thin gelatinous aril.

The flavor is rather sweet, with a slight musky tang which is sometimes objectionable to the novice, and which varies greatly in degree; the best types being of a bland agreeable taste which is almost sure to be relished. In Brazil the flavor is believed to be improved if the fruit is lightly scored when taken from the tree, and then allowed to stand for a day so that the milky juice may run out.

The native home of the papaya is known to be in tropical America, but the exact area in which it originated has not been determined. Jacques Huber, after reviewing the evidence presented by Alphonse DeCandolle and others, reached the conclusion that the species originally came from Mexico. Count Solms-Laubach, who monographed the Caricaceæ, believes that the cultivated papaya may have originated as a cross between some of the species of Carica native to Mexico.

The plant is now widely distributed. In nearly all parts of tropical America it is one of the common fruits. It is abundant in India, Ceylon, and the Malay Archipelago. In Hawaii it probably attains greater comparative importance than in any other region. It is common in Australia, where it is cultivated as far south as Sydney.

In the United States it has been planted in Florida and California. It is entirely successful in the southern part of Florida, but in California its cultivation is limited to the most protected situations, and even there the fruit produced is not of good quality.

The name papaya is held to be a corruption of the Carib *ababai*. In one form or another it has been carried around the world; *papaia*, *papeya*, and *papia* are some of the corruptions which are in use. The English name papaw is widely employed, but in the southern United States its use has the disadvantage of confusing this fruit with *Asimina triloba*. The

Portuguese name, current in Brazil, is *mamão* (probably referring to the mammiform apex of the fruit); in French the fruit is called *papaye*, in German *papaja*, and in Italian *papaia*. Several other names are used in tropical America, notably *fruta de bomba* in Cuba, *lechosa* in Porto Rico, *melon zapote* in parts of Mexico, and tree-melon in English-speaking countries. Botanically the species is *Carica Papaya*, L.

While most commonly used, perhaps, as a breakfast-fruit, like the muskmelon or cantaloupe in northern countries, the papaya can be prepared in numerous ways. In Brazil it is served as a dessert, sliced, with the addition of a little sugar and whipped cream. As a salad, in combination with lettuce, it is excellent. As a crystallized fruit it is good, but has not much character. When green it is sometimes boiled and served as a vegetable, much as summer-squash is in the North. It can also be made into pickles, preserves, jellies, pies, and sherbets. When used as a breakfast-fruit, it is cut in halves longitudinally, and after the seeds are removed served with the addition of lemon juice, salt and pepper, or sugar, according to taste.

The fruit of the papaya, as well as all other parts of the plant, contains a milky juice in which an active principle known as papain is present. This enzyme, which was first separated by Theodore Peckholt, greatly resembles animal pepsin in its digestive action, and in recent years has become an article of commerce. Aside from its value as a remedy in dyspepsia and kindred ailments, it has been utilized for the clarification of beer. Its digestive action has long been recognized in the tropics, as is evidenced by the common practice of the natives, who rub the juice over meat to make it tender, or, in preparing a fowl, wrap it in papaya leaves and let it remain overnight before cooking it.

Much has been written concerning the preparation and properties of papain. Lengthy accounts will be found in the Philippine Journal of Science, Section A, January, 1915; Agriculture (Habana, Cuba), April, 1917; the Tropical Agriculturist (Colombo, Ceylon) No. 3, 1915; and the American Journal of Pharmacy, 1901.

Alice R. Thompson of Hawaii has published the following analyses of several different seedling strains grown at Honolulu:

TABLE IV. COMPOSITION OF THE PAPAYA

STRAIN	TOTAL SOLIDS	ASH	ACIDS	PROTEIN	TOTAL SUGARS	FAT	FIBER
	%	%	%	%	%	%	%
Trinidad . .	12.14	.53	.06	.43	9.72	.06	.78
South Africa	13.00	.54	.09	.68	10.73	.07	.81
Honolulu . .	12.20	.56	.07	.50	10.29	.05	.66
Barbados .	11.72	.48	.06	.46	8.05	.06	.76
Panama . .	14.41	.90	.14	.50	11.12	.25	1.09

The sugar found in the papaya is principally invert sugar, only traces of sucrose being present.

Cultivation.

The papaya is tropical in its requirements, but it can be grown in regions where light frosts are experienced. It prefers a warm climate and rich, loamy, well-drained soil. In southern Florida it grows best on hammock soils, but it is successful on "high pine" lands if properly fertilized. On the Florida Keys the plant has become thoroughly naturalized and springs up wherever a clearing is made, the seeds being scattered by birds and other agencies. While commercial papaya-culture probably should not be attempted north of Palm Beach, good fruits are occasionally produced in the central part of the state when a mild winter allows the plants to reach fruiting age without injury.

In California the cool nights do not permit the fruit to mature

Plate XII. A plate of fine loquats.

perfectly. It has been observed in the tropics that papayas ripened in cool weather are insipid or squash-like in flavor. The best situations in southern California are the protected foothill regions, where the heat during the summer months is more intense than on the seacoast. An old tree at Hollywood, near Los Angeles, bore fruit several years, but finally succumbed to the cold rains of winter which cause the plants to rot off at the base, especially if the drainage is in the least defective.

Higgins and Holt, whose bulletin "The Papaya in Hawaii" [1] is the most valuable contribution yet made to the literature of papaya-growing, have the following to say concerning climate and soil:

> "In regard to rainfall and moisture requirements, the plant is able to adapt itself to a wide range of conditions, and when established suffers much less from a shortage of water than the orange or the avocado, but makes beneficial use of a large amount if supplied. Yet, withal, it is one of the most insistent plants in the matter of drainage. In waterlogged soils the papaya makes a spindling growth and drops its lower leaves prematurely, while the remaining foliage becomes yellow, the whole plant indicating an unhealthy condition.
>
> "There are few, if any, soils in which the papaya will not grow if aëration and drainage are adequately supplied. Most of the plantings at this station are upon soils regarded as unsuitable for other fruit trees and upon which the avocado is a failure. . . . They are very porous, permitting perfect drainage and aëration. Rich soils give correspondingly better and more permanent results if they permit of the free passage of water and entrance of air."

For a permanent orchard, the plants should be set not less than 10 feet apart. The papaya is short-lived, and will not usually remain in profitable bearing more than three to five years. That it is extremely simple of culture is proved by the ease with which it becomes naturalized in tropical regions, and by the thriftiness of the wild plants which spring up everywhere along the roadsides.

[1] No. 32 of the Hawaii Exp. Sta.

P. J. Wester writes as follows regarding the planting and care of papayas:

"When the plants have attained a height of about 3 to 4 inches, they are ready to be transplanted to the place where they are intended to grow.

"Unless the transplanting has been preceded by a good rain, the plants should be thoroughly watered before they are removed from the seed-bed. In order to reduce the evaporation of water from the plants until they are well established in their new quarters, about three-fourths of the leafblades should be trimmed off.

"In transplanting, take up the plants with so large a ball of earth that as few roots are cut or disturbed as possible. Do not set out the young plant deeper in the new place than it grew in the nursery; firm the soil well around the roots, making a slight depression around the plant, and water it thoroughly.

"In order to protect the tender plant from the sun until it is established, it is well to place around it a few leafy twigs at the time of planting. It is well to set out three plants to each hill, and as the plants grow up and fruit, to dig out the males or the two poorest fruiting plants.

"If the plants cannot be set out in the field at the time indicated, transplant them from the seed-bed to a nursery, setting out the plants about 8 to 12 inches apart in rows a yard apart, or more, to suit the convenience of the planter. While the best plan is to set out the plants in the field before they are more than 12 inches tall, the plants may be transplanted to the field from the nursery with safety after they are more than 5 feet high, provided that all except young and tender leaf-blades are removed, leaving the entire petiole, or leafstalk, attached to the plant; if the petiole be cut close to the main stem, decay rapidly enters it. If the entire petiole is left it withers and drops and a good leaf scar has formed before the fungi have had time to work their way from the petiole into the stem of the plant.

"When a plant has grown so tall that it is difficult to gather the fruit, which also at this time grows small, cut off the trunk about 30 inches above the ground. A number of buds will then sprout from the stump, and will form several trunks that will bear fruit like the mother-plant in a short time. These sprouts, except two or three, should be cut off, for if all are permitted to grow the fruit produced will be small."

When first set out in the field, the young plants should be watered every day or two; after a few weeks have elapsed and they have become established, waterings may be less frequent.

Mature plants should be irrigated liberally unless rainfall is abundant. Since they are gross feeders, stable manure or commercial fertilizers should be supplied liberally. This is particularly true of plants which are grown on the sandy lands of southeastern Florida. Organic nitrogen is especially desirable.

Propagation.

The papaya is usually propagated by seeds, which in Florida should be sown as early in the year as possible, preferably in January, in order to have the plants in bearing by the following winter. If seeds are washed and dried after removal from the fruit, and stored in glass bottles, they will retain their viability for several years. Higgins and Holt say:

"It is best to plant the seeds in a well-drained, porous soil in flats or boxes, covering them about half an inch deep. In from two to six weeks the seedlings should appear, germination being hastened by heat. In the open in cool weather the time will not be less than a month, but in a warm greenhouse it may be shortened to two weeks. In about a month after germination the seedlings should be large enough to be transferred to pots, in which they should remain for another month before being placed in the orchard or garden."

Wester advocates planting in seed-beds and transferring the young seedlings directly into the open ground. He writes:

"The seed-bed should be prepared by thoroughly pulverizing the soil by spading or hoeing the ground well, and the clearing away of all weeds and trash. Sow the seed thinly, about 1 to 2 centimeters apart, and cover the seed not more than 1 centimeter with soil, then water the bed thoroughly. In the dry season it is well to make the seed-bed where it is shaded from the hot midday rays of the sun, under a tree; or it may be shaded by the erection of a small bamboo frame on the top of which is placed grass or palm leaves. If the seed is planted during the rainy season a shed of palm leaves should always be put up over the seed-bed to protect the seed from being washed out and the plants from being beaten down by the heavy rains."

Vegetative propagation of the papaya by two means has been shown to be possible, but it is not yet demonstrated that either of these methods produces satisfactory plants. Cuttings are readily grown, but they develop more slowly than seedlings. Grafted plants are more rapid in growth and come into fruit early, and it was thought at one time that this method offered great possibilities; but later experience has shown that when propagated by this means in Florida, a given variety degenerates rapidly, and in the third or fourth generation from the parent seedling the grafted plants make very little growth and their fruits are small and practically worthless. The explanation of this behavior has not been found, nor is it known whether it will occur in other regions; but its effect in Florida has been to do away with grafting and cause all growers to return to seed-propagation.

In order that those who are interested in the subject may experiment for themselves, a brief extract is given here from "The Grafted Papaya as an Annual Fruit Tree," by Fairchild and Simmonds.[1] These investigators found that seeds of the papaya, when planted in the greenhouse in February, produce young seedlings large enough to graft some time in March; that these grafted trees, which can be grown in pots, when set out in the open ground in May or the latter part of April, make an astonishing growth and come into bearing (in Florida) in November or December; that they continue bearing throughout the following spring and summer, and if it is advisable, can be left to bear fruit into the following autumn.

"After a seedling begins to fruit, it does not normally produce side-shoots which can be used for grafting. It has been observed for some time, however, that if the top of a bearing tree is cut or broken

[1] Circ. 119, Bur. Plant Industry.

off accidentally, a large number of shoots begin to form, one from the upper part of each leaf scar; that is, the axil of the leaf. This takes place three or four weeks after the tree is decapitated. It is these small shoots, of which as many as 50 or more may be produced by a single tree, that are used in grafting the papaya. One of these shoots is taken when a few inches long and about the diameter of a lead pencil, is sharpened to a wedge point, the leaf surface reduced, and inserted in a cleft in a young seedling papaya plant which has been decapitated when 6 to 10 inches high and split with an unusually sharp, thin grafting knife. At this age the trunk of the young seedling has not yet formed the hollow space in the center. It is not necessary for the stock and the cion to be of equal size; the cion should not, however, be larger than the stock. After inserting the cion, the stock is tied firmly, but not tightly, with a short piece of soft twine. The grafted plant should be shaded for a few days after the grafting has been done and the twine should be removed on the sixth or seventh day. The best success has been secured in these experiments by grafting potted seedlings in the greenhouse, or under the shade of a lath-house, presumably because the stock can be kept in good growing condition under these circumstances."

One of the most remarkable features of the papaya is the irregularity which it presents in the distribution of the sexes. Normally it is diœcious, with staminate and pistillate (male and female) flowers produced on different plants. Cross-pollination is necessary to enable the pistillate flowers to develop fruits. This is effected by insects. Among seedling plants the number of staminates is usually greater than that of pistillates. Only a few of the former being necessary as pollinizers (certainly not more than one in ten), this excess of staminates is, from the grower's standpoint, an objectionable feature.

In addition to the staminate and pistillate forms, many intermediates have been observed in which both sexes are combined in one plant. Staminate flowers may occur with rudimentary stigmas and ovaries which give rise to small worthless fruits; and there is a hermaphrodite type which regularly produces perfect flowers, is self-pollinated, and yields excellent fruits. Numerous other forms have been described (see the bulletin by Higgins and Holt), but the importance of these is lessened by the fact that during the lifetime of a plant it may change from one form to another.

In general, it may be said that plants which develop from the seed as pure pistillates will retain their sex without modification, but plants which commence life as pure staminates may undergo a change of sex. It has been asserted that a change of sex may be induced by topping the male tree or breaking its roots. M. J. Iorns, who studied this question in Porto Rico, reached the conclusion that other conditions than the loss of the terminal bud must be present to induce a change of sex, and he suggested that the trees may pass through definitely recurring cycles of development, and be subject to the change only at certain periods. L. B. Kulkarni,[1] who investigated the matter in India, came to the belief that change of sex is not in any way connected with the removal of the terminal bud. He found that male plants, in the course of their development, may present a number of different sex-combinations, as follows:

First stage: Staminate flowers only.
Second: Staminate, with a few hermaphrodite flowers.
Third: A few staminate, with many hermaphrodite flowers.
Fourth: A few staminate, with many hermaphrodite, and a few pistillate flowers.
Fifth: Hermaphrodite flowers only.
Sixth: Hermaphrodite, with a few pistillate flowers.
Seventh: A few hermaphrodite, with many pistillate flowers.
Eighth: Pistillate flowers only.

Thus the plant in the course of its life history may change from a staminate to a hermaphrodite and then to a pure pistillate.

At the Hawaii Experiment Station, much attention has been devoted to breeding papayas. Some of the objects in view have been hermaphroditism (in order to eliminate the necessity

[1] Poona Agrl. College Magazine, 1, 1915.

of male trees to act as pollinizers), fruit of suitable size and shape for market purposes, uniformity in ripening, good keeping qualities, and good color and flavor of flesh. The diœcious type has not been satisfactory in breeding, principally because the staminates do not show the characters which are inherent in them and which will appear in the fruits of their progeny. "The hope, therefore," says J. E. Higgins,[1] "must lie in the use of a hermaphrodite type. Here it is possible to select an individual of known qualities. This may be used as the sole parent stock or may be combined with another parent of known qualities. What mixtures there may be in the individual at the start may not be known; but through repeated selections and elimination of undesirable characters, it should be possible to produce a reasonably pure strain, provided, of course, that the stock is kept pure by constantly avoiding cross-pollination with plants of different characters."

Some excellent hermaphrodite forms have already been produced, and, although they do not breed true, a sufficient number of the seedlings are hermaphrodites and produce fruit of good quality for it to be felt that a definite advance has been made. Breeding work should be continued until a strain has been purified to a point that it will breed true and retain its fruit characteristics as closely as do cultivated varieties of eggplant, tomato, and other vegetables.

Yield and market.

In the tropics papayas are in season during a large part of the year and the yield is enormous, a single plant bearing in the course of its life (not more than a few years) a hundred or more immense fruits. In Florida the season extends from December to June, with a few fruits ripening at other times. Higgins and Holt say: "The first ripe fruits may be expected (in Hawaii) in about a year from the time when the plants are

[1] Journal of Heredity, May, 1916.

set in the orchard or garden, and thereafter fruits and flowers in all stages of development may be in evidence at all times of the year. In the cool season the fruits are slow in ripening, thus causing a short crop and high prices for a month or two."

Sometimes the fruits are produced in such abundance that it is necessary to thin them in order to avoid their remaining small in size or becoming malformed by the pressure of neighboring fruits. Thinning should be done when the fruits are rather small.

If the fruits are to be sent to market they should be picked as soon as the surface begins to turn yellow. "Certain varieties become ripe enough for serving while showing little yellow coloring." It is difficult to ship the fully ripe fruit, since it is large, heavy, and has no firm outer covering, but only a thin membranous skin, to protect it. For this reason papayas must be shipped before they are fully ripe, and even then great care is necessary. Shipments have been made from Hawaii to San Francisco in cold storage with good results. When shipped from southern Florida to New York by express, the percentage of loss is usually large, unless the fruit is picked while still green; and in the latter case it does not ripen properly after reaching the market. It is advised to encase the fruits in cylinders of corrugated strawboard, and pack them in single-tier cases holding four to six fruits.

Pests and diseases.

Two pests have become sufficiently troublesome in south Florida to require attention. One, the papaya fruit-fly (*Toxotrypana curvicauda* Gerst.) threatened at one time to become serious. This insect occurs in several parts of tropical America. The female inserts her eggs into the immature papaya by means of a long ovipositor, and the larvæ first feed in the central seed-mass, but later work into the flesh of the fruit, frequently rendering it unfit for human consumption. The only means of control which have been suggested are the destruction of wild plants and infested fruits, and the production of varieties having very thick flesh, so that the ovipositor will not reach to the seed-cavity (the young larvæ are unable to live in the flesh). A fungous disease known as papaya leaf-spot (*Pucciniopsis caricæ* Earle) frequently attacks the foliage in the winter season, forming small black masses on the under-surfaces of the leaves. It is not very destructive and is easily controlled by spraying with bordeaux mixture.

In Hawaii a red mite (Tetranychus sp.) sometimes occurs in scattered colonies on the lower surfaces of the leaves and on the fruits. The larvæ of a moth (*Cryptoblades aliena* Swezey) feeds under a web on the floral stems and beneath the flower-clusters. Neither of these pests is said to be serious. The Mediterranean fruit-fly (*Ceratitis capitata* Wied.) attacks the fruit; its presence in Hawaii has made necessary a quarantine order prohibiting the shipment of papayas from that territory to the mainland of the United States. Two scale insects, *Aspidiotus destructor* Sign. and *Pseudoparlatoria ostriata* Ckll., are reported on the plant in Africa and Cuba respectively.

Seedling races.

With the introduction of grafting as a means of propagating choice papayas in Florida, one named variety, the Simmonds, was established, but the stock has degenerated and it is no longer grown. Grafted plants of the third and fourth generation from the original seedling developed to a height of 3 or 4 feet only, produced a few small fruits, and were always yellowish and sickly in appearance.

There are marked differences in the size, shape, and quality of the fruits produced by different seedlings, and the papayas of certain regions in the tropics are uniformly superior to those of other regions. In Bahia, Brazil, there are two distinct types, one with small nearly spherical fruits not over 6 inches in diameter, and a very superior type called *mamão da India* which produces fruits 18 inches long, cylindrical in form, and of excellent flavor. The hermaphrodite seedlings produce some of the sweetest fruits, and they usually have thick flesh. Some papayas are very sweet, while others are insipid. The production of seedling races which will produce fruits of good quality and breed fairly true is much to be desired.

The Mountain Papaya

(*Carica candamarcensis*, Hook. f.)

Since it comes from elevations of 8000 or 9000 feet in the mountains of Colombia and Ecuador, this species is more frost-resistant than its near relative the papaya, and in this characteristic lies its greatest interest. It has been suggested that hybridization of the two species might result in a plant which would be sufficiently hardy for regions like southern California and the shores of the Mediterranean, and yet would produce fruit nearly as good as that of the papaya. Such a hybrid has not yet been produced.

The mountain papaya resembles its more tropical relative in habit and general appearance, but it is smaller in all its parts; it grows only 8 or 10 feet high, its leaves are smaller (and deeply lobed), and its fruits are only 3 or 4 inches in length. H. F. Macmillan says: "The tree has been introduced at Hakgala Gardens, Ceylon, in 1880, and is now commonly grown in hill gardens for the sake of its fruit, being often found in a semi-naturalized state about up-country bungalows." A. Robertson-Proschowsky of Nice, France, writes, in the Petite Revue Agricole et Horticole: "It is a handsome plant, growing a few meters high, and often without branches, though the latter are developed when the top is killed by frost. For several years I have grown this species and I find it to produce good fruits, of a sweetish, acidulous, perfumed taste. They are suitable, as I have had occasion to learn from experience, for persons with weak stomachs, who cannot eat other fruits. They are particularly good for dyspeptics." Macmillan notes that the fruit, which ripens in Ceylon throughout the year, is too acid to be used for dessert, but is very agreeable when stewed and can be made into jam and preserves.

The requirements of the plant are much the same as those of the papaya, except in regard to climate. It withstands 28° above zero without serious injury. The seeds are sown in the same manner as those of the papaya.

There are other species of Carica in tropical America, many of them as yet little known, which may be of value in connection with papaya breeding. *C. quercifolia*, Benth. and Hook., with leaves like those of the English oak, is even hardier than the mountain papaya, but its fruit, the size of a date, is worthless. There appear to be in Ecuador several species closely resembling *C. candamarcensis*, but some of them may be nothing more than varieties of the latter.

The Purple Granadilla (Plate X)

(*Passiflora edulis*, Sims)

The passifloras are known in the Temperate Zone as flowering plants, but the species commonly grown in the tropics are cultivated principally for their edible fruits. The most important one is the purple granadilla, *P. edulis*, known in Australia, where its culture is extensive, as passion-fruit.

The plant is a strong-growing, somewhat woody climber, with deeply three-lobed, serrate leaves. The flower, which is white and purple, is attractive but not so handsome as that of some other members of the genus. The fruit is oval, 2 to 3 inches long, deep purple in color when fully ripe. Within the brittle outer shell are numerous small seeds, each surrounded by yellowish, aromatic, juicy pulp, the flavor of which is rather acid.

From its native home in Brazil the purple granadilla has been carried to all parts of the world. It attains its greatest importance as an economic plant in Australia, but it is grown also in Ceylon, the Mediterranean region, in the southern United States, and elsewhere. The fruit is used for flavoring sherbets, for confectionery, for icing cakes, for "trifles," — a dish composed of sponge-cake, fruits, cream, and white of egg, — and for other table purposes. The pulp is also eaten directly from the fruit, after adding a little sugar, or it may be used to prepare a refreshing drink by beating it up in a glass of ice-water and adding a pinch of bicarbonate of soda.

The term passion-fruit, which is often applied to this species, confuses it with other members of the same genus, many of which are known by the same common name. In order to distinguish between these different species, it is well to adopt a different name for each. *P. edulis* is called *lilikoi* in Hawaii.

In California this fruit is easily grown, but it has not yet reached a position of importance in the markets; indeed, it is rarely seen in them, — a condition which contrasts strikingly with its prominence in Australia. It withstands light frosts, but when young is injured by temperatures more than one or two degrees below the freezing-point. While it bears abundantly in California, plants grown in Florida have in some instances failed to produce fruits. The reason for this is not definitely known, but it may be due to defective pollination. The pollination of this and other edible-fruited passifloras deserves investigation, for it is probable that the secret of many failures in their cultivation lies in this detail. Paul Knuth, in his "Handbook of Flower Pollination," states that the passifloras are protandrous (the anthers shedding their pollen before the stigmas are in condition to receive it) and adapted to cross-pollination by humble-bees and humming-birds. In describing the pollination of *P. cœrulea* he says: "In the first stage of anthesis, a large insect (such as a humble-bee) when sucking the nectar, receives pollen on its back from the downwardly dehiscing anthers. In the second stage the styles have curved downwards to such an extent that the now receptive stigmas are lower than the empty anthers. It follows that older flowers are fertilized by pollen from younger ones."

The passifloras are easily propagated by seeds or cuttings, the latter method being preferable in most cases. Seeds should be removed from the fruit, dried in a shady place, and planted in flats of light soil. They do not germinate quickly, but the young plants are easily raised, and may be set out in the open ground when six months to a year old. Cuttings should be taken from fairly well-matured shoots, and should be about 6 inches in length. They are easily rooted in sand, no bottom-heat being required. Cuttings of the purple granadilla will often fruit in pots at the age of two years.

Directions for the commercial cultivation of this fruit, based on American experience, cannot be given, since no commercial plantings, with the exception of a few small ones on an experimental scale, have yet been made in this country. The following extracts are taken from an article by W. J. Allen in the Agricultural Gazette of New South Wales for November 2, 1912:

"Although this fruit is not grown so extensively as it should be throughout the many districts on the coast where it will do well, it nevertheless plays quite an important part in some of the young citrus orchards in the County of Cumberland, on the Penang Mountain, and around the Gosford district, where it is frequently planted among the trees. As it begins to bear very early, growers are enabled to make considerably more from this crop than pays for the working of the orchard until the young trees begin to produce crops of fruit, which they invariably do after the third or fourth year.

"Generally speaking, the vines are most productive before having attained to four or five years of age. After that period they begin to lose vigor and gradually die out, or cease to be very profitable, and are in consequence removed.

"The passion-vine is found to thrive well on many classes of soil, — some so poor that one is led to wonder how anything could profitably be grown on it. On the light sandstone and poorer coastal country there is no other fruit which will give the same return as this, and with proper working and heavy manuring, it is wonderful the amount of fruit that can be taken from an acre of such vines. The area planted is comparatively small, and, in consequence, the fruit usually commands very high prices. As an addition to a fruit salad there is no flavor that can surpass it, and when eaten with cream it rivals the most delicious of strawberries. If this fruit were known in Great Britain and America, I venture to say that there would be an unlimited demand for it, if once we were successful in landing it in those countries in large quantities.

"In selecting a site for the planting of a vineyard, one of the important points to keep in view is to avoid a district or situation where frosts are at all severe or of frequent occurrence in the winter. There is one thing which this vine will not stand, and that is severe frosts; and the Easter, winter, and spring crops are those which are in most demand. During the summer time there is a superabundance of other fruits, and hence the consumption of the passion-fruit is not so great; from Easter until Christmas time there is a splendid market for all well grown fruit. It is during part of this time that we have our coldest weather, and a severe frost or two would destroy the whole crop, and in all probability kill the vine back to the root.

"The chief feature about the passion-vine, however, is its habit of producing two crops per annum. The summer crop comes in about February or March, and prices are necessarily low. The winter crop is ready for pulling when other fruits are not so plentiful on the market. The practice of the growers, has, therefore, been to secure a heavy winter crop by pruning away the summer crop when about half grown; or generally speaking, about the month of November. This stimulates the vines to throw out fresh fruiting laterals for the winter.

"The next point of importance is to put the land in thorough order before planting, and in places where it is very sour and deficient in lime, which it mostly is on our coastal country where the passion-fruit is grown, it would be advantageous to give the land at least half a ton of good lime to the acre.

"The vines should be planted out about August or September, when the ground is in good condition.

"The seed is sown in February. The rows should be 30 inches to three feet apart, and the seed every inch or so in the row, afterwards thinning out to three inches apart to make good stocky plants.

"In erecting the trellis, the posts should be six feet and a half long, firmly set in the ground to a depth of 18 inches, and placed at distances of about 24 feet apart, or at farthest 32 feet in the row. On the tops of these posts are tightly stretched, at a distance of six inches apart, two strong No. 8 galvanized iron wires. The rows should run north and south, so that they get the sunlight on both sides. The rows are placed in the center of the tree-rows, or when alone, 10 feet apart, with the vines 12 feet in the row, thus requiring about 362 plants to the acre.

"The young vine is trained with a single stem up the stakes until it reaches the wires, when it is allowed to throw out from two to four leaders, which are trained to run either way on the wires. As the vine puts forth further growth, the main leaders and laterals are trained along the wires.

"Without judicious manuring there are very few districts where the growing of this fruit would prove highly satisfactory, while, on the other hand, those growers who are giving the most attention to this important adjunct are the ones who are making the greatest profits out of the industry. It has become a recognized fact that liberal dressings of manure must be used from the time of planting until the plants cease to be productive.

"On making inquiry among the different growers, I found that scarcely any two of them were using the same mixture. Some, on the lighter soils, were using considerable quantities of blood and bone with a little potash; others were using bone, superphosphate, and potash; while others were using a mixture of nitrate of soda, dried blood, and superphosphate and sulphate of potash, etc., etc.; and judging from the appearance of the different vines, all with very gratifying results.

"When the fruit begins to ripen it should be picked at least twice a week. It will keep well in a cool dry place, but I would recommend marketing every week.

"All badly formed and inferior fruit is discarded, and the better fruit is mostly packed in layers, so that when opened at the markets it presents a good appearance. In grading, color as well as size is taken into consideration, any badly colored fruits being sorted out and packed separately."

The Sweet Granadilla (Fig. 30)

(*Passiflora ligularis*, A. Juss.)

Next in importance to the purple granadilla or passion-fruit comes the sweet granadilla, a species extensively used by the inhabitants of the mountainous parts of Mexico and Central America. In flavor it is perhaps the best of the genus, and it certainly merits a wider distribution than it enjoys at present.

Henry Pittier speaks of this fruit as "neither a food nor a beverage." Its white pulp is almost liquid, acidulous, and perfumed in taste. Among the Indians of Central America it is a favorite, and figures prominently in many of the markets.

The plant is a vigorous climber, scrambling over buildings and trees of considerable size. The leaves are cordate and acuminate, and commonly about 6 inches long. The flowers are solitary, with the petals and sepals greenish, and the corona white with zones of red-purple. The fruit is somewhat larger than that of *P. edulis*, oval or slightly elliptic in form, and orange to orange-brown, sometimes purplish, in color. The shell is strong, so that the fruit can be transported long distances without injury. The seeds are numerous and each surrounded by translucent whitish pulp. The Indians eat the fruit out of hand.

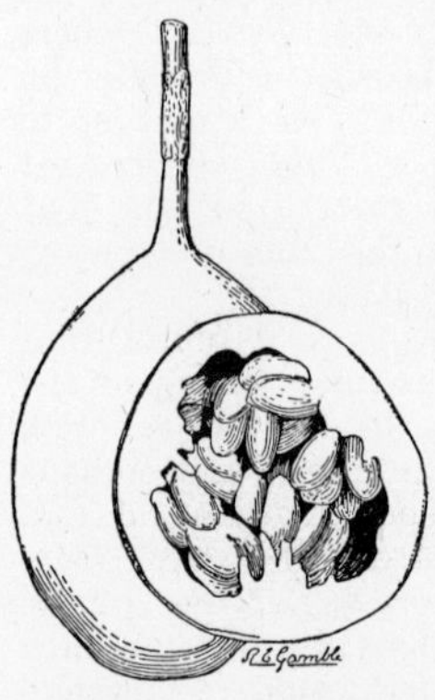

FIG. 30. The sweet granadilla (*Passiflora ligularis*), one of the best-flavored fruits of its genus. (× ⅓)

The species is a native of tropical America and does not seem to be known in other regions. Recently it has been introduced into California and Florida by the United States Department of Agriculture, but so far as is known, it has not yet fruited in either state. Since it grows in Central America at elevations of 6000 to 7000 feet, it should be sufficiently cold-resistant to withstand light frosts, although it is doubtful whether it will survive temperatures more than two or three degrees below freezing-point.

Its requirements in regard to soil and cultural attention are probably about the same as those of *P. edulis*. It does not fruit quite so abundantly as the latter, nor has it been observed to produce more than one crop a year in Central America. Propagation is usually by seed.

THE GIANT GRANADILLA (Fig. 31)

(*Passiflora quadrangularis*, L.)

While this is the largest-fruited species of the genus, and one of the most widely distributed, it is not the best in quality. From its native home in tropical America it has been carried to the eastern tropics, where it is now grown in many places. It is common in the West Indies, but nowhere is it cultivated on a commercial scale.

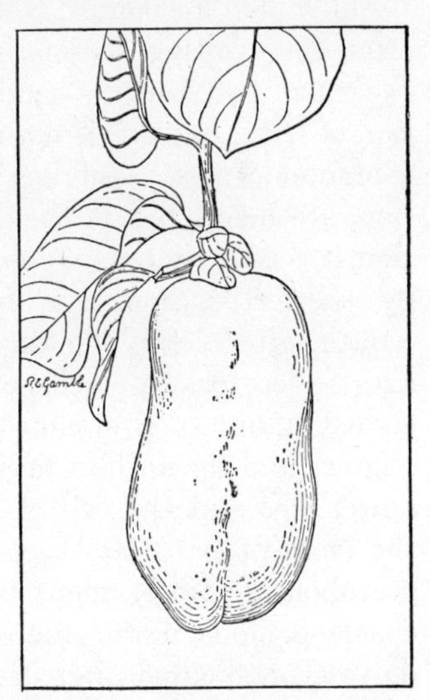

FIG. 31. The giant granadilla (*Passiflora quadrangularis*). (× ¼)

The plant is somewhat coarse and is a strong climber. The stems are four-angled, as indicated by the specific name, and the leaves are ovate or round-ovate, cordate at the base and mucronate at the apex, entire, and 6 or 8 inches long. The flowers, which are about 3 inches in diameter, are white and purple in color. The fruits are oblong, up to 10 inches in length. H. F. Macmillan says: "Its large, oblong, greenish-yellow fruit is not unlike a short and thick vegetable-marrow, and contains in its hollow center a mass of purple, sweet-acid pulp mixed with flat seeds." A horticultural form exists which has leaves variegated with yellow.

This species is more tropical in its requirements than *P. ligularis* and *P. edulis.* It will grow in southern Florida, but is not successful in California. A. Robertson-Proschowsky reports, however, that it has fruited on the French Riviera at Golfe-Juan and perhaps elsewhere, and in his own garden at Nice was only killed after surviving several winters.

The fruit is known in French as *barbadine,* in Portuguese as *maracujá melão,* and in Spanish as *granadilla* or *granadilla real.* The name granadilla is applied, in different parts of the tropics, to several species of Passiflora, and in order to distinguish them it is necessary to append a qualifying word. It is derived from *granada,* and means "small pomegranate."

Macmillan recommends that the shoots be well cut back after the fruiting season is past. It is commonly believed necessary to resort to hand-pollination to insure the production of fruit, but this is not always the case. The protandrous character of the passifloras, and the necessity of cross-pollination, are mentioned in the discussion of the purple granadilla; that it is sometimes possible, however, for fruits to be produced by self-fertilization, has been shown by experience. Paul Knuth, after describing the character of the passiflora flower, says: "Autogamy (self-pollination) would seem to be excluded under such circumstances, yet it is possible that the stigmas and the anthers may be brought into contact when the flower closes at the end of the single day's anthesis. This is the more probable as Warnstorf saw a fully formed fruit in a greenhouse. Here, then, is a case in which an obviously chasmogamous flower (one in which the perianth opens) is only self-pollinated after it has closed." If *P. quadrangularis* is self-sterile, however, it would do no good to have the flowers self-pollinated. If insects are lacking to do the work, cross-pollination must be effected by hand.

Propagation is by seed or by cuttings, which should be 10 to 12 inches long and from well-matured stems, and should be inserted in sand.

Several other species of Passiflora are cultivated in the tropics for their fruit. *P. laurifolia,* known as yellow granadilla, water-lemon, Jamaica honeysuckle, sweet-cup, bell-apple, and *pomme d'or,* is cultivated in the West Indies, and to a limited extent in other regions. H. F. Macmillan states that it is not fruitful in the eastern tropics. *P. maliformis,* L. is grown in the West Indies, and in the mountains of Colombia, where it is called *curuba* or *kuruba.*

CHAPTER VIII

THE LOQUAT AND ITS RELATIVES

HEREIN are grouped the few fruits of the Rose family that are cultivated to any extent in the tropics and subtropics. In temperate regions, this family supplies the leading tree-fruits, as apple, pear, quince, stone-fruits, and also such small-fruits as raspberry, dewberry, blackberry, and strawberry.

THE LOQUAT (Plate XII)

(*Eriobotrya japonica*, Lindl.)

The production of loquats in Japan is estimated at twenty million pounds annually. From one small village in the Chekiang Province of China, twenty thousand dollars' worth have been shipped in a single year. In the Occident this excellent fruit has not attained the commercial prominence which it deserves, nor has it been improved through cultivation and selection to any such extent as have many other Asiatic fruits now grown in Europe and America.

To northern residents and travelers in tropical and subtropical countries, the loquat should possess an especial attraction, inasmuch as it recalls in flavor and character the fruits of the North. As a matter of fact, it is a close relative of the apple and the pear, while its flavor distinctly suggests the cherry. Those whose palates have been educated to demand the subacid sprightly flavored fruits of the Temperate Zone often criticize tropical fruits as being too sweet and rich. The loquat is not open to this objection, and it can be grown throughout

PLATE XIII. *Left*, the wild cherry of Central America (*Prunus salicifolia*); *right*, the manzanilla (*Cratægus stipulosa*).

the tropics wherever there are elevations of a few thousand feet.

To reach its greatest perfection, the loquat requires particular climatic conditions. Quite satisfactory results are obtained with it, however, in situations where the plant cannot realize its best possibilities. The tree is simple of culture, and has become widely distributed throughout the tropics and subtropics.

Not until rather recently has it been planted in regions where systematic attention is given to the improvement of fruits; hence its development to meet the ideals of European and American pomologists, while accomplished in part, is still far from complete. The progress made during the last twenty years is highly encouraging, and several varieties now available are sufficiently good to merit extensive cultivation.

Because of its ornamental appearance alone, the loquat is often planted in parks and gardens. It is a small tree, rarely more than 30 feet high and commonly not exceeding 20 or 25 feet. It has a short trunk, usually branching two or three feet from the ground to form a crown round or oval in form, and normally compact and dense. The leaves, which are somewhat crowded towards the ends of the stout woolly branchlets, are elliptic-lanceolate to obovate-lanceolate in outline, 6 to 10 inches long, remotely toothed, deep green in color, and woolly below. The fragrant white flowers are ½ inch broad and are borne in terminal woolly panicles 4 to 8 inches long. The calyx is composed of five small, imbricate, acute teeth; the corolla has five oblong-ovate clawed petals, white in color and delicate in texture. The stamens are twenty, the pistils five, joined toward the base. The fruits, which are borne in loose clusters, are commonly round, oval, or pyriform, 1 to 3 inches in length, pale yellow to orange in color, and somewhat downy on the surface. The skin is about as thick as that of a peach, but slightly tougher; the flesh firm and meaty in some varieties, melting in others, ranging from almost white to deep orange in color, juicy, and of a sprightly, subacid flavor. The seeds may be as many as ten, since there are five cells in the ovary and two ovules in each cell; but usually several of the ovules are aborted, and not more than three to five seeds develop. They are ovate in form, flattened on the sides, light brown in color, and about $\frac{3}{4}$ of an inch long. Sometimes fruits with only one seed are found, and varieties constantly one-seeded have been reported.

Although formerly considered indigenous to Japan and China, it is now believed that the loquat was originally limited to the latter country. The late Frank N. Meyer considered the species to be "in all probability indigenous to the hills of the mild-wintered, moist regions of central-eastern China." He found it in a semi-wild state near Tangsi, in Chekiang Province, a region in which loquats are extensively cultivated for market. The Chinese graft superior varieties on seedling stocks, but according to Meyer [1] they are not very skillful in this work. Their finest variety is said to be the *pai bibaw* or white loquat.

The loquat has been cultivated in Japan since antiquity, and is at present one of the important fruits of that country. It is grown in the same regions as the citrus fruits, or even farther north than the latter. T. Ikeda [2] points out that localities noted for unusually fine loquats always lie close to the sea. Numerous varieties have originated in Japan, the best of which have been introduced into the United States and a few other countries. While there are commercial orchards in many places, the total number of trees growing in Japan is said to be less than one million; hence it would seem that the industry there should be capable of extension, for the fruit is popular and the territory adapted to its production is large.

In northern India the loquat is a fruit-tree of considerable importance. A. C. Hartless, superintendent of the Government

[1] Bull. 204, Bur. Plant Industry.
[2] Fruit Culture in Japan.

Botanical Gardens at Saharanpur, observes that certain localities have been much more favorable than others, and that the best results are obtained where the soil is sandy loam and where abundant water is supplied: and reports that "In the plains the loquat is in season in April, but in the colder climate of the hills it fruits in the autumn." Most of the trees in India are seedlings, but several grafted varieties have been distributed from Saharanpur.

Throughout a large part of the Mediterranean region the loquat is highly successful; it is said, in fact, to have become naturalized in several places. In southern France it is a common tree, but there are no large commercial plantations. In Italy and Sicily it is abundant. David Fairchild states it to be one of the principal fruits of the island of Malta, but the trees are seedlings and practically none of them worth propagating. L. Trabut says of the loquat in Algeria: "The Horticultural Society, the Botanical Service, and a certain number of amateurs have collaborated in producing superior varieties which are now propagated by grafting. The Botanical Service has introduced the best varieties obtainable in Japan, and public opinion is undergoing a change regarding this fruit. Formerly it was not esteemed." The tree is common in the gardens of Algiers, and during early spring the fruit is abundant in the markets.

Regarding its behavior in England, the Gardener's Chronicle (May 3, 1913), referring to it under an alternative name, says: "The Japanese Medlar is an old garden favorite, grown in this country for its handsome evergreen foliage, and in warmer regions for the sake of its edible fruits. Messrs. Sander have obtained from some source a variegated sport of it, which is likely to become a popular garden plant, the variegation being particularly pleasing, some of the leaves being more milk-white than green. It is not generally known that the Japanese Medlar is quite happy when grown under the shelter of a wall in the neighborhood of London; in other words, it is much hardier than is supposed."

According to Paul Hubert, the loquat is grown in Madagascar and in some islands of French Oceania. It is also cultivated in Indo-China. In Hawaii it is fairly common as a garden tree. In Australia its cultivation is limited to Queensland, but Albert H. Benson says that it can be grown in the more southerly coast districts, in the foothills of the Coast Range, and on the coast tablelands. It is not extensively cultivated in any of these regions. Grafted varieties are offered by nurserymen in Brisbane.

The loquat has become widely distributed throughout America, where its cultivation extends from California and Florida to Chile and Argentina. In Mexico, Central America, and northern South America, it is grown usually in mountain valleys and on plateaux at elevations of 3000 to 7000 feet. In those situations it succeeds well, and merits more attention than is now given it, especially the introduction of superior varieties, propagated by grafting.

California is probably the most favorable region for loquat culture in the United States. There are many areas in the southern end of the state which are admirably adapted to the production of choice fruit, and the commercial development of loquat culture in these localities is slowly but steadily progressing. Already there are several orchards ten to twenty acres in size, and many budded trees of superior varieties have been planted in dooryards and home gardens.

Throughout the Gulf states the tree grows well, but in many regions frosts interfere with the production of fruit. Several small orchards have been started in Florida, and while these have not been altogether successful in most instances, there are certain districts in the southern part of the state which seem well adapted to its culture. W. J. Krome has had signally good results with this fruit at Homestead. At Miami it has not done so well, probably because the soil is too light for it and not sufficiently moist.

While the name loquat is universally recognized among English-speaking peoples as the correct one for this fruit, it is sometimes called Japanese medlar and Japan-plum. The Spanish name is *níspero del Japón*, the Italian *nespola giapponese;* both of these mean Japanese medlar, and have been applied because of the resemblance of this fruit to the European medlar, *Mespilus germanica*. The French use this same term, as *néflier du Japon;* they also use the name *bibace*. Yule and Burnell say of the word loquat: "The name is that used in S. China, lu-küh, pronounced at Canton lukwat, and meaning 'rush orange.' Elsewhere in China it is called pi-pa." This later suggests *biwa*, which is the common name in Japan.

The botanical name of the loquat is *Eriobotrya japonica*, Lindl., of which *Photinia japonica*, Gray, is a synonym. The latter name is retained by those who prefer not to separate the two genera, for the generic name Photinia is older than Eriobotrya.

Although most commonly eaten as a fresh fruit, the loquat can be utilized in several ways. For culinary purposes it is nearly as useful as its temperate-zone relative the apple; it may be stewed and served as a sauce, or it may be made into excellent jelly. Loquat pie, if made from fruit which is not fully ripe, can scarcely be distinguished from the renowned article made from cherries. The seeds are usually removed from the fruit before it is cooked, as otherwise they impart a bitter flavor to it.

The following analyses of two California varieties, made by M. E. Jaffa, have been published by I. J. Condit in his bulletin "The Loquat"[1] unquestionably the most thorough treatise on this fruit which has appeared up to the present:

[1] Bull. 250, Calif. Agr. Exp. Sta.

TABLE V. COMPOSITION OF THE LOQUAT

VARIETY	WATER	PROTEIN	FAT	SUGAR		FIBER	ASH
				Dextrose	Sucrose		
	%	%	%	%	%	%	%
Thales	89.0	0.35	0.06	8.95	0.94	0.30	0.29
Champagne . . .	84.0	0.32	0.03	11.96	0.83	0.37	0.36

Cultivation.

The climatic requirements of the loquat, except as an ornamental plant, are distinctly subtropical. It is not successful in the hot tropical lowlands, nor can it be grown for fruiting purposes in regions subject to more than a few degrees of frost. Cool weather during part of the year and a rainfall of 15 to 50 inches (with artificial irrigation where the dry season is severe) suit it best. These conditions are found in southern Japan, in parts of southern California, along the shores of the Mediterranean, and in several other regions. It has been noted in Japan that the best loquat situations always lie close to the sea; and in California much finer fruit has been produced near the coast than in the foothill tracts twenty to thirty miles inland. Thus it seems that the mild climate of the seacoast is peculiarly favorable to the development of the fruit.

While mature trees have withstood temperatures as low as 10° above zero without serious injury, the flowers and young fruits may be killed by temperatures only a few degrees below freezing; hence loquats cannot be produced successfully where heavy frosts may occur at the time of flowering. Condit notes: "Frost coming when the fruit is less than half grown may result in killing the seeds only, while the flesh continues to develop, so that seedless fruits mature. On the other hand, frost may have somewhat the same effect as sunburn, injuring the tissues and causing them to shrink or to develop irregularly."

When grown in regions where the weather during the ripening season is extremely hot and dry, the fruit is subject to sun-scald or sunburn. The exposed surface withers and turns brown, and the product is rendered unfit for market. If, on the other hand, the weather is cool and foggy during the ripening season, the fruit lacks sweetness and flavor.

Sandy loam is considered the ideal loquat soil, and it should be of good depth. Several other types of soil have proved satisfactory; thus, in southern California good orchards have been produced on heavy clay of the adobe type, and in Florida the shallow rocky soils of the Homestead region on the lower east coast have given excellent results. Deep sandy soils, when of little fertility, are not suitable. Frank N. Meyer points out that the best loquat orchards in China are situated on low, rich, moist land.

In California orchards, loquat trees are planted 12 to 24 feet apart. When planted on the square system, they should not be nearer than 20 feet. Close planting has been practiced in Orange County, where the rows are set 24 feet apart and the trees 12 feet apart in the row. This is believed to result in greater regularity and uniformity of production than wider planting. March and April are good months for planting in California; late September and October are also suitable. In southern Florida the best time is probably in the autumn.

The amount of tillage given the orchard varies in different regions. Condit says: "Clean culture may be practiced throughout the season, but the growth either of a winter or a summer leguminous cover-crop is much more advisable." For a winter cover-crop, the natural vegetation which springs up in California with the arrival of the rains may be allowed to grow until it reaches its maximum development, when it should be cut with a mowing-machine and plowed under after the fruit is harvested. Following this the ground should be cultivated and a summer cover-crop such as buckwheat or the whip-poor-will cowpea should be planted. "Winter cover-crops may be planted as early as September, in which case they may have made sufficient growth to be turned under before the harvest begins. This is not always possible, especially if an early variety of loquat is grown; in fact, it is a question whether it is advisable to plow or work the ground deeply or at all during the setting and maturing of the fruit." In Florida and other regions different methods of cultivation may be required, but the liberal use of green cover-crops seems universally desirable.

In addition to cover-crops, stable manure is often used to enrich the land in California orchards. Bearing loquat trees exhaust the fertility of the soil rapidly and it is necessary to replenish the supply of plant-food annually if fruit of large size is to be expected. Condit observes: "When the average California soil begins to fail from heavy production, nitrogen is likely to be the first crop limiter; after nitrogen, phosphoric acid, and after phosphoric acid, potash." Particular care should be taken, therefore, to see that the supply of nitrogen is sufficient to meet the demands of the tree. C. P. Taft, of Orange, California, has found the green cover-crops of great value in this connection. E. Pillans, Government Horticulturist at the Cape of Good Hope, says that a yearly application of well-rotted stable manure is amply repaid by larger crops and increased size of fruit. The loquat groves of Japan are said to be fertilized with litter, weeds from the roadsides, and, recently, with commercial fertilizers. Condit advises the application of 15 cubic feet of stable manure biennially to each bearing tree.

It is ordinarily considered that the amount of water required by loquat trees corresponds closely to that needed by citrus fruits. Probably it would be more accurate to say that the loquat is more drought-resistant than any of the citrus fruits, but that the best results are obtained when the orchard is irrigated as liberally as the citrus orchard. In California there is usually abundant rainfall at the time the fruits are approaching maturity; in other regions, or in California if the season is abnormally dry, it may be desirable to supply water at this time, since the fruits only develop to large size when there is abundant moisture in the soil. In southern France the tree is said not to do well on soils which are over-moist in winter.

The young tree should be headed 24 to 30 inches above the ground, and three to five main branches forced to develop. The loquat is a compact grower, and the mature tree requires much less pruning than most of the temperate-zone fruits. It has been found by C. P. Taft, however, that a certain number of branches must be cut out from time to time, in order to limit the amount of fruiting wood and to admit light to the center of the tree. It must be remembered that the tendency of the loquat is to overbear, and for the production of commercially valuable fruit this must be checked by pruning and thinning. The best time for pruning is soon after the crop has been harvested.

Propagation.

In many countries it is still the custom to propagate the loquat by seed, but in regions where the commercial cultivation of this fruit has received serious attention, this method has been replaced by budding and grafting. Seedling loquats are no more dependable than seedlings of other tree-fruits. As ornamental trees for parks and dooryards they can be recommended, but they will not serve when commercially marketable fruit is required.

Choice named varieties are budded or grafted on seedling loquat stocks or on the quince. Other plants have been used as stock-plants, but have not proved altogether satisfactory.

When budded on quince the tree is dwarfed. This stock is easy to bud; and it is believed to produce a tree which bears at an early age, while its fibrous root-system readily permits of transplanting. In spite of these advantages it is considered unsatisfactory in Florida, and in California it is commonly held that the seedling loquat is preferable. To produce stock-plants, loquat seeds may be planted singly in four-inch pots; they may be sown in flats of light soil and later transplanted; or they may be germinated in moist sand or sawdust and potted off as soon as they are 3 or 4 inches high. Potting soil should be light and loamy. After the young plants are 8 inches high, they may be planted in the field in nursery rows. When the stems are about ½ inch in diameter at the base, the plants are ready for budding or grafting.

In California, budding is best done in October or November. Budwood should be of young smooth wood, preferably that which has turned brown and lost its pubescence and from which the leaves have dropped. Shield-budding is the method used (a description of the operation will be found in the chapter on the avocado). The buds should be cut at least 1½ inches long. After inserting them in T-incisions made in the stocks at a convenient point not far above the ground, they are tied with raffia, soft cotton string, or waxed tape. Three or four weeks later the wraps should be loosened to keep them from cutting into the stock, and the eye should be left exposed. The wraps should not be finally removed until the bud has made several inches' growth. In California the stock-plant is cut off 2 or 3 inches above the bud in early spring. This usually forces the bud to grow, but sometimes it shows a tendency to lie dormant, and many adventitious buds develop around the top of the stock. These must be removed as fast as they make their appearance.

In Florida it has been found that buds unite readily with the stock-plant, but that it is difficult to force them into growth.

For this reason grafting has superseded budding in that state. The stocks should be of the same size as for budding, and the cion should be of well-matured wood. Cleft-grafting is the method commonly employed.

The young trees should be stake-trained in the nursery, and headed 24 to 30 inches above the ground. In a year from the time of budding or grafting they should be ready for transplanting.

In California, budded or grafted trees begin to bear the second or third year after they are planted in the orchard, but they cannot be expected to produce commercial crops until four or five years old. According to Condit, a ten-year-old tree should produce 200 pounds of fruit. Early in the season, the latter part of February and all of March, prices are high. Fancy fruit will bring 25 to 35 cents a pound at this time. Later, in May and June, the average price drops to 5 cents and occasionally lower, but fancy fruit rarely sells for less than 8 to 10 cents a pound. It is the opinion of experienced loquat-growers that the gross returns from an orchard should be $300 to $500 an acre; more than this has been obtained in some instances. The advisability of planting early varieties, in order to place the crop on the market while prices are high, is emphasized by all growers. If late fruit is to be produced, it should be of large-fruited varieties which ship well; otherwise the profits will be small.

Yield and picking.

The loquat tree is productive, and a regular bearer. Barring crop failures due to severe frosts at flowering time, the trees rarely fail to produce well every year. Their tendency is to overbear, with the result that the fruits are apt to be undersized. It has been profitable to thin the crop, since the increased size of the fruits remaining on the tree more than compensates for the loss of those removed. The practice of experienced loquat-growers in California is to clip out the ends of the fruit-clusters with a pair of thinning-shears: this should be done as soon as the young fruits have formed. Where choice varieties are grown, and where birds and insects are troublesome, it has been profitable, in a small way, to protect the fruit by inclosing each cluster in a cloth or paper bag. The Japanese, who practice bagging in connection with the production of fancy loquats, find that it results in larger fruit and a greater degree of uniformity in ripening.

The season during which loquats are marketed in California extends from the latter part of February to June. A given variety may ripen several weeks earlier in one locality than in another. In Florida the season is considerably earlier than in California. The fruits should be left on the tree until they are fully ripe, unless it is desired to use them for jelly or for cooking. Unripe the loquat is decidedly acid, whereas the fully ripe fruit is sweet and delicious. Clippers such as are used by orange-pickers are employed in gathering the fruit. Sometimes whole clusters can be picked, and again it may be necessary to clip off two or three ripe fruits and leave the remaining ones to mature.

The fruit is sorted and graded by hand. For shipping to near-by markets it is packed in thirty-pound wooden boxes ("lug boxes") without the use of excelsior, straw, or other soft material to prevent bruising. For distant markets smaller packages and considerable care will be required, since the fruit is bruised rather easily.

Pests and diseases.

The principal enemies of the loquat in California are pear-blight (*Bacillus amylovorus* Trev.) and loquat-scab (*Fusicladium dendriticum* var. *eriobotryæ* Scalia). Condit says of the former: "The pear blight is a serious enemy of the loquat at times, blossom blight often being especially abundant on trees during the spring months. Infected twigs should be cut off well back of the diseased area and burned, care being taken to sterilize the pruning shears in alcohol or formalin after each cut so as to reduce the danger of further infection. Occasionally entire trees are killed by the blight, which gradually extends downward from the branches into the trunk, although in most cases the disease does not seem to progress much beyond the branches. Some varieties are more susceptible than others. For example, the Advance is quite resistant and the trees of the Victor, which were very susceptible when young, have in later years become more or less immune; the Champagne showed considerable blossom blight in the spring of 1914, but to no greater extent than young trees of other varieties. The trees seem to gain resistance as they grow older."

In regard to the scab he says: "This is reported to be a serious disease of the loquat in Australia. The fruit is attacked when half grown by brownish black spots, which soon extend, stop its further development, and disfigure its appearance. The fleshy part of the fruit becomes desiccated and the skin seems to cling to the stones. A large proportion of the crop may in a short space of time be rendered absolutely unsalable. It is also well known in Italy upon the leaves. In California the scab is quite common both on nursery and bearing trees, attacking both leaves and fruit. . . . Spraying with Bordeaux mixture after the blossoms have fallen and the fruit is setting should prove an effective remedy."

In Florida the flowers are sometimes blighted by the anthracnose fungus (*Colletotrichum glœosporioides* Penz.). Bordeaux mixture, prepared according to a 3–3–50 formula, should be used to combat this disease.

E. O. Essig[1] mentions four insects which occasionally attack the loquat in California. One of these is the well-known codlin-moth (*Cydia pomonella* L.). Another is the green apple

[1] Injurious and Beneficial Insects of California.

aphis (*Aphis pomi* DeGeer), and the remaining two are scale insects, one the San José scale (*Aspidiotus perniciosus* Comstock), and the other the Florida wax scale (*Ceroplastes floridensis* Comstock). None of these insects is a serious pest at present. In other countries the fruit is sometimes attacked by the Mediterranean fruit-fly (*Ceratitis capitata* Wied.) and the Queensland fruit-fly (*Bactrocera tryoni* Froggatt). In India the anar caterpillar (*Virachola isocrates* Fabr.) bores in the fruit.

Varieties.

The regions in which named varieties of the loquat have been developed are China, Japan, Queensland, India, Sicily, Algeria, and California.

Little is known of the Chinese varieties. Frank N. Meyer observed several in his travels in China, but mentioned specifically only one, the *pai-bibaw*, or white loquat. T. Ikeda lists forty-six varieties which are cultivated in Japan, but only nine of them are important. One of them, Tanaka, has been introduced into the United States by David Fairchild and into Algeria by L. Trabut. Four sorts are listed by the Government Botanical Garden at Saharanpur, India, but only one, the Golden Yellow, is recommended by A. C. Hartless, Superintendent of the Garden. The Queensland varieties are not extensively planted, and probably are not so good as those of California. Out of five or six named forms which have originated in Italy (including Sicily), not one has been planted extensively. More than fifteen varieties have been described from Algeria, but most of them have already been discarded. One, named Taza, which Trabut produced by crossing Tanaka and one of the best Algerian loquats, is considered meritorious.

Most of the improved sorts at present cultivated in California and Florida have been produced by C. P. Taft of Orange, California. Taft has done more than any other man in the

United States to improve the loquat. His method of procedure has been to grow a large number of seedlings and select the most desirable ones. In this way he has established eight named varieties, of which Champagne, Advance, Early Red, Premier, and Victor are the best.

Little attention has been devoted to the classification of loquat varieties. Takeo Kusano, professor in the Imperial College of Agriculture and Forestry at Kagoshima, states that the Japanese classify them into two groups, called Chinese and Japanese. The Chinese type is large, pyriform, and deep orange-colored, while the Japanese is smaller, lighter colored, and sometimes slender in form. This classification may correspond to one suggested in 1908 by L. Trabut of Algiers. Trabut's two groups were defined, one as having crisp white flesh and the other orange or yellow flesh.

The Chinese group, so far as is known at present, includes only late-ripening varieties. The flesh differs in texture from that of loquats belonging to the Japanese group, while the flavor is very sweet. Kusano states that Tanaka belongs to this class. The variety known in California as Thales, which is thought by some to be identical with Tanaka or very close to it, appears also to belong to the Chinese list.

The Japanese group includes the loquats of California origin, such as Champagne and Premier. These fruits have not the firm meaty flesh of the Chinese group, but are more juicy, and also are distinct in flavor. The flesh is whitish or light-colored, except in the variety Early Red.

The varieties described below are the important ones cultivated in the United States at the present time. For others of minor value, the reader is referred to Condit's bulletin and to the articles by Trabut in the Revue Horticole de l'Algérie.

Advance. — Shape pyriform; size large, weight 2½ ounces, length 2½ inches, breadth 1⅜ inches; base somewhat tapering; apex narrow, the basin medium deep, narrow, abrupt, corrugated; the calyx-segments short, converging, the eye closed; fruit-cluster large, compact; surface downy, deep yellow in color; skin thick and tough; flesh whitish, translucent, melting and very juicy; flavor subacid, very pleasant; quality good; seeds commonly 4 or 5, the seed cavity not large. Season March to June at Orange, California.

This variety was originated by C. P. Taft of Orange, California, in 1897. It is a productive variety, and the fruit-clusters are large and handsome.

Champagne. — Shape oval to pyriform; size large, weight 2 ounces, length $2\frac{1}{2}$ inches, breadth $1\frac{1}{2}$ inches; base tapering, slender; apex flattened, rather narrow, the basin shallow, narrow, flaring, and the calyx-segments broad, short, the eye small, open; fruit-cluster large, loose; surface deep yellow in color with a grayish bloom; skin thick, tough, somewhat astringent; flesh whitish, translucent, melting, and very juicy, flavor mildly subacid, sprightly and pleasant; quality very good; seeds 3 or 4, the seed cavity not large. Season late April and May at Orange, California.

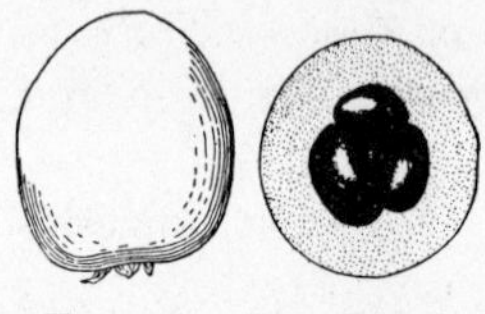

Fig. 32. The Premier loquat, of California origin which has been planted commercially. (× $\frac{1}{3}$)

Originated by C. P. Taft at Orange, California, in 1908. Taft considers it superior to his other varieties in flavor. It is precocious and productive.

Early Red. — Shape oval pyriform to oblong pyriform; size medium large, weight 2 ounces, length $2\frac{1}{2}$ inches, breadth $1\frac{3}{4}$ inches; base tapering slightly; apex broad, flattened, with the basin shallow, narrow, abrupt, the calyx-segments short, broad, the eye small and closed; fruit-cluster compact; surface yellowish orange, tinged with red in the fully ripe fruit; skin thick, tough, acid; flesh pale orange, translucent, melting and very juicy; flavor very sweet, pleasant; quality good; seeds 2 or 3, the seed cavity not large. Season February to April at Orange, California.

The Early Red loquat was originated by C. P. Taft of Orange, California, in 1909. This is the earliest variety known in California. It is valuable for commercial cultivation in regions that are free from severe frosts.

Premier (Fig. 32). — Shape oval to oblong-pyriform; size large, weight $2\frac{1}{2}$ ounces, length $2\frac{1}{2}$ inches, breadth $1\frac{3}{4}$ inches; base tapering slightly; apex flattened, the basin shallow, moderately broad, rounded, the calyx-segments short, the eye large, nearly open; surface orange-yellow to salmon-orange in color, downy; skin moderately thick and tough; flesh whitish, translucent, melting and juicy; flavor subacid, pleasant; quality good; seeds 4 or 5, the seed cavity not large. Season April and May at Orange, California.

Originated by C. P. Taft of Orange, California, in 1899. It is a good variety for home use, but not a good shipper.

Tanaka. — Shape commonly obovoid, weight 2 to 3 ounces. L. Trabut says of it: "Tanaka is characterized by a beautiful color, remarkable size, firm flesh of rich color, agreeable perfume, and little acidity. The proportion of flesh to seeds is large. This loquat owes to the consistence of its flesh unusual keeping quality, — it can be handled without turning black. Left for a week it wrinkles and dries but does not rot. Among the plants, grafted on quince, which were introduced from Japan, two subvarieties can be distinguished; one with pear-shaped fruits, the other subspherical. Tanaka is vigorous, the leaf a little narrower than in our loquats. The tree is productive." Tanaka is famed as the largest loquat in Japan, and one of the best. It has been planted in Algeria and in California.

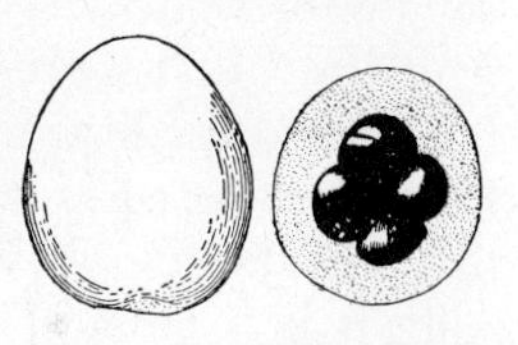

FIG. 33. Thales loquat, late-ripening, large, and of excellent quality. (× about $\frac{1}{3}$)

Thales (Fig. 33). — Shape round to pyriform; size large, weight $2\frac{1}{4}$ to $2\frac{3}{4}$ ounces, length $2\frac{5}{8}$ inches, breadth $1\frac{3}{4}$ to 2 inches; base rounded; apex flattened, the basin shallow and flaring, the calyx-segments broad and short, eye open or closed; surface yellow-orange to orange in color; skin not thick, tender; flesh orange-colored, firm and meaty, juicy; flavor sweet, suggesting the apricot; quality good; seeds 4 or 5, the seed cavity not large. Season April to June at Placentia, California.

Syns. *Placentia Giant, Gold Nugget.* Introduced into California, without name, from Japan betwen 1880 and 1890. It is a large, handsome fruit, and possesses unusually good shipping qualities. It is considered to be very close to Tanaka, if not synonymous with that variety.

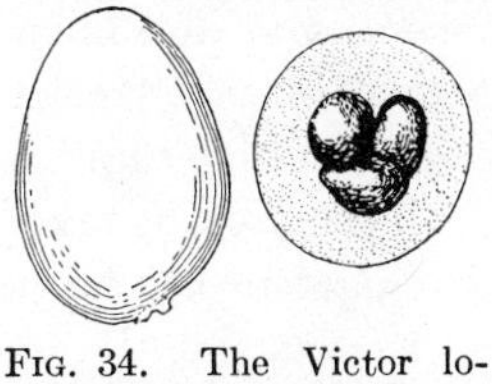

FIG. 34. The Victor loquat. (× about $\frac{1}{3}$)

Victor (Fig. 34). — Shape oblong-pyriform; size large, weight $2\frac{1}{2}$ ounces, length $2\frac{1}{4}$ inches, breadth $1\frac{3}{4}$ inches; base tapering slightly; apex slightly flattened, with a shallow, flaring basin; fruit-cluster large, loose; surface deep yellow in color; skin moderately thick and tough; flesh whitish, translucent, melting, very juicy; flavor sweet, not very rich; quality good; seeds 3 to 5, the seed cavity medium-sized. The season of this variety is May and June at Orange, California.

Originated by C. P. Taft of Orange, California, in 1899. A large and showy fruit, but not considered valuable in California because it ripens late in the season. It is considered especially good for canning.

The Capulin (Plate XIII)

(*Prunus salicifolia,* HBK.)

One of the best rosaceous fruits of tropical countries is the capulin or wild cherry of Central America and northern South America. In its present wild and semi-wild state a fruit of fairly good quality, it would seem that with a little attention from plant-breeders it might become a valuable addition to the list of fruits suitable for moist subtropical regions. Geographically it is a tropical fruit, but in climatic requirements it is distinctly subtropical, for it does not thrive upon the tropical littoral, but grows in cool mountain regions at elevations of 4000 to 9000 feet. It should, therefore, be sufficiently hardy to permit of cultivation as far north as California, Florida, and the Gulf states, and it may also be of value for northern India, southern Brazil, and similar regions.

The botany of this species is confused. It seems to differ very little from the *Prunus Capollin,* Zucc., of Mexico (*P. Capuli,* Cav., *Cerasus Capollin,* DC); possibly the two are identical. *Prunus Capollin* is abundant in the Mexican highlands, where it is an important fruit. *Prunus salicifolia* is supposed to be found only in South America, but specimens collected in Guatemala have been identified as of this species. Horticulturally there is little difference between the capulins of Mexico and those of Central America. The name is taken from the Nahuatl language of Mexico. In Spanish the fruit is often termed *cereza* (cherry).

The tree is erect, often somewhat slender, and reaches a height of 30 feet. The trunk is stout, reaching as much as 3 feet in thickness, with bark rough and grayish. The leaves, which are borne upon slender petioles $\frac{3}{4}$ inch long, are commonly $4\frac{1}{2}$ inches in length, oblong-lanceolate in outline, with a long slender tip, and are deep green on the upper surface, glaucous below, with margins finely serrate. The flowers, which in Guatemala are produced from January to May, are white, about $\frac{3}{4}$ of an inch broad, very numerous, on slender racemes 2 to 4 inches long. As many as fifteen or twenty fruits sometimes develop on a raceme, but half or more fall before reaching maturity. The ripening season in Guatemala is May to September. The fruits resemble northern cherries in appearance; they are $\frac{1}{2}$ to $\frac{3}{4}$ inch in diameter, and deep, glossy, maroon-purple in color. The skin is thin and tender, though sufficiently firm for the fruit not to be easily injured by handling. The flesh is pale green, meaty, and full of juice, and the flavor sweet, suggestive of the Bigarreau type of cherry, with a trace of bitterness in the skin. The stone is rather large in proportion to the size of the fruit.

Pleasant to eat out of hand, this cherry can also be used in various other ways, — stewed, preserved whole, or made into jam. In the highlands of Guatemala, where it is abundant, it is usually eaten as a fresh fruit or made into a sweet preserve. While not equal to the cultivated cherries of the North, — fruits which have been produced by generations of selection and vegetative propagation, — the capulin is a fruit of remarkably good quality for one which has never had the benefit of intelligent cultivation and has been propagated only by seed. Naturally, some trees produce much better fruit than others, and it will be worth while to select the best seedling forms now existing in tropical America and propagate them by budding or grafting.

The Manzanilla (Plate XIII)

(Cratægus spp.)

The manzanilla of Guatemala and the tejocote of Mexico are fruits so similar in character that they may perhaps belong to one species; the former is considered at present to be *Cratægus stipulosa*, Steud., and the latter *C. mexicana*, Moç. &

Sessé. It may be allowable to use the common name manzanilla (the diminutive of the Spanish *manzana,* hence little apple) for both, since it is better adapted to the English language than the Mexican *tejocote* (from the Nahuatl *texocotl,* meaning stone-plum). According to Gabriel Alcocer, *Cratægus stipulosa* is found in Mexico as well as in Guatemala.

The manzanilla closely resembles some of the northern haws in appearance, but it is a larger fruit than most of the latter. It occurs only in the highlands, at elevations of 3000 to 9000 feet. It withstands heavy frosts unharmed, and should be suitable for cultivation in subtropical regions with rather dry climates. It has done well in southern California, where it was introduced some years ago by F. Franceschi under the name *Cratægus guatemalensis.*

The plant is variable in habit, in some cases shrubby, in others becoming a small tree, with a short thick trunk. Commonly it is seen as an erect slender tree about 20 feet high. In spring it produces white flowers resembling those of the apple. In early fall, beginning about October, the yellow fruits ripen and remain abundant in the markets of Mexico and Guatemala until Christmas. They resemble small apples in appearance. The largest specimens are nearly 2 inches in diameter, but the average size is not over 1 inch. The flesh is mealy in texture, and not so juicy nor so sprightly in flavor as that of a good apple. The seeds, commonly three in number, are rather large.

The fruits, which are much used for decorative purposes, are eaten in the form of jelly and preserves. For stewing, they are first boiled with wood-ashes, by which means the skin is easily removed; they are then placed in hot sirup and boiled for a short time. The flavor of the cooked fruit suggests that of stewed apples.

The plant is simple of culture. It grows most commonly on heavy soil and does not require a large amount of water. Propagation is usually by seed, but it would be an easy matter to bud or graft superior varieties. Both in Mexico and in Guatemala the European pear is sometimes top-worked on the manzanilla by cleft-grafting.

The Icaco

(*Chrysobalanus Icaco*, L.)

Although not a fruit of great value, the icaco is extensively used in the tropics, especially by the poorer classes. It is abundant along the seacoasts of tropical America as a wild plant, and is frequently planted in gardens. In southern Florida, where it is known as coco-plum, it is not considered valuable. In Cuba, where the Spanish name icaco (often spelled *hicaco* and *jicaco*) is current, the wild fruit is gathered and made into a sweet preserve, which is served in Habana restaurants as a *sobremesa* or dessert. In Brazil, where it is called *uajurú,* its use is limited. It is said to occur in Africa as well as in America.

The icaco is a large shrub or small tree, attaining a maximum height of 25 or 30 feet. When grown as a shrub it is rather ornamental and it is sometimes planted for this reason. The leaves are obovate or obcordate in outline, about 2 inches long, thick, glossy, and deep green in color. The flowers are small and white, in axillary racemes or cymes. The fruit resembles a large plum in appearance, being oval, 1½ inches long, and pinkish white, magenta-red, or almost black in color. The skin is thin, and the white flesh, which is cottony and of insipid taste, adheres closely to the large oblong seed.

Jacques Huber says that the icaco grows wild in the Amazon region on dry sandy soils. In other parts of tropical America it is often found on moist rich ground. It is propagated only by seed. While there is hope of improving the quality of the fruit through selection, it is doubtful, in view of the abundance of more promising subjects, whether the species would repay attention. The plant is easily grown and withstands light frosts.

CHAPTER IX

FRUITS OF THE MYRTLE FAMILY

The myrtaceous fruits comprise an interesting lot of aromatic things, and with blossoms bearing many long and conspicuous stamens. The Myrtaceæ include many of the spices, as clove, cinnamon, allspice, nutmeg. The eucalyptus belongs here; also such ornamental plants as myrtus, callistemon, metrosideros, melaleuca. The guava is the most important pomological fruit of the family. Most of the guavas belong to what is usually considered to be a single species, although several Latin names have been applied in the group.

The Guava (Fig. 35)

(*Psidium Guajava*, L.)

The guava, while useful in many ways, is preëminently a fruit for jelly-making and other culinary purposes. To the horticulturist the species is admirable as being one of the least exacting of all tropical fruits in cultural requirements, for it grows and fruits under such unfavorable conditions, and spreads so rapidly by means of its seeds, that it has in truth become a pest in some regions. It is a fruit of commercial importance in many countries, and one whose culture promises to become even more extensive than it is at present, for guava jelly is generally agreed to be *facile princeps* of its kind, and is certain to find increasing appreciation in the Temperate Zone.

The first account of the guava was written in 1526 by Gonzalo Hernandez de Oviedo, and published in his "Natural History of the Indies." Oviedo says:

"The guayabo is a handsome tree, with a leaf like that of the mulberry, but smaller, and the flowers are fragrant, especially those of a certain kind of these guayabos; it bears an apple more substantial than those of Spain, and of greater weight even when of the same size, and it contains many seeds, or more properly speaking, it is full of small hard stones, and to those who are not used to eating the fruit these stones are sometimes troublesome; but to those familiar with it, the fruit is beautiful and appetizing, and some are red within, others white; and I have seen the best ones in the Isthmus of Darien and nearby on the mainland; those of the islands are not so good, and persons who are accustomed to it esteem it as a very good fruit, much better than the apple."

FIG. 35. The common guava of the tropics (*Psidium Guajava*), an American plant which has become naturalized in southern Asia and elsewhere. (× ½)

The guava is an arborescent shrub or small tree, sometimes growing to 25 or 30 feet. The trunk is slender, with greenish-brown scaly bark. The young branchlets are quadrangular. The leaves are oblong-elliptic to oval in outline, 3 to 6 inches long, acute to rounded at the apex, finely pubescent below, with the venation conspicuously impressed on the upper surface. Flowers are produced on branchlets of recent growth, and are an inch broad, white, solitary, or several together upon a slender peduncle. The calyx splits into irregular segments; the four petals are oval, delicate in texture. In the center of the flower is a brush-like cluster of long stamens. The fruit is round, ovoid, or pyriform, 1 to 4 inches in length, commonly yellow in color, with flesh varying from white to deep pink or salmon-red. Numerous small, reniform, hard seeds are embedded in the soft flesh toward the center of the fruit. The flavor is sweet, musky, and very distinctive in character, and the ripe fruit is aromatic in a high degree.

The native home of the guava is in tropical America. The exact extent of its distribution in pre-Columbian days is not known. In the opinion of Alphonse DeCandolle, it occurred from Mexico to Peru. In the former country the Aztec name for it was *xalxocotl*, meaning sand-plum, probably a reference to the gritty character of the flesh. The name *guayaba* (whence the English guava) is believed to be of Haitian origin. The plant was carried at an early day to India, where it has become naturalized in several places. It is now cultivated throughout the Orient. In Hawaii it has become thoroughly naturalized. Occasional specimens are said to be found along the Mediterranean coast of France, and in Algeria. In short, the guava is well distributed throughout the tropics and subtropics.

In the United States, the two regions in which guavas can be grown are Florida and southern California. The plant is said by P. W. Reasoner to have been introduced into the former state from Cuba in 1847. It is now naturalized there in many places and cultivated in many gardens. It is successful as far north as the Pinellas peninsula on the west coast and

Cape Canaveral on the east, but has been grown even farther north. If frozen down to the ground, the plant sends up sprouts which make rapid growth and produce fruit in two years. In California the species has not become common, as it has in Florida, nor is it suited to so wide a range of territory in the former as in the latter state. Accordingly it can only be grown successfully in California in protected situations. At Hollywood, at Santa Barbara, at Orange, and in other localities it grows and fruits well, although occasional severe frosts may kill the young branches.

Guayaba is the common name of *Psidium Guajava* throughout the Spanish-speaking parts of tropical America. The French have adopted this in the form *goyave*, the Germans as *guajava*, and the Portuguese as *goiaba*. The latter name is used in Brazil, where the indigenous name (Tupi language) is *araçá guaçú* (large araçú). In the Orient there are many local names, some of them derived from the American *guayaba*. The commonest Hindustani name, *amrud*, means "pear." The term *safari am*, meaning "journey mango," is also current in Hindustani.

The two species *Psidium pyriferum* and *P. pomiferum* of Linnæus are now considered to be the pear-shaped and round varieties of *P. Guajava*. They represent two of the many variations which occur in this species. The pear-shaped forms are often called pear-guava, and the round ones apple-guava. A large white-fleshed kind was formerly sold by Florida nurserymen under the name *Psidium guineense*, and in California as *P. guianense;* but it is now known to be a horticultural form of *P. Guajava*, as is also a round, red-fleshed variety introduced into California under the name *P. aromaticum*. The true *P. guineense*, Sw. (see below) has been itself confused with *P. Guajava*, but can be distinguished from it by its branchlets, which are compressed-cylindrical in place of quadrangular, and by the number of the transverse veins, which is less than in the latter-named species.

The fruit is eaten in many ways, out of hand, sliced with cream, stewed, preserved, and in shortcakes and pies. Commercially it is used to make the well-known guava jelly and other products. When well made, guava jelly is deep wine-colored, clear, of very firm consistency, and retains something of the pungent musky flavor which characterizes the fresh fruit. In Brazil a thick jam, known as *goiabada,* is manufactured and sold extensively. A similar product is made in Florida and the West Indies under the name of guava cheese or guava paste. An analysis at the University of California showed the ripe fruit to contain: Water 84.08 per cent, ash 0.67, protein 0.76, fiber 5.57, total sugars 5.45 (sucrose none), starch, etc., 2.54, fat 0.95.

The guava succeeds on nearly every type of soil. In Cuba it does well on red clay, in California it has been grown on adobe, and in Florida it thrives on soils which are very light and sandy. While not strictly tropical in its requirements, it can scarcely be called subtropical. It is found in the tropics at all elevations from sea-level to 5000 feet, and it withstands light frosts in California and Florida. Mature plants have been injured by temperatures of 28° or 29°, but the vitality of the guava is so great that it quickly recovers from frosts which may seem to have damaged it severely. Young plants, however, may be killed by temperatures of only one or two degrees below freezing. As regards moisture, writers in India report that the guava prefers a rather dry climate.

The plants may be set from 10 to 15 feet apart, the latter distance being preferable. They should be mulched with weeds, grass, or other loose material immediately after planting. In certain parts of India, where guava cultivation is conducted commercially on an extensive scale, it is the custom to set the plants 18 to 24 feet apart. Holes 2 feet wide and deep are prepared to receive the trees. Occasionally the soil is tilled and once a year each plant is given about 20 pounds of barnyard manure. During the dry season the orchard is irrigated every ten days. Very little pruning is done.

Seedling guavas do not necessarily produce fruit identical with that from which they sprang. It is the custom in most regions to propagate the guava only by seed, but choice varieties which originate as chance seedlings can be perpetuated only by some vegetative means of propagation, such as budding or grafting.

Although the seeds retain their viability for many months, they should be planted as soon after their removal from the fruit as possible. They may be sown in flats or pans of light sandy loam and covered to the depth of $\frac{1}{4}$ inch. When the young plants appear they should not be watered too liberally. After they have made their second leaves, they may be transferred into small pots. Since they are somewhat difficult to transplant from the open ground, they had better be carried along in pots until ready to be planted in the orchard. The proper season for planting varies in different regions; in India it is said to be July or August; in California it is April and May; while in Florida October and March are good months.

Both shield-budding and patch-budding are successful with the guava. Shield-budding is the better method of the two. P. J. Wester, who says that the guava was first budded, so far as known, in 1894 by H. J. Webber at Bradentown, Florida, describes the method in the Philippine Agricultural Review for September, 1914. He states that budding should be performed in winter. While it has been done successfully as late as May, the months from November to April are the best (in the southern hemisphere the season would, of course, be at the opposite time of year). The stock-plants should be young; it is best to use them just as soon as they are large enough to receive the bud. When inserted in old stocks the buds do not sprout readily. The method of budding is the same as that described for the avocado and mango. The budwood should be so far mature that the green color shall have disappeared from the bark. The buds should be cut 1 to 1½ inches long.

Patch-budding has been successful in California when large stock-plants have been used. They should have stems 1 inch in diameter, and the buds should be cut 1½ inches in length, square or oblong in form. Propagation by cuttings is also possible if half-ripened wood is used and bottom-heat is available.

A simple method of propagation, which may be employed when it is desired to obtain a limited number of plants from a bush producing fruits of particularly choice quality, is as follows: With a sharp spade cut into the soil two or three feet from the tree, severing the roots which extend outward from the trunk. Sprouts will soon make their appearance. When they are of suitable size they may be transplanted to permanent positions. They will, of course, reproduce the parent variety as faithfully as a bud or graft.

The guava is a heavy bearer and ripens its fruit during a long season. In some regions guavas are obtainable throughout the year, though not always in large quantities. Seedlings come into bearing at three or four years of age; budded plants may bear fruit the second year after they are planted in the orchard. Indian horticulturists state that the plants bear heavily for fifteen to twenty-five years, and thereafter gradually decline in production. The guava is not a long-lived plant, but may live and bear fruit for forty years or more. The season of ripening in India is November to January; in Florida and the West Indies it is in late summer and autumn.

The guava is subject to the attacks of numerous insect and fungous enemies. The list of scale insects injurious to it is a particularly long one, including numerous species belonging to the genera Aspidiotus, Ceroplastes, Icerya, Pseudococcus, Pulvinaria, and Saissetia. All of these can be held in check by the usual means, *i.e.*, spraying with kerosene emulsion or some other insecticide, but little attention is given to this matter in most tropical countries. The fruit-flies, including species of Anastrepha, Ceratitis, and Dacus, cause serious trouble in many regions. It is said that 80 per cent of the guavas produced in Hawaii have in some seasons been infested with the larvæ of the Mediterranean fruit-fly (*Ceratitis capitata* Wied.). The guava fruit-rot, a species of Glomerella, is a common fungous disease in some places. There are other pests, some of them serious, which the guava-grower may have to combat.

Within the species there evidently exist more or less well-defined races, each of which includes many seedling variations. Of true horticultural varieties, propagated by cutting or grafting, there are as yet practically none. The so-called varieties listed in different regions are presumably seedling races. Indian nurserymen distinguish a number of forms, such as "smooth green," "red-fleshed," Karalia, Mirzapuri, and Allahabad. In the United States, seedlings are offered of the Allahabad guava, and of forms termed Brazilian, Peruvian, lemon, pear, smooth green, snow-white, sour, Perico, and Guinea. The number of such forms which could be listed is considerable. The Guinea variety, a white-fleshed, sweet-fruited guava with few seeds, has been propagated in California by budding, but it has not been planted extensively.

The Strawberry Guava (Fig. 36)

(*Psidium Cattleianum*, Sabine)

Unlike the preceding species, the strawberry guava is subtropical in its requirements, and can be grown wherever the orange succeeds. It is ornamental in appearance, and for this reason has become a favorite garden-shrub in many regions. Though somewhat less valuable than the tropical guava for the commercial production of guava jelly, the fruit is popular with housewives and is put to several uses.

The strawberry guava is ordinarily a bushy shrub, but sometimes becomes a small tree up to 25 feet high. The bark is smooth, gray-brown in color, and the young branchlets are cylindrical (not quadrangular as in *P. Guajava*). The leaves are elliptic to obovate in outline, acute, 2 to 3 inches long, thick and leathery in texture, somewhat glossy, and deep green in color. The flowers, which are produced singly upon axillary peduncles, are white, and nearly an inch broad. The calyx is obscurely lobed; the corolla is composed of four orbicular petals. The numerous stamens are clustered at the bases of the calyx lobes. The fruit is obovate to round in form, 1 to 1½ inches in diameter, purplish red in color, with a thin skin; the soft flesh, which is white toward the center, contains numerous hard seeds. The flavor is sweet and aromatic, suggesting that of the strawberry (whence the common name). It has not the pronounced muskiness of *P. Guajava*, and for this reason is preferred by some.

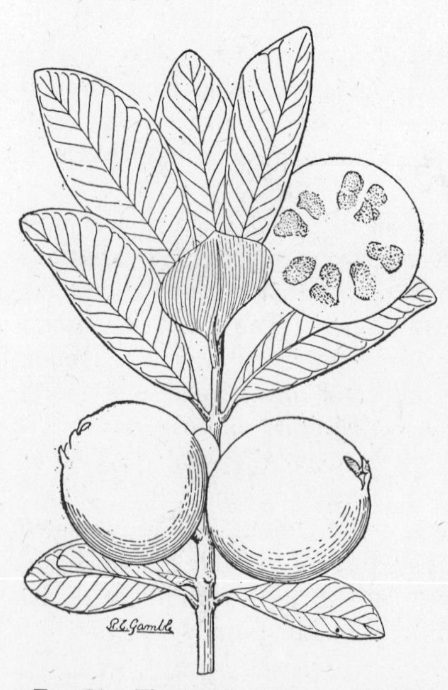

FIG. 36. The strawberry guava (*Psidium Cattleianum*), a hardier species than the common guava. (× ½)

The strawberry guava is a native of Brazil, whence it was

Plate XIV. Feijoas ready for gathering.

carried to southern China at an early period, presumably by the Portuguese. From China it was taken to Europe, where it was for some time considered to be native to China and hence was called Chinese guava. It has been planted in India, but is not widely known in that country. According to H. F. Macmillan, in Ceylon it thrives at elevations of 2000 to 4000 feet. In the Mediterranean region it succeeds in southern France, Spain, Algeria, and elsewhere. It is grown in Hawaii, but is not common there. In Mexico, Central America, and the West Indies it is occasionally seen in gardens. In Brazil, its native home, it is nowhere extensively cultivated. In the United States it thrives in both Florida and California.

Besides the name strawberry guava, the terms Cattley guava and Chinese guava are sometimes applied to this species. In Brazil it is known as *araçá da praia* and *araçá vermelho*. Nurserymen in the United States have sent out seedling races or horticultural forms of this species under several botanical names; the yellow-fruited form (variety *lucidum* of horticulture) has been listed as *Psidium lucidum, P. chinense,* and *P. sinense.* A form distributed in California under the name *P. acre,* Ten., has proved to be identical with the yellow-fruited strawberry guava, except in its elongated fruits. Otto Berg considered *P. Cattleianum* and *P. littorale,* Raddi, synonyms of *P. variabile,* Berg; more recently, however, the two last-named species have been included under *P. Cattleianum.* A large-fruited guava which appears to be nothing more than a form of this species has been distributed in California under the name *Psidium Araçá.* It is similar to the yellow-fruited strawberry guava, except in its larger and broader leaves, larger fruit, and fewer seeds.

The fruit is used principally for jelly-making, but is sometimes eaten out of hand. An analysis made at the University of California showed the ripe fruit to contain: Water 79.42 per cent, ash 0.77, protein 0.88, fiber 6.58, total sugars 5.06, starch, etc., 6.49, and fat 0.80.

Like other species of Psidium, the strawberry guava succeeds on a wide variety of soils. A rich sandy loam seems to suit it best, but it grows well on red clay and on adobe. It is not quite so successful on shallow sandy soils. The mature plant withstands severe frosts without serious injury; temperatures of 22° above zero have not killed it. A dry climate suits it better than a moist one, if the soil is irrigated with reasonable frequency. It is fairly drought-resistant.

Young plants 12 inches high may be set in the open ground in the positions they are to occupy permanently. They may be spaced 10 × 10 feet in California, but in the tropics, and on rich soil, they should be somewhat farther apart. They require the same cultural treatment as the tropical guava. During the first few years the plants retain a compact bushy form; later they may develop trunks and become small trees. They need very little pruning.

Propagation is usually by seed, which method is more satisfactory with the strawberry guava than it is with many other of the fruits here discussed, since there is less variation among seedlings. Particularly choice varieties cannot, however, be propagated by this means. Cuttings are sometimes grown, and the species may be budded in the same manner as *Psidium Guajava*. Seeds are germinated as in that species.

The plants come into bearing very early and should produce a few fruits the second or third year after planting. Their growth is slow; hence good crops cannot be expected until the plants are five or six years old. The season of ripening in Florida and California is from August to October.

The strawberry guava suffers much less from the attacks of insect parasites than does the tropical guava. As a garden plant in California and Florida it has thus far been subject to few pests.

No horticultural varieties have as yet been established. The seedling race or botanical variety *lucidum*, known as the Chinese or yellow Cattley guava, differs from the typical strawberry guava in its sulfur-yellow color and more delicate flavor. It is very productive, and so far as is known, comes true when grown from seed.

Other Guavas

Costa Rican guava (*Psidium Friedrichsthalianum*, Ndz.).

This is a species from Central America which recently has been introduced into California, Florida, and a few other regions. In the countries where it is native it is found occasionally in gardens, but nowhere is it cultivated extensively. Its fruit is highly acid and is valued for jelly-making.

The tree is erect, about 25 feet high, with slender trunk and branches. The young branchlets are wiry, quadrangular and reddish in color. The leaves are elliptic, oblong-elliptic, or oval in form, 1½ to 3 inches long, acuminate at the apex, almost glossy on the upper surface and puberulent on the lower. The flowers are produced singly on slender peduncles; they are white, fragrant, and about an inch broad. The calyx is closed, but splits into irregular segments when the flower expands. The petals, five in number, are waxy in appearance. The fruit is round or oval in form, and 1½ to 2½ inches long, sulfur-yellow in color, with comparatively few seeds, and soft white flesh of acid flavor with none of the musky aroma which characterizes some of the other guavas.

In Costa Rica the indigenous name for this fruit is *cás*. A plant which has been introduced into the United States from the island of Trinidad under the name *Psidium laurifolium* is evidently *P. Friedrichsthalianum*. When planted in southern Florida it has grown well, but in southern California it has usually been killed by frost. Plants in Florida have not borne heavy crops, and the species does not seem to possess great promise for that state.

Guisaro (*Psidium molle,* Bertol.).

This shrub from southern Mexico and Central America is now cultivated in a few gardens in southern California and southern Florida. The acid fruits, smaller than those of the Costa Rican guava, are used only for jelly-making.

The plant is of slender habit, and rarely grows more than 10 feet high. The young branchlets, peduncles, and lower surfaces of the leaves are reddish-velvety, which makes it easy to distinguish the species from *P. Guajava.* The leaves are oblong-oval, 3 to 5 inches long, obtuse at the apex, and rather stiff. The flowers, of which three are commonly borne upon each peduncle, resemble those of the common guava (*P. Guajava*). The fruit is round, about 1 inch in diameter, yellowish green to pale yellow in color, with whitish flesh containing numerous small hard seeds. The flavor is acid with little of the muskiness which characterizes some other guavas.

This is the *chamach* of northern Guatemala, often called *guayaba acida* in Spanish. In California it has proved to be hardier than *P. Guajava* and of simple culture. In Florida some plants have not borne good crops while others have been productive. It cannot be considered a valuable species.

Brazilian guava (*Psidium guineense,* Sw.).

While this species is scarcely known horticulturally, so much confusion has existed regarding its identity that it seems desirable to include it here. As was stated on a former page, the guava which has been disseminated in Florida under this name is properly a horticultural form of *P. Guajava;* the true *P. guineense* may have been planted in a few Florida gardens, but it is not well known in that state. It is grown in Cuba, although not widely, so far as is known.

The shrub is of slender habit. The young branchlets are compressed-cylindrical and finely hairy. The leaves are oblong-oval, acute or obtuse, 3 to 5 inches long, with the lower surfaces pubescent. The flowers, of which one to three are borne upon a single peduncle, resemble those of *P. Guajava*. The fruit is round or nearly so, 1 to $1\frac{1}{2}$ inches in diameter, greenish-yellow and rather hard when ripe, with whitish flesh containing numerous small seeds. The flavor is subacid, and not so musky as that of *P. Guajava*.

This guava was considered by Swartz, who first described it, to be indigenous to Africa, but more recent knowledge shows this to be improbable. *P. Araçá*, Raddi, is a synonym of this species. In Brazil many wild guavas are known by the indigenous (Tupi) name *araçá*, a fact which has led North American nurserymen, who have obtained seeds from that country, to apply the name *P. Araçá* erroneously to several species of Psidium. *P. guineense* is easily distinguished from *P. Guajava* by its compressed-cylindrical branchlets; by the upper surfaces of the leaves not having the venation impressed as in the latter species, and by the number of the lateral veins, which are 7 to 12 (commonly 8 or 9) pairs, in place of 12 to 18 (commonly 14 to 16) pairs.

The quality of the fruit is not sufficiently good to make the species of great horticultural value.

Pará guava (*Britoa acida*, Berg).

Since it does not belong to the genus Psidium this fruit is not properly entitled to be called a guava, but its similarity to the true guavas in nearly every respect makes it horticulturally permissible to include it with them. In Brazil it is known as *Araçá do Pará*. It is indigenous there, and is occasionally seen in cultivation in several parts of the country. It has been introduced into Florida, where it has been distributed under the name *Psidium Araçá*. This is a good fruit, worthy of wider dissemination in tropical countries. Its requirements appear to resemble closely those of the common guava, *P. Guajava*.

The plant becomes an erect shrub or small tree up to 25 or 30 feet in height, with slender branches and quadrangular winged branchlets. The leaves, which are borne upon very short petioles, are oblong-ovate or oblong-lanceolate in outline, 2 to 4 inches long, acuminate, glabrous, and somewhat glossy. The flowers, which are solitary in the leaf-axils, are borne upon slender quadrangular peduncles. The calyx is closed, splitting when the flower expands, forming several irregular segments. The petals are white and five in number. The fruit is oval or roundish in form, 2 to 3 inches long, sulfur-yellow in color, with soft whitish pulp containing a few seeds of larger size than those of the true guavas. The flavor is acid but pleasant, similar to that of the guava but with little of that fruit's musky aroma.

The Pitanga (Fig. 37)

(*Eugenia uniflora*, L.)

The pitanga is the best of the Eugenias. Outside of Brazil it is not appreciated as it deserves to be, although it is commonly grown in several countries. In its native home it is a popular favorite. Father Tavares observes: "Surely Brazil does not need to envy Europe her cherry trees, bending in May under the weight of their ruby fruits. Our pitangas surpass them both in beauty and taste."

In the United States the pitanga is usually seen as a broad compact shrub, but in Brazil it sometimes becomes a small tree up to 25 feet in height. The foliage is deep green and somewhat glossy, the new growth being of rich wine-color. The branchlets are thin and wiry, the leaves subsessile, ovate in outline, bluntly acuminate at the apex and rounded at the base, 1 to 2 inches long, and glabrous. When crushed, the leaves emit a pungent agreeable odor, for which reason they are sometimes scattered over the floors of Brazilian houses.

The fragrance they give off when trampled under foot is doubly appreciated as being thought efficacious in driving away flies. The white slightly fragrant flowers are $\frac{1}{2}$ inch broad, and are borne in the axils of the leaves. They have four oblong cupped petals, with a prominent cluster of stamens in the center. The fruit is oblate in form, conspicuously eight-ribbed, up to one

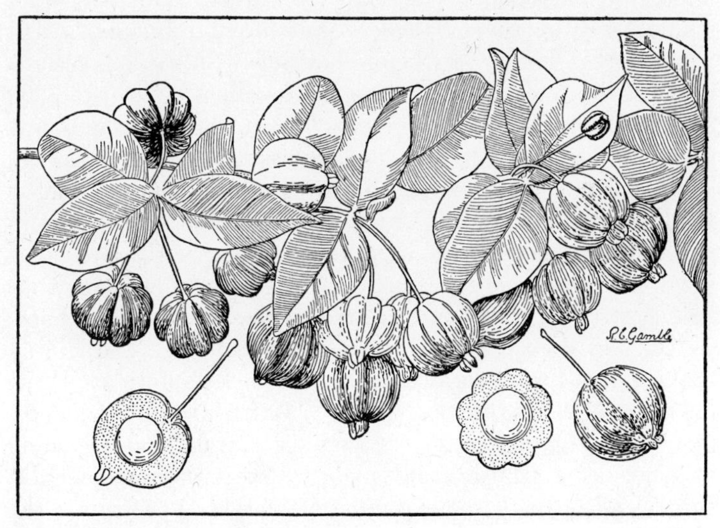

FIG. 37. The pitanga (*Eugenia uniflora*), an excellent fruit from Brazil which should be more extensively cultivated in tropical and subtropical regions. ($\times \frac{3}{5}$)

inch in diameter, deep crimson in color when fully ripe, and crowned at the apex with the persistent calyx-lobes. The flesh is soft, juicy, concolorous with the skin, and of aromatic sub-acid flavor. Usually there is one large round seed, but sometimes two hemispherical ones.

The pitanga is indigenous in Brazil, extending over a wide area. Father Tavares[1] reports that it occurs in the states

[1] Broteria, No. 5, 1912.

of Rio de Janeiro, Paraná, Santa Catharina, and Rio Grande do Sul. In wild form it usually grows along the banks of streams and in the edge of the forest, but it is also common in cultivation throughout many parts of Brazil. At Bahia it is particularly abundant; in fact, it takes an important place among the cultivated fruits of the region.

From its native home it was carried to India at an early date, undoubtedly by the Portuguese, but is not commonly cultivated in that country at the present time. In Ceylon, according to H. F. Macmillan, it thrives at elevations of 1000 to 3000 feet. It has been reported from southern China, where it was probably introduced by the Portuguese. In Hawaii it has become a common garden-shrub. L. Trabut[1] believes that it would rapidly become more popular in Algeria if it fruited more abundantly, since it has been found quite hardy along the coast. It has also proved hardy in the garden of A. Robertson-Proschowsky at Nice, on the Mediterranean coast of France. In Cuba it is occasionally seen in gardens. In the United States its culture is limited to Florida and California. Writing in 1887, P. W. Reasoner said: "The tree is quite frequently met with in Orange county and middle Florida, and is gaining in favor as a fruit-bearing plant." It is now common in gardens along the lower east coast of Florida (especially in the vicinity of Miami, where the fruit has begun to appear in the markets) and on the west coast from Fort Myers northward to Tampa Bay. After the plants have attained the requisite age, they fruit abundantly, sometimes producing two crops a year.

In California the pitanga has not become so common as in Florida, owing perhaps to the fact that many of the plants which have been tested in various parts of the state have not fruited well. F. Franceschi reported in 1895 that it was growing at Montecito, near Santa Barbara, but it still is rare in

[1] Revue Horticole de l'Algérie, p. 161, 1908.

California gardens. If it is found that the plants, after attaining sufficient age, will fruit abundantly, as has been indicated by the behavior of specimens at Santa Barbara and Orange, the pitanga should certainly be planted extensively in California. Up to four or five years of age it does not produce much fruit.

The local names of this fruit are several. In the United States it is known as Surinam-cherry, and less commonly Cayenne-cherry and Florida-cherry. The name pitanga which is used throughout Brazil was applied to this fruit by the Tupi Indians, who inhabited Brazil at the time of its discovery by Europeans. According to Martius, the word is derived from the Tupi *piter*, to drink, and *anga*, odor or scent. In India it is called Brazil-cherry, and in Ceylon, *goraka-jambo*. The common names in French are *cerise de Cayenne* and *cerise carrée*. In Spanish it is sometimes called *cereza de Cayena*. While most commonly known botanically as *Eugenia uniflora*, L., several synonyms have been used by botanists: *E. Michelii*, Lam., is one which is frequently seen. *Stenocalyx Michelii*, Berg, was used by Barbosa Rodrigues in Brazil, and *S. brasiliensis*, Berg, by M. Pio Correa. A plant introduced into California as *E. Pitanga*, Kiaersk., seems to be of the uniflora species; the true *E. Pitanga* has narrow leaves acute at the base.

The uses of the pitanga are numerous. As a fresh fruit it is delicious, when fully ripe, although the novice sometimes finds the strongly aromatic flavor slightly disagreeable. Before full ripeness, the flavor is resinous and pungent. As the fruits ripen they lose their green color, becoming yellow, then orange, and finally scarlet or crimson. They should never be eaten until quite mature. Jelly made from the pitanga possesses a distinctive flavor, and vies in popularity with guava jelly among the inhabitants of Bahia, Brazil. Pitanga sherbet is a favorite refreshment in Bahia, and is regularly served in the cafés. It is salmon-pink in color and delicious in flavor. A liqueur is sometimes prepared from the fruit, and also sirups and wines which are considered by the Brazilians to have medicinal value.

Alice R. Thompson, who has analyzed the fruit in Hawaii, finds that it contains: Total solids 9.30 per cent, ash 0.34, acids 1.44, protein 1.01, total sugars 6.06, fat 0.66, and fiber 0.34.

In Brazil the plant is commonly used to form hedges, for which purpose it is admirably adapted since it withstands heavy pruning, and is evergreen, with foliage of rich green color. Plants in hedgerows, however, produce little fruit compared with those which are allowed to develop naturally. The foliage is often used for decorative purposes, in the same manner as holly is employed in northern countries.

The pitanga thrives in both the tropical and subtropical zones, its culture extending as far north as southern California and central Florida in this hemisphere, and the Mediterranean region in Europe. Mature plants withstand temperatures of 27° or 28° above zero without serious injury. They are more at home and fruit more profusely in a warm moist climate such as that of southern Florida than in a semi-arid region. On the dry plains of northern India, on the Algerian littoral, and in southern California, the complaint is made that they do not bear well, although in Florida and in the moist tropical regions they are heavily productive. It is not known, however, just what is the limiting factor.

Father Tavares states that the plant prefers a light sandy soil. It grows very well in southern Florida on shallow sand overlying soft limestone, and equally well in California on sandy loam. At Bahia, Brazil, it is commonly found on stiff clay. It can thus be seen that it is very adaptable in regard to soil and apparently does not object to a large amount of lime, as is indicated by its growth in Florida.

Unless trained, the plants usually assume a bushy compact form, branching close to the ground. They may be planted in the open when they are a foot high, and require no unusual care. In California they have proved to be fairly drought-resistant, but they succeed best when watered liberally. Their growth is not rapid under any circumstances, and several years are required for them to reach fruiting size. In the tropics they come into bearing the third or fourth year.

In Florida no serious enemies of the plant have been noted. E. A. Back has found in Bermuda, however, that the pitanga is one of the principal hosts of the Mediterranean fruit-fly (*Ceratitis capitata* Wied.), a widely distributed pest in tropical regions.

Seed propagation is the only means of multiplication of the pitanga in common use. Whip-grafting has been reported as successful but has not been employed extensively. Seedlings sometimes spring up beneath the bushes from fruits allowed to fall to the ground; these can readily be transplanted and saved. Seeds should be planted while fresh, though they may be kept for a month or more if they are washed immediately after being removed from the fruit and then dried. They may be germinated in two-inch pots, or may be planted in flats and potted off when they are 2 to 3 inches high. Germination usually takes place within two or three weeks. Unlike the rose-apple (*Eugenia Jambos*), which is polyembryonic and produces four or five plants from a single seed, the pitanga produces only one plant from each seed. The young plants grow slowly and do not require frequent shifting into larger pots. Light sandy loam, which need not be very rich, is the best potting soil.

In Florida and the West Indies the main crop ripens in early spring. The plants flower in February, and the fruits develop very rapidly, maturing five or six weeks after the flowers have fallen. The main crop, which is usually a heavy one, ripens at one time, extending over a period of about two weeks; following this the plants sometimes produce scattering flowers, and begin to ripen a second crop about a month after the first, extending through early summer. In the second crop only a limited number of fruits ripen at one time.

In Brazil the plants bloom in September and ripen their first crop in October, flowering again for the second crop in December and January. Father Tavares says that the fruits ripen at Bahia within three weeks from the appearance of the flowers. In California the season is late summer.

Under favorable conditions the pitanga is one of the most prolific of fruits. The flowers, which are very fragrant, are pollinated by bees and probably by other insects. The plants must be watered liberally when the fruits begin to color, otherwise the latter will remain small.

Since the pitanga is rarely propagated vegetatively, no horticultural varieties have been established. Nurserymen in Florida have disseminated a seedling race under the name of "black-fruited" which differs from the common form in being deeper crimson in color and having a distinctive flavor. There is considerable variation among seedlings of the common type although they come sufficiently true from seed for this method of propagation to be satisfactory. The size of the seed is not always the same in proportion to the size of the fruit, and plants have been observed in Brazil which normally produce larger fruits than the average. Differences in productiveness have also been noticed. It will be worth while, therefore, to perfect means of grafting or budding this species so that the best seedling forms can be propagated.

The Feijoa (Plate XIV, Fig. 38)

(*Feijoa Sellowiana*, Berg)

Edouard André, one of the greatest French horticulturists of the past century, took home with him when he returned from a voyage to South America in 1890 plants of *Feijoa Sellowiana*,

PLATE XV. A fruiting jaboticaba tree.

a fruit at that time unknown save as a wild species upon the campos of southern Brazil, Uruguay, Paraguay, and parts of Argentina. He tried them in his garden on the Riviera, and they succeeded remarkably well. In 1898, by means of an article in the *Revue Horticole,* he brought the stranger to the attention of horticulturists, and it was soon planted experimentally all along the Riviera. About 1900 it was introduced into California, where its cultivation has attracted much attention in the past few years. Its prompt dissemination in that state was due largely to the efforts of F. Franceschi of Santa Barbara.

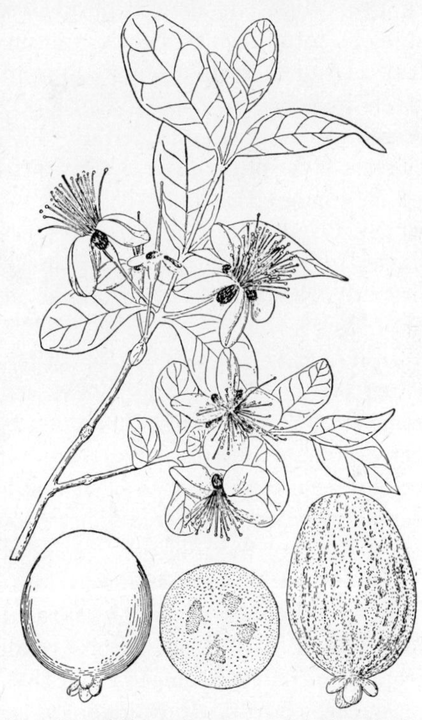

FIG. 38. Foliage, flowers, and fruits of the feijoa (*Feijoa Sellowiana*). (× $\frac{1}{3}$)

As a rule wild fruits, or those which have not been improved by cultivation, are seedy or have scanty flesh. The feijoa, taken directly from the wild, is remarkable for the minute size of its seeds, its abundance of flesh, and its delicious perfumed flavor.

The plant reaches an ultimate height of 15 or 18 feet. There are several types in cultivation; one may be compact, low-growing, while another will be tall, open, and inclined to be straggling in habit. The leaves are similar in form and appearance to those of the olive, but usually larger. The upper surface is glossy green, the lower silver-gray. The flowers are $1\frac{1}{2}$ inches broad and strikingly handsome. They are peculiar in that the fleshy petals are good to eat. The four petals are cupped, white outside and purplish within; and the long stiff stamens form a conspicuous crimson tuft in the center. The fruit is round, oval, or oblong in shape, 1 to 3 inches long, dull green in color, overspread with a thick whitish bloom, and sometimes blushed dull red on one side. The thin skin incloses a layer of granular flesh, whitish and about $\frac{1}{4}$ inch thick, which surrounds a quantity of translucent, jelly-like pulp in which twenty to thirty minute seeds are embedded. The flavor is suggestive of pineapple and strawberry, and when properly ripened the fruit has a penetrating and delightful aroma.

In its native country the feijoa is scarcely known as a cultivated plant. It is a wild species, called *guayabo del pais*. In southern France it is found in a number of gardens, but it is not yet commercially cultivated there, although the desirability of extending its culture has been pointed out by several prominent horticulturists. It has been found to succeed in Algeria and L. Trabut recommends it as a promising new fruit for that country. Although introduced into Cuba, southern Florida, and several other tropical regions, it has not been successful in any of them. It has become evident that the plant is subtropical in its requirements, and that it cannot be expected to produce good fruit in moist tropical regions. In the dry climate of California it is eminently successful. Numerous small commercial plantings have been made in various parts of the state, and the fruit has begun to appear regularly in the markets.

The feijoa may be eaten as a fresh fruit, or it may be stewed, or made into jam or jelly. Different opinions have been expressed regarding its value as a fresh fruit; those who have eaten perfectly ripened specimens of a good variety have invariably praised it, while others who have been less fortunate and have chanced to try improperly ripened ones or those of an inferior variety, have considered that the feijoa does not merit the praise which has been bestowed on it. An analysis of the ripe fruit made at the University of California shows it to contain: Water 84.88 per cent, ash 0.56, protein 0.82, fat 0.24, carbohydrates 4.24 (invert sugar 2.66, sucrose 1.58), and crude fiber 3.35.

The feijoa is hardier than many other subtropical fruits. It has withstood with little injury temperatures as low as 15° above zero. It delights in a dry climate but one free from extremely high temperatures. As was mentioned above, it has not proved successful in moist tropical regions. It is so drought-resistant that it has been grown successfully at Santa Barbara, California, with no artificial irrigation; yet it must be irrigated as liberally as the citrus fruits if the best results are to be obtained. In the extremely hot desert valleys of California, such as the Coachella, it has not been fully successful. Edouard André pointed out that the native home of the feijoa is the region of *Cocos australis;* it is probable, therefore, that the climate to which the plant is naturally adapted is a mild one, free from extremes of temperature, and having a yearly rainfall of 30 to 40 inches.

A sandy loam, rich in humus, is considered to be the ideal soil for the feijoa. In California it has been grown successfully on adobe, red clay, and sandy loam. French horticulturists consider that the plant will not tolerate much lime. It is not known whether its failure to produce good fruit in Florida is due solely to unfavorable climatic conditions, or whether the light sandy soils, often containing much lime, are partly responsible.

The plants should be spaced 15 to 18 feet apart if they are not to crowd one another when mature. While young they should be watered liberally, and it is desirable to keep a heavy mulch around them to prevent evaporation. In California it is customary to form a basin around each plant; after the mulch is added there is still room for water, of which one or two buckets should be given weekly during the dry season. After the plants reach fruiting age, they should be irrigated every two or three weeks. When a mulch is not used, the ground should be cultivated after each irrigation.

The amount of manure which can be used advantageously has not been determined. It has been the general practice in California to give the young plants an abundance of stable manure, and the effect of this seems to be highly beneficial. There has been a suspicion that large amounts of manure, if applied to bearing plants, would decrease the production of fruit, but the evidence is not convincing. Lack of pollination is probably the cause of many crop failures which are attributed to excessive soil fertilization.

Plants of the compact low-growing type require almost no pruning. Those of tall straggling form often need cutting back in order to keep the branches from developing to such great length that they cannot support their own weight.

Seedling feijoas do not reproduce the parent variety and are less satisfactory than plants propagated by some vegetative means. Layering is used in France. In the United States many plants have been grown from cuttings, and not a few by whip-grafting.

When seedlings are grown, they should be from plants which produce good fruits in abundance. If kept dry, feijoa seeds will retain their viability a year or more. One of the best mediums for germinating them is a mixture of silver-sand and well-rotted redwood sawdust. They are small and delicate, and should not be planted in heavy soil. A light sandy loam, containing much humus, is satisfactory. The seeds should be sown in pans or flats, covering to the depth of $\frac{1}{4}$ inch. Germination usually takes place within three weeks. A glasshouse is not necessary, but the flats containing the seeds should be kept in a frame with lath or slat covering to provide half-shade. As soon as the young plants have made their second leaves they should be pricked off into two-inch pots; after attaining a height of 4 inches they should be shifted into three-inch pots, from which they can later be transplanted into the open ground.

Layering is somewhat tedious, but with the feijoa is more successful than any other vegetative means of propagation. Those branches which are nearest the ground are bent down and covered with soil for the space of 3 to 6 inches. They require no care except keeping the soil moist. They will root in about six months, after which time they may be severed from the parent and set in their permanent positions.

Cuttings are successfully rooted under glass, and occasionally in the slat-house or lath-house. They should be of young wood from the ends of branches, and about 4 inches in length. Inserted in clear sand over bottom-heat, they will strike roots in a month or two; without bottom-heat they root very slowly. It is sometimes advised to keep them covered with a bell-jar. In Florida good results have been obtained by using as cuttings the young sprouts which appear around the base of the plant; these are removed with a heel when still quite small, and are planted in sand. Although they are slow to form roots, the percentage of loss is lower than when branch-tips are used.

Whip-grafting has given good results in some instances, and is probably one of the best methods of propagating the feijoa. The stock-plants should be of the diameter of a lead-pencil, the cions slightly smaller and of firm wood. Grafting has been successful both under glass and in the open ground.

Many feijoa plants which have been grown in California have borne little or no fruit. It has commonly been thought that wrong cultural practices were the cause of this, but the investigations of K. A. Ryerson and the author indicate that self-sterility may be to blame in many instances.

In its native home, the feijoa is believed to be pollinated by certain birds that visit the flowers in order to eat the fleshy sweet petals. The stamens and style project to a considerable height in the center of the flower; they brush against the breast of the visiting bird and pollen-grains adhere to its feathers. When it visits the next plant some of these pollen-grains are likely to come in contact with the stigmas of other flowers and remain upon them. Cross-pollination is thus effected.

In the United States the birds which do this work in the habitat of the feijoa are not present; consequently the plant must depend on other pollinating agencies. In some instances feijoa plants are self-fertile, and abundant fruits are produced when the flowers are self-pollinated. In other instances, it has been found that they are self-sterile, and can develop fruits only when pollen is brought from a different plant. The pollen of self-sterile feijoas has been found potent, when applied to flowers of other individuals.

To avoid the dissemination of self-sterile feijoas, varieties known to be self-fertile should be propagated by vegetative means. Seedlings, even if grown from a self-fertile variety, may nevertheless be self-sterile.

Grafted or layered plants begin bearing two or three years after they are planted. Seedlings may not bear until the fourth or fifth year. Self-fertile varieties often yield regularly and abundantly. The ripening season in California is October to December. The fruits fall to the ground when mature, and must be laid in a cool place until they are in condition for eating, — which can be known by their becoming slightly soft, and by their perfumed aroma. They spoil quickly in a hot, humid atmosphere, but if stored in a cool place they may be kept a month in good condition. They can be shipped long distances without difficulty. Feijoas are usually packed for market in fruit-baskets holding about two quarts.

To be appreciated, this fruit must be eaten at the proper degree of ripeness. M. Viviand-Morel says, "Everyone knows that the finest pears are only turnips if eaten a trifle too soon or a trifle too late." The observation is applicable also to the feijoa.

The plant is attacked by few insect pests. The black scale (*Saissetia oleæ* Bernard) is the principal enemy which has been noted. No fungous parasites have yet become troublesome.

In the Pomona College Journal of Economic Botany (February, 1912), the writer has described three varieties of the feijoa, the André, the Besson, and the Hehre. The André, described below, is the only one which has been widely disseminated. Other varieties which have originated in California as seedlings have been propagated to a limited extent, but they are little known as yet.

André. — Form oblong to oval; size medium, length 2 to 2½ inches, breadth 1½ inches; base rounded, the stem inserted without depression; apex rounded, the calyx-segments cupped; surface roughened, light green in color, overspread with a thick whitish bloom; flesh whitish, juicy, of spicy, aromatic flavor suggesting the pineapple and the strawberry; seeds few, small. Season November and December on the French Riviera and in southern California.

This variety is of unknown origin. It was brought to France from Uruguay in 1890 by Edouard André, and was planted in his garden at Golfe-Juan, on the Riviera. Layered plants were later sent from France to California. It is self-fertile, and fruits profusely. The shrub is sometimes erect and open in habit, and in other instances low, compact, and broad.

The Jaboticaba (Plate XV)

(Myrciaria spp.)

In southern Brazil there are a number of indigenous fruits of genuine merit. The jaboticaba is one of the best, but like many of the others it has until recently received little attention outside its native home.

Among the fruit-trees cultivated in Rio de Janeiro and its vicinity, the jaboticaba is one of the commonest and certainly the one which first attracts the attention of the newcomer. Its habit of producing the fruit directly upon the trunk and larger limbs, together with the unusual beauty of its symmetrical and umbrageous head of pale green foliage, makes this a peculiarly striking tree. The fruit is popular and highly esteemed by all classes of Brazilians, and occupies an important position in the markets.

When grown on rich soil, the tree reaches a height of 35 or 40 feet. The leaves are ovate-elliptic to lanceolate, acute to acuminate at the apex, usually glabrous, and vary from $\frac{3}{4}$ inch to 3 inches in length. The flowers are small, white, with four petals and a prominent cluster of stamens. They are produced singly or in clusters on the bark of the trunk and limbs. The fruit is round, $\frac{1}{2}$ to $1\frac{1}{2}$ inches in diameter, maroon-purple in color, and crowned with a small disk at the apex. The skin is thicker and tougher than that of a grape. The translucent juicy pulp, whitish or tinged with rose, is of agreeable vinous flavor. The seeds, one to four in number, are oval to round in outline and compressed laterally.

The jaboticaba is usually listed as *Myrciaria cauliflora,* Berg. There are several closely related species, however, whose fruits are all known under the same common name. *M. trunciflora,* Berg, and *M. jaboticaba,* Berg, probably furnish many of the fruits sold as jaboticabas in the markets of Rio de Janeiro. Father Tavares considers that the cultivated forms are in some instances the result of hybridization.

As a wild plant the jaboticaba is limited to southern Brazil, from Rio Grande do Sul to Minas Geraes. It is cultivated in the same area, as well as in a few other parts of Brazil. It has been introduced into the United States and a few other countries, but has not yet become established in any of them.

The uses of the jaboticaba are several. As a fresh fruit it is as popular in southern Brazil as the grape is in the eastern United States. A wine can be made from it, and also an excellent jelly.

While the tree is said to succeed on any soil, it prefers one that is rich and deep. Its growth is slow, six to eight years being required for it to come into bearing. In Brazilian orchards this tree is nearly always planted too closely; the distance apart should be 30 feet at least. Though rarely grown in those parts of Brazil which are subject to severe frosts, the jaboticaba has shown in the United States that it resists comparatively low temperatures. At Miami, Florida, it has passed successfully through a freeze of 26° above zero. So far as can be judged from the limited experience which has been gained, the soils of southern Florida are not well adapted to it. Those of southern California are more suitable, but the climate has proved to be too cold in all but the most protected spots in that state. The jaboticaba appears to demand for full success a deep rich soil and a moist, equable, rather cool climate with temperatures preferably never below the freezing-point.

Little attention is given in Brazil to the culture of this tree. Father Tavares says that the *fazendeiros* (planters) of São Paulo, who irrigate their trees at times when there is a scarcity of rain, succeed in having ripe jaboticabas throughout the year. Without irrigation, fruit is produced usually during the warmest months of the year. When heavily laden with fruit, the tree is a curious sight. Not only is the trunk covered with clusters of glistening jaboticabas, but the fruiting extends to the limbs and out to the tips of the smallest branches.

Propagation is usually by seed. It is said, however, that young plants can be inarched successfully: if so, choice varieties could well be propagated in this manner. Other methods of propagation will doubtless be developed when the jaboticaba becomes more widely grown.

The Brazilians cultivate as named varieties a number of forms which must either be distinct species or seedling races. The name jaboticaba, without any qualifying word, is considered to be applied properly only to *Myrciaria cauliflora.* The closely allied *M. jaboticaba* is known as *jaboticaba de São Paulo, jaboticaba de cabinho,* and *jaboticaba do matto.* According to Father Tavares, *M. tenella,* Berg, is known as *jaboticaba macia.* The fruits of the various species are very much alike. The form *coroa,* which is one of the commonest named "varieties" recognized in Rio de Janeiro and Minas Geraes, can probably be referred to *M. cauliflora.* The form *murta* has small leaves; it is, perhaps, another form of the same species. The variety *branca* (white) is listed by nurserymen in Rio de Janeiro, also *roxa* (red); both are said to be distinct from the ordinary jaboticaba in color.

Other Myrtaceous Fruits

Grumichama (*Eugenia Dombeyi,* Skeels) (Fig. 39).

This is a better fruit than several other species of Eugenia which are much more widely grown. It is found both wild and cultivated in southern Brazil, particularly in the states of Paraná and Santa Catharina. Elsewhere, with the exception of Hawaii, it is scarcely known.

The tree, which grows to the same size as the orange, is shapely and attractive in appearance, with ovate-elliptic, glossy, deep green leaves 2 to 3 inches long. The small white flowers are followed by pendent fruits, round or slightly flattened, the size of a cherry, and deep crimson in color. The persistent green sepals which crown the apex are a distinguishing characteristic. The skin is thin and delicate; the flesh soft, melting, of a mild subacid flavor suggesting that of a Bigarreau cherry.

The seeds are round or hemispherical when one or two in number; sometimes there are three or more, in which case the size is reduced and they are angular.

The rapidity with which the fruits develop is surprising; within a month from the time of flowering they have reached maturity and are falling to the ground. Father Tavares states that all the trees do not ripen their crops at the same time, some blooming later than others and thus extending the fruiting season from November to February (in Brazil). Three varieties are distinguished by him, one with dark red flesh, another with vermilion, and the third with white. All three are said to be equally good in quality. The fruit is usually eaten fresh, but may also be used to make jams and preserves.

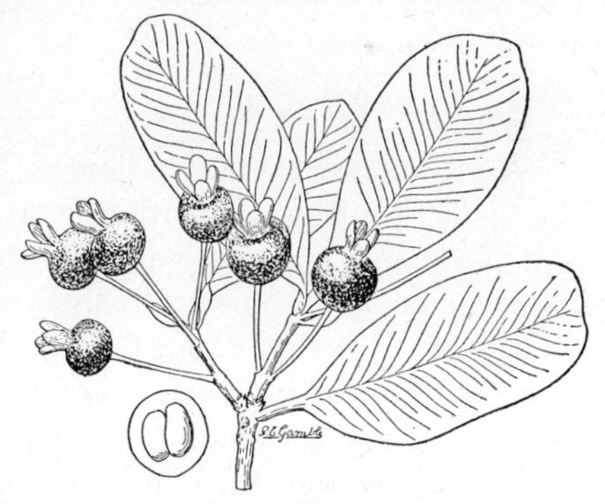

FIG. 39. The grumichama (*Eugenia Dombeyi*), a little-known fruit from southeastern Brazil. ($\times \frac{3}{8}$)

The grumichama (sometimes *grumixama*, to conform to old Portuguese orthography) has recently been planted in California and Florida. In the latter state it has withstood a temperature of 26° without injury, which indicates that it is subtropical, rather than strictly tropical, in character. It prefers a deep sandy loam, but succeeds in Florida on shallow sandy soils. Vaughan MacCaughey says: "In the Hawaiian islands it is usually about 20 feet high. It requires considerable moisture for its best development, as do all the Eugenias in our flora; the largest crops are borne by plants at the lower levels, up to 300 feet . . . flowering and fruiting continues from July until December, the main crop coming in the fall. . . . The first plants in Hawaii were probably introduced by the Spaniard, Don Francisco de Paula Marin, who came to the islands in 1791." The grumichama is sometimes listed as *Eugenia brasiliensis,* Lam. *Stenocalyx brasiliensis,* Berg, is another synonym.

Seedlings are said to commence fruiting when four or five years old. They grow rather slowly. No one appears to have budded or grafted the species as yet. For its value as an ornamental plant as well as for its pleasant fruit, the grumichama deserves cultivation throughout the tropics and subtropics.

Jambolan (*Eugenia jambolana,* Lam.).

This species, whose native home is in the East Indies, is of little value in comparison with several of its congeners. It is a small tree, with large, oblong, apiculate, glossy leaves, white flowers, and oval, purplish red fruits commonly ½ inch in length. It is said that forms with large fruits of good quality are known in the Orient, but those which have been grown in the United States are scarcely worth cultivating. The plant is slightly less hardy than the rose-apple (see below), but can be grown successfully in southern Florida and in protected situations in southern California. It is said to succeed in Algiers, where it is known by the French name *jamelongue.* In English it is sometimes called Java-plum, while in India it is commonly known as *jambu* and *jaman.* According to Watt's "Dictionary of the Economic Products of India," it is "A small evergreen tree met with throughout India and Burma, ascending the hills to about 6000 feet. It is chiefly found along river beds and is specially cultivated for its fruit in gardens (*topes*) and in avenues. There are several varieties that yield much better flavored fruit than others, but as a rule it is astringent, and only serviceable when cooked in tarts and puddings."

The propagation of the plant is usually by seed. Its botanical synonymy is rather extensive; *Syzygium jambolana*, DC., *Syzygium Cumini*, Skeels, and *Eugenia Cumini*, Druce, are names under which it is sometimes listed.

Rose-apple (*Eugenia Jambos*, L.) (Plate XVI).

As an ornamental tree, the rose-apple is of value for all tropical and subtropical regions. As a fruit it is beautiful and interesting, but is not much used except for making preserves.

The tree grows to 25 or 30 feet in height, and is shapely and attractive in appearance. The leaves are oblong-lanceolate, acuminate, 5 to 8 inches long, thick and glossy, with the new growth wine-colored. The flowers are produced upon the young branchlets in short terminal racemes. They are greenish white in color and have a conspicuous tuft of long stamens which almost hide the other floral parts from view. The calyx-tube is turbinate, and the corolla composed of four obovate concave petals. The fruit is round or oval, 1 to 2 inches in length, and crowned at the apex with the calyx-segments. In color it is whitish green to apricot-yellow; it is perfumed with the odor of the rose, and is attractive in appearance. The flesh is crisp, juicy, and sweet. The single, round seed (or sometimes two hemispherical ones) is loose in the large hollow seed-cavity.

The rose-apple is indigenous in the East Indies, whence it has been carried to all parts of the tropics. It has become naturalized in the West Indies, in Hawaii, and in other regions. In India, where it is very abundant, it is usually known as *gulab-jaman* (rose jaman). Yule and Burnell state that the Sanskrit name *jambu* is applied in the Malay language, with distinguishing adjectives, to several species of Eugenia. *Jambo* and *yambo* are sometimes used in English for the rose-apple. In French it is called *pomme-rose*, in Spanish *poma-rosa*. Botanically it is sometimes listed as *Jambosa vulgaris*, DC., sometimes as *Caryophyllus Jambos*, Stokes.

The tree is hardy in southern California and throughout the southern and central parts of Florida. It succeeds equally well in warm, moist, tropical regions and in the cool and dry subtropics. In Florida it is esteemed as an ornamental plant. The fresh fruit is fragrant and attractive, but owing to its peculiar character it is not pleasant to eat unless in small quantities; yet as a preserve or crystallized it is delicious. On account of its beauty it is often used for table decoration. Its enticing perfume, strikingly similar to that of rose-water, makes it unique among fruits.

According to an analysis made in Hawaii by Alice R. Thompson, the ripe fruit contains: Total solids 15.85 per cent, ash 0.29, acids 0.03, protein 0.79, total sugars 11.73, fat 0.18, and fiber 0.98 per cent.

The plant thrives on soils of diverse types. While a rich loam perhaps best suits it, the shallow sandy soils of southeastern Florida have proved altogether satisfactory. It is of slow growth, and comes into bearing when four or five years old. When in bloom it is highly ornamental as it is also when the yellow fruits are ripe. It does not bear heavily, but fruiting extends over a long season.

Propagation of the rose-apple is usually by seed. Like the mango it is peculiar in that its seeds are polyembryonic; thus a single seed may give rise to seven or eight plants. P. J. Wester has found that the species lends itself to bud-propagation. The method is the same as that used with the avocado and mango. Wester says: "Use greenish to brownish and roughish, well-matured budwood; cut the buds an inch and a quarter long. The age of the stock at the point of insertion is unimportant." Large-fruited varieties, or those otherwise desirable, may be propagated by this means.

No named varieties of this very interesting fruit have been disseminated.

PLATE XVI. Flowers and fruits of the rose-apple (*Eugenia Jambos*).

Pera do campo (*Eugenia Klotzschiana*, Berg) (Fig. 40).

This is a rare eugenia from the campos or rolling plains of central Brazil (Minas Geraes), which has recently been introduced into the United States. It is slender in habit and grows not more than 4 or 5 feet high. The leaves are lanceolate, 3 to 5 inches long, hard and brittle in texture and silvery pubescent on the lower surface. The pear-shaped, downy, golden-yellow fruits, 2 to 4 inches in length, ripen in Brazil from November to January. They have soft, juicy, acid flesh, and are highly aromatic in odor and flavor. The seeds, one to four in number, are irregularly oval in form and small in size.

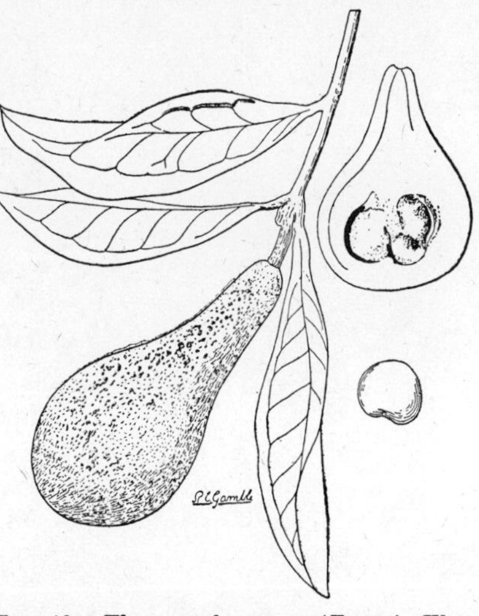

FIG. 40. The pera do campo (*Eugenia Klotzschiana*). (× 3/7)

Pitomba (*Eugenia Iuschnathiana*, Berg).

This is a fruit-tree found wild and cultivated in the state of Bahia, Brazil. It attains a height of 25 to 30 feet, and is of handsome appearance. The leaves are lanceolate, 3 inches long, glossy and deep green on the upper surface, light green below. The fruits are broadly obovate in form, 1 inch long, with the apex crowned by four or five green sepals ½ inch long. The color is bright orange-yellow. The skin is thin, and the flesh soft, melting, juicy, with an acid, highly aromatic flavor. The seeds, commonly one but sometimes as many as four, are rounded or angular according to number, and attached to one side of the seed cavity. Propagation is usually by seed. The botanical name *Phyllocalyx Luschnathianus*, Berg, is sometimes used.

Ohia (*Eugenia malaccensis*, L.) (Fig. 41).

This species is a native of the Malayan Archipelago, whence it has been introduced into other tropical regions. It is now the most important eugenia in the Hawaiian flora. Vaughan MacCaughey[1] says of it: "This beautiful tree was introduced by the primitive Hawaiians and is now abundant in the humid valleys and ravines on all the islands. It is distinctly a tree of the lower forest zone, where it forms pure stands, some of which, on the broad valley floors, cover areas of several hundred acres."

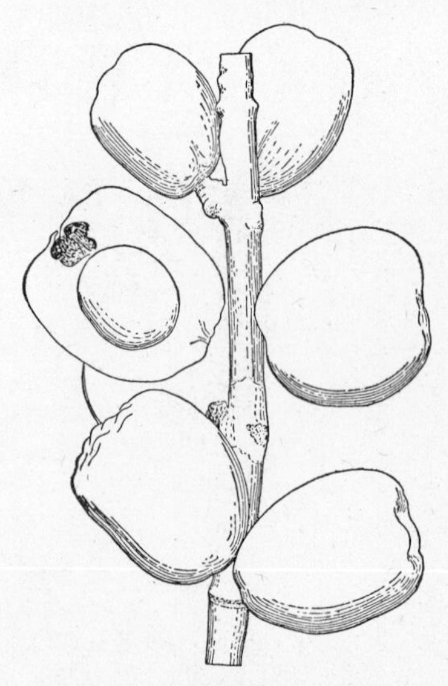

FIG. 41. The ohia (*Eugenia malaccensis*), a Malayan fruit little known in the American tropics. The tree is handsome, but the fruit is not of very good quality. (× about $\frac{1}{3}$)

The tree sometimes reaches 60 feet in height. The leaves are elliptic-oblong to oblong-obovate, acute, 6 or 7 inches in length, thick, glossy, and dark green in color. To quote again MacCaughey: "The flowers are showy clusters of long, spreading, bright red stamens, that contrast charmingly with the rich foliage. During the flowering season, in early summer, the shady interior of the tree seems to be filled with a delicate scarlet haze." The fruits are oval to obovate in form, 2 to 3

[1] Torreya, Dec., 1916.

inches long, and white to crimson in color. The skin is thin, the flesh crisp, "apple-like," white, and juicy, with refreshing subacid flavor. In Hawaii this species is often called mountain-apple. In the Orient it is known as Malay-apple, and in French, *jamelac*. In the British West Indies, where it was introduced from Tahiti in 1793, it is often termed Otaheite-apple. According to W. Harris, it is now common in the wet warm districts of Jamaica. Botanically it is sometimes listed as *Jambosa malaccensis*, DC., and also as *Caryophyllus malaccensis*, Stokes.

The fruit is not especially esteemed. It is somewhat pithy in texture and its flavor is not rich. Alice R. Thompson finds it to contain: Total solids 8.61 per cent, ash 0.13, acids .06, protein 0.21, total sugars 6.88, fat 0.03, and fiber 0.56.

The ohia is tropical in its requirements and cannot be grown in regions subject to frost. So far as is known, it has never been brought to fruiting age either in California or Florida. It is probable, however, that few attempts have been made to grow it in either state, and that it might succeed in the extreme southern part of Florida. In Ceylon it is said by H. F. Macmillan to thrive at elevations up to 2000 feet. It is grown successfully at Rio de Janeiro, Brazil.

It is propagated by seed, and also, according to Thomas Firminger, by layers. It cannot be recommended, however, for extensive cultivation as a fruit-producing tree. As an ornamental plant for the tropics it is distinctly valuable. In Hawaii a variety with white flowers and fruits is found.

Uvalha (*Eugenia Uvalha*, Cambess.).

This shrub or small tree is found both wild and cultivated in southern Brazil. The leaves are oblong, obtuse, and aromatic when crushed; the fruits are round or oblate in form, 1 inch in diameter, yellow or orange in color, and crowned at the apex with the remnants of the calyx. The skin is thin, the flesh soft and juicy, with an intense and agreeable aroma. The acid flavor causes the fruit to be used principally for making refreshing drinks. It is little-known outside of Brazil.

Cabelluda (*Eugenia tomentosa*, Cambess.).

This myrtaceous fruit is found both indigenous and cultivated in the vicinity of Rio de Janeiro, Brazil. When well grown, the tree is handsome and of value as an ornamental plant. It reaches a height of 15 to 25 feet. The leaves are oblong-lanceolate, 2 to 4 inches long, bright green and tomentose above, dull green and tomentose below. The fruits, which ripen in October and November at Rio de Janeiro, resemble large gooseberries in appearance. They are yellow when fully ripe and nearly 1 inch in diameter. The skin is firm and tough, downy externally; the flesh is juicy and of pleasant subacid flavor. The one or two large seeds are surrounded by coarse short fibers. The cabelluda, sometimes listed as *Phyllocalyx tomentosus*, Berg, is scarcely known outside of Brazil. It has been introduced recently into the United States, where it should succeed in California and Florida. It is not a fruit of much merit.

Guabiroba (*Abbevillea Fenzliana*, Berg).

This is another small tree found both wild and cultivated in southern Brazil, especially in the vicinity of Rio de Janeiro. It grows to 30 or 40 feet in height, and has foliage which resembles that of some of the European oaks. The leaves are elliptic-ovate in form, about 2 inches long, with the venation depressed above and salient below. The flowers resemble those of the guava. The fruits are oblate in form, nearly 1 inch in diameter, orange-yellow in color, and crowned with the large disk and persistent calyx-segments. The surface is somewhat wrinkled, and the thin skin surrounds a soft, yellow flesh in which numerous seeds are embedded. The flavor is similar to that of the guava but less pleasant. This plant, sometimes listed as *Campomanesia Fenzliana*, Glaziou, has been introduced into Florida, where it grows well and has withstood a temperature of 26° above zero without injury. The fruit is used in the same manner as the guava, but is not of great value.

Downy myrtle (*Rhodomyrtus tomentosa*, Wight).

This myrtle is a small, handsome shrub, valuable as an ornamental plant as well as for its fruit. The leaves are elliptic or obovate, obtuse, 1 to 2½ inches long. The rose-pink flowers are followed by round fruits somewhat resembling a large black currant in size and character. The downy myrtle (sometimes called hill-gooseberry) is probably best known in southern India, where it occurs commonly in the mountains. It is said by H. F. Macmillan to succeed in Ceylon only at high elevations. It is grown also in southern China, and to a very limited extent in Florida and California. It withstands several degrees of frost. The fruits are said to make excellent pies, and they may also be eaten out of hand. Sir Joseph Hooker says that they are used in India to prepare a jam called *theonti*. The plant is not particular regarding soil, and is readily propagated by means of seeds, which should be sown in flats of light soil and covered to a depth of ⅛ inch. The botanical name *Myrtus tomentosa*, Aiton, is sometimes given it. Everything considered, the downy myrtle should repay wider cultivation than it receives at present.

CHAPTER X

THE LITCHI AND ITS RELATIVES

THE Sapindaceæ or Soapberry family comprises a number of fruits prized in the tropics, which may be brought together in one chapter. In temperate climates the family yields no important edible fruits. Some botanists place the maples and buckeyes in this family, but these plants are now commonly separated in other closely related families.

THE LITCHI (Plate XVII, Fig. 42)

(*Litchi chinensis*, Sonn.)

While living in exile at Canton, the poet Su Tung-po declared that litchis would reconcile one to eternal banishment. Yet he did not allow his enthusiasm to draw him into gastronomic indiscretions, for he limited himself to a modest three hundred a day, while other men (so he says) did not stop short of a thousand.

Chang Chow-ling, an illustrious statesman of the eighth century of our era, composed a poem on the litchi in which he praised it as the most luscious of all fruits. Modern Chinese critics fully concur in this opinion. Neither the orange nor the peach, two of the finest fruits of southern China, is held to equal it in quality.

Nor is the litchi one of those rare and delicate fruits known only to the favored few. In southern Asia, where its cultivation dates back at least two thousand years, it is grown extensively and millions are familiar with it. That it should still be unknown in most parts of the western tropics is probably due to the perishable nature of the seeds. Before the days of steam navigation, it was difficult to transport them successfully from one continent to another.

"An orchard of litchis," wrote the eminent E. Bonavia of India, "say of a few hundred trees, and with ordinary care, would give a handsome and almost certain annual return for not improbably a hundred years." While it has been considered that the litchi is somewhat exacting in its cultural requirements, it can be grown successfully in many parts of the tropics and subtropics. Now that it has been established in tropical America, there is no reason why it should not there become one of the common fruits, nor why fresh litchis should not be found on fruit-stands of northern cities at least as abundantly as are the dried ones at present.

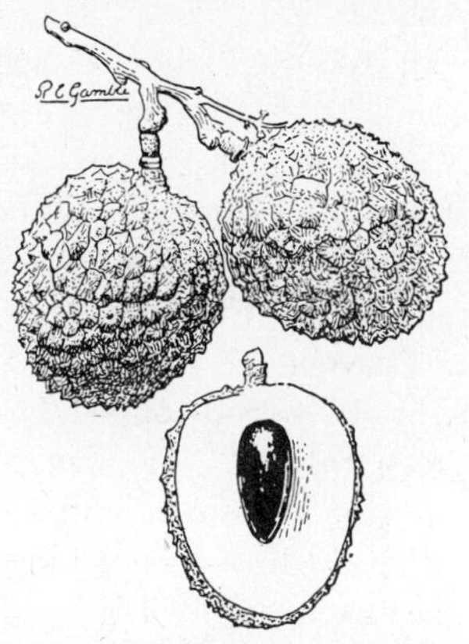

FIG. 42. Fruits of a good variety of the litchi. Kinds which are altogether seedless have been reported, but in the best-known sorts the seed is about the size of the one here shown. (× ½)

It is in the form of dried litchis, "litchi nuts," that North Americans are usually acquainted with this fruit. The Chinese who live in the United States import them in large quantities, and are particularly prone to indulging in them at the time of their New Year celebrations. But the dried litchi resembles the fresh one even less than the dried apple of the grocery store resembles a Gravenstein just picked from the tree. To appreciate its excellence, one must taste the fresh litchi; although a fairly true estimate of it may be acquired from the canned or preserved product, which much resembles preserved Muscat grapes in flavor.

Judging by the experience of the past few years, it should be possible to produce litchis commercially in southwestern Florida (the Fort Myers region), where there is relative freedom from frost and where the soils are deep and moist. It is doubtful whether there are any localities in southern California adapted to commercial litchi culture, but trees have been grown at Santa Barbara and in the foothill region near Los Angeles (Monrovia, Glendora). While the dry climate and cool winter weather of California are unfavorable, it seems probable that litchis may be grown on a small scale in this state, if planted in sheltered situations and given protection from frost for the first few years.

Because of its value as an ornamental tree, the litchi is recommended for planting in parks and gardens. It grows to an ultimate height of 35 or 40 feet (less in some regions), and forms a broad round-topped crown well supplied with glossy light green foliage. The leaves are compound, with two to four pairs of elliptic-oblong to lanceolate, sharply acute, glabrous leaflets 2 to 3 inches long. The flowers, which are small and unattractive, are borne in terminal panicles sometimes a foot in length. They are said to appear in northern India in February and in China during April. The fruits, which are produced in loose clusters of two or three to twenty or even more, have been likened to strawberries in appearance. In shape they are oval to ovate, in diameter 1½ inches in the better varieties, and in color deep rose when fully ripe, changing to dull brown as the fruit dries. The outer covering is hard and brittle, rough on the surface and divided into small scale-like areas. The seed is small, shriveled, and not viable in some of the grafted varieties; in seedlings it is as large as a good-sized castor-bean, and glossy dark brown in color. Surrounding it and separating from it readily is the flesh (technically aril), which is white, translucent, firm, and juicy. The flavor is subacid, suggestive of the Bigarreau cherry or (according to some) the Muscat grape.

Regarding the origin and early history of the fruit Alphonse DeCandolle says: "Chinese authors living at Pekin only knew the litchi late in the third century of our era. Its introduction into Bengal took place at the end of the eighteenth century. Every one admits that the species is a native of the south of China, and, Blume adds, of Cochin-China and the Philippine Islands, but it does not seem that any botanist has found it in a truly wild state. This is probably because the southern part of China towards Siam has been little visited. In Cochin-China and in Burma and at Chittagong the litchi is only cultivated."

Macgowan [1] recounts that litchis were first sent as tribute to the emperor Kao Tsu about 200 B.C. These were dried fruits, however; later fresh ones were forwarded by relays of men, and one is happy to learn that though the cost in human life was frightful they reached the emperor in good condition. The Emperor Wu Ti (140–87 B.C.) made several attempts to bring trees from Annam and plant them in his garden at Chang-an, but he was not successful in raising them.

According to Walter T. Swingle, the first published work devoted exclusively to fruit-culture was written by a Chinese scholar in 1056 A.D. on the varieties of the litchi.

The principal provinces of China in which litchis are grown are Fukien, Kwantung, and Szechwan. In Kwangtung Province alone the annual crop is said to be twenty million to thirty million pounds, worth $1,000,000 to $1,500,000. The region around Canton is considered the most favorable part of China for litchi culture. North of Foochow the tree is not successful.

While litchis are by no means so extensively grown in India as they are in southern China, there are several districts in which they are produced commercially. The most important are said to be in Bengal; about Muzaffarpur (in Bihar); and at Saharanpur (United Provinces of Agra and Oudh). E.

[1] Journal of the Agri-Horticultural Society of India, 1884, p. 195.

Bonavia says: "The tree does admirably in Lucknow, and should do as well all over the northwestern provinces, but it flourishes best, I believe, in Bengal. Who knows what untold litchi wealth there may be in the fine black soil of the central provinces, so centrally situated for fruit trade?"

In Cochin-China, in Madagascar, and in a few other countries of the East, the tree is cultivated on a limited scale. In Hawaii, where it is believed to have been introduced about 1873, it has succeeded remarkably well, and much attention has lately been given to its commercial cultivation, without, however, any large orchards having been established as yet.

According to William Harris, it was introduced into Jamaica in 1775, but it is still rare in that island. A tree at Santa Barbara, California, which produced a few fruits in 1914, was the first to come into bearing in the United States. While the litchi is believed to have been planted in Florida as early as 1886, it was not until 1916 that the first fruits were produced in that state. These were from plants introduced from China in 1906. A few trees have borne in Cuba, Brazil, and other parts of tropical America.

The common name of this fruit is variously spelled, — *litchi, lichee, lychee, leechee, lichi, laichi,* and so on. Yule and Burnell state that the pronunciation in northern China is *lee-chee,* while in the southern part of the country it is *ly-chee.* Since the form *litchi* has been fixed as a part of the botanical name of the species, and since it is employed extensively as the common name, it may be well to retain it in preference to others. The pronunciation *ly-chee,* which is used in the region where the fruit is grown, is generally preferred to *leechee.* Botanically the plant is *Litchi chinensis,* Sonn. *Nephelium Litchi,* Cambess., is a synonym.

While the litchi is probably best as a fresh fruit, Frank N. Meyer says that it is considered by some to be more delicious when preserved (canned) than when fresh; and he adds: "No good dinner, even in northern China (where the litchi is not grown) is really complete without some of these delicious little fruits." The dried litchi tastes something like the raisin. Consul P. R. Josselyn of Canton writes: "There are two ways of drying litchis, — by sun and by fire. The sun dried litchi has a finer flavor and commands a better price than the fire dried fruit." Only two or three varieties are considered suitable for drying. Regarding the preserving industry, Josselyn remarks: "It is estimated by dealers that the annual export of tinned litchis from Canton is about 3000 boxes, or 192,000 pounds. Each box of preserved litchis contains 48 tins, weighing 1 catty each. Each tin contains about 28 litchis. There are five large dealers in Canton who make a business of preserving these litchis. In addition to the preserved litchis exported from Canton large quantities of the fresh fruit are shipped from the producing districts surrounding Canton to Hongkong and are there preserved in tin."

An analysis of the fresh fruit, made in Hawaii by Alice R. Thompson, shows it to contain: Total solids 20.92 per cent, ash 0.54, acids 1.16, protein 1.15, and total sugars 15.3.

Cultivation.

In general it must be considered that the litchi is tropical in its requirements. It likes a moist atmosphere, abundant rainfall, and freedom from frosts. It can be grown in subtropical regions, however, where the climate is moist or if abundant water is supplied, and where severe frosts are not commonly experienced.

Young plants will not withstand temperatures below the freezing point. In regions subject to frost they should, therefore, be given careful protection during the winter. The mature tree is not seriously injured by several degrees of frost, but at Miami, Florida, plants six feet high were killed by a temperature of 26° above zero.

Rev. William N. Brewster of Hinghua, Fukien, China, describing the conditions under which the trees are cultivated in that country, says: "They will not flourish north of the frost line. They are particularly sensitive to cold when young. It is the custom here to wrap the trees with straw to protect them from the cold. After the trees are five or six years old they are less sensitive, and it takes quite a heavy frost to injure them."

Regarding soil, G. W. Groff of the Canton Christian College writes: "The litchi seems to do best on dykes of low land where its roots can always secure all the water needed, and where they are even subjected to periods of immersion. In some places they grow on high land but not nearly so successfully." The Rev. Mr. Brewster says on this subject: "The trees flourish in a soft, moist black soil; alluvium seems best. Near by or on the bank of a stream or irrigation canal is best, though this is not essential. Where there is no stream the trees should be watered so frequently that the ground below the surface is always moist; about twice a week when rain is not abundant should be enough. After the young trees are well started, about two or three years old, the irrigations may be less frequent."

These authorities are quoted to show the conditions under which the litchi is grown in China. Experience in other countries has shown the tree to be reasonably adaptable in regard to both climate and soil. While it prefers a humid atmosphere, it has succeeded in the relatively dry climate of Santa Barbara, California, without more frequent irrigation than other fruit-trees. On the plains of northern India, where the atmosphere is comparatively dry and the annual rainfall about 40 inches, it is cultivated on a commercial scale. Although the best soil may be a rich alluvial loam, it has done well in Florida on light sandy loam. It has not been successful, however, on the rocky lands of southeastern Florida. Whether these lands are too dry, or whether the litchi dislikes the large amounts of lime which they contain, cannot be stated definitely. In undertaking to grow this tree, four desiderata should be kept in mind: first, freedom from injurious frosts; second, a humid atmosphere; third, a deep loamy soil; and fourth, an abundance of soil-moisture. When one or more of these is naturally lacking, efforts must be made to correct the deficiency in so far as possible. Frost-injury can be lessened by protecting the trees; low atmospheric humidity is not badly prejudicial if the soil is abundantly moist; sandy soils may be made more suitable by adding humus-forming material; and a soil naturally dry may be irrigated regularly and frequently.

In regions where the litchi tree grows to large size, it is not advisable to space the plants closer than 30 feet apart, and 40 feet is considered better. In Florida they can be set more closely without harm; 25 feet will probably be a suitable distance. In localities where frost protection must be given, it may be desirable to plant the trees under sheds, and in this case economy will demand that they be crowded as much as possible. At Oneco, near Bradentown, Florida, E. N. Reasoner has fruited the litchi very successfully in a region usually considered too cold for it, by growing it in a shed covered during the winter with thin muslin to keep off frost, and opened in the summer. If it is commercially profitable to erect sheds over pineapple-fields, — and it has proved so in certain parts of Florida, — there seems to be no reason why it should not be much more profitable to grow the litchi in this way, in regions where protection from frost is necessary.

The trees should be planted in holes previously prepared by excavating to a depth of several feet, and incorporating with the soil a liberal amount of leaf-mold, well-rotted manure, rich loam, or other material which will increase the amount of humus. This is, of course, more important where the soil is light and sandy, as it is in many parts of Florida, than where the humus-content is high. Basins may be formed around the trees to hold water.

Bonavia writes: "As the trees grow, their thalas or water-saucers should be enlarged and on no account should the fallen leaves be removed from them, but allowed to decay there and form a surface layer of leaf-mold. . . . Every hot weather thin layers of about two or three inches of any other dried leaves should be spread over the thalas, and allowed to decay there, to be renewed when they crumple up and decay." This corresponds to the mulching generally practiced in western countries. It has been remarked by several writers that the litchi is a shallow-rooted tree, with most of its feeding roots close to the surface. If this really is the case, mulching will probably be an essential practice, and deep tilling of the soil will have to be avoided.

Rev. Mr. Brewster says: "Fertilization is important. Guano is probably as good as anything. The Chinese use night soil. They dig a shallow trench around the tree at the end of the roots and fill it with liquid manure of some sort. This is done about once in three months." J. E. Higgins,[1] in his bulletin "The Litchi in Hawaii," notes that "Some growers prefer to put the manure on as a top dressing and cover it with a heavy mulch because of the tendency of the litchi to form surface roots."

The tree requires little pruning. Higgins says: "The customary manner of gathering the fruit, by breaking with it branches 10 to 12 inches long, provides in itself a form of pruning which some growers insist is necessary for the continued productivity of the tree." But a thorough study has yet to be made of this subject in the Occident.

Hand-in-hand with the development of litchi-growing in the American tropics and subtropics will come the development of new cultural methods. The information at present available

[1] Bull. 44, Hawaii Agri. Exp. Sta., 1917.

is meager, and too apt to be characterized by the generalities of the Hindu horticulturist: "Too much manure should not be applied to newly planted or small trees. As the tree flourishes, more and more manure should be applied," writes one of them, in a treatise on litchi-culture. The literature of tropical pomology is burdened with information of this nature, and the need is for more specific data based on experience.

Propagation.

Propagation of the litchi is commonly effected by two means: seed, and air-layering (known in India as *guti*). Higgins writes on this subject:

"As seeds do not reproduce the variety from which they have been taken, and as the seedlings are of rather slow growth and require many years to come into bearing, it has for many years been the custom in China, the land of the litchi, to propagate the best varieties by layering or by air-layering, a process which has come to be known as 'Chinese layering' and is applied to many kinds of plants. In air-layering, a branch is surrounded with soil until roots have formed, after which it is removed, and established as a new tree. In applying the method to the litchi, a branch from ¾ to 1½ inches in diameter is wounded by the complete removal of a ring of bark just below a bud, where it is desired to have the roots start. The cut is usually surrounded by soil held in place by a heavy wrapping of burlap or similar material, although sometimes a box is elevated into the tree for this purpose. Several ingenious devices have been made to supply the soil with constant moisture. Sometimes a can with a very small opening in the bottom is suspended above the soil and filled with water which passes out drop by drop into the soil. Again, sometimes the water is conducted, from a can or other vessel placed above the soil, by means of a loosely woven rope, one end of which is placed in the water, the other on the soil, the water passing over by capillarity.

"Air-layering is commenced at about the beginning of the season of most active growth, and several months are required for the establishment of a root system sufficient to support an independent tree. When a good ball of roots has formed, the branch is cut off below the soil, or the box, after which it is generally placed in a larger box or tub to become more firmly established before being set out permanently. At first it is well to provide some shade and protection from the wind, and it is often necessary to cut back the top of the branch severely, so as to secure a proper proportion of stem to root."

Regarding methods of propagation employed in China, Groff says: "I have never seen a budded or grafted litchi tree, and I understand it is never done. Litchi trees are either inarched or layered, the latter being the most common and most successful. If inarched it is on litchi stock. The common practice in inarching is to use the Loh Mai Chi variety for cion and the San Chi for stock." The method of layering mentioned by Groff is that described above. Inarching is treated in this volume in connection with the propagation of the mango. It is a tedious process of grafting little used in America, but more certain than budding and other methods.

Litchi seeds are short-lived. If removed from the fruit and dried, they retain their viability not more than four or five days. If they remain in the fruit, however, and the latter is not allowed to dry, they can be kept for three or four weeks. In this way they can be shipped to great distances, or they may be removed from the fruit, packed in moist sphagnum moss, and allowed to germinate en route. Some of the choice grafted varieties, such as the Bedana of India, do not produce viable seeds.

Higgins recommends that the seeds be sown in pots sunk in well-drained soil. They should be placed hortizontally about ½ inch below the surface of the soil, and after they have germinated the seedlings should be kept in half-shade.

Attention has recently been given to the possibility of grafting or budding the litchi on the longan (*Euphoria Longana*) and other relatives (see below). Higgins has successfully crown-grafted the litchi on large longan stocks. He says, "Repeated experiments with this method have shown that there is no great difficulty in securing a union of the litchi with the longan. A noteworthy influence of the stock on the cion should be mentioned here. The growth produced is very much more rapid than that of the litchi on its own roots, and in some cases the character of the foliage seems to undergo a change." Additional experience is required, however, to show the practical value of the longan and other stocks. The field is an interesting one, and important results are likely to be secured.

Yield and season.

Seedling litchis have been known to bear fruit at five years of age. It is commonly held that they should bear when seven to nine years old. In some instances, however, trees twenty years old have failed to produce fruit. Higgins remarks, "Wide variability in the age of coming into bearing has been noted with seedlings of other tropical fruits, especially the avocado, but the litchi appears most extreme in this respect."

Layered plants tend to bear when very young. Sometimes they will flower a year after planting, and mature a few fruits when two years old, but three to five years is the age at which they normally come into bearing.

The litchi is famed as a long-lived tree. An early Chinese account (not necessarily to be credited) mentions one which was cut down when it was 800 years old. Bonavia considered that litchis should remain in profitable bearing for a century at least. Mature trees have been found in Hawaii to yield 200 to 300 pounds of fruit yearly, and crops of 1000 pounds have been reported. Under good cultural conditions, the tree can be expected to produce a crop every year. Again quoting Bonavia, it may be said that the tree "bears annually an abundant crop of fine, well-flavored and aromatic fruits, which can readily be sent to distant markets. Instead of being planted by ones or twos, it should be planted by the thousand."

In picking the fruit, entire clusters are usually broken off, with several inches of stem attached. If the individual fruits are pulled off the stems, they are said not to keep well. After they are picked the fruits soon lose their attractive red color, but they can be kept for two or three weeks without deteriorating in flavor. The Chinese sometimes sprinkle them with a salt solution and pack them in joints of bamboo for shipment to distant markets. At the Hawaii Experiment Station it was found that "refrigeration, where it is available, furnishes the best means of preserving the litchi for a limited period in its natural state. . . . There is no doubt that refrigeration will provide a very satisfactory method for placing upon American markets the litchi crop grown in Florida, California, Hawaii, Porto Rico, or Cuba."

The season of ripening in southern China is from May to July. In northern India it is slightly earlier. In Honolulu fresh litchis sell for 50 to 75 cents a pound.

Pests and diseases.

Little is known regarding the enemies of the litchi in China. Brewster says: "There is a worm which makes a ring around the trunk under the bark. When the circle is complete the tree dies; but the bark is broken by it, and by careful watching this can be prevented before the worm does serious harm. There is also a sort of mildew upon the leaves in certain years that does much harm, and the Chinese do not seem to have any way of dealing with it."

Several insect pests are reported from India. A small brown weevil (*Amblyrrhinus poricollis* Boh.), the larvæ of a gray-brown moth (*Plotheia celtis* Mo.), and the larvæ of *Thalassodes quadraria* Guen. feed on the leaves. The larvæ of *Cryptophlebia carpophaga* Wlsm. attack the fruits. Several species of Arbela (notably *A. tetraonis* Mo.) occur as borers on the tree.

It has been found in Hawaii that the dreaded Mediterranean fruit-fly does not attack the litchi fruit, except when the shell has been broken and the pulp exposed. The litchi fruit-worm, the larva of a tortricid moth (*Cryptophlebia illepida* Btl.), is said to have caused much damage to the fruit crop at times. The hemispherical scale (*Saissetia hemispherica* Targ.) occa-

PLATE XVII. The litchi, favorite fruit of the Chinese.

sionally attacks weak trees. The larvæ of a moth (*Archips postvittanus* Walker) sometimes injure the foliage and flowers.

A disease which has been termed erinose, caused by mites of the genus Eriophyes, has been reported from Hawaii, where it has become serious on certain litchi trees. Spraying with a solution of 10 ounces nicotin sulfate and 1¾ pounds whale-oil soap in 50 gallons of water was found to eradicate the mites.

Varieties.

Since the litchi has been propagated vegetatively from ancient times, it is natural that many horticultural varieties should be grown at the present day. Most of these, however, are unknown to the western world. Recently they have been studied by Groff, and it is to be hoped that the best will be brought to light, and their successful introduction into the American tropics realized.

The variety Loh mai chi is said to be one of the best in the world. It is grown in the vicinity of Canton. Haak ip is another Canton litchi said to be choice. All together thirty or forty kinds are reported from this region, some of them being particularly adapted for drying, others for eating fresh, and so on.

The varieties cultivated in India are not in all instances clearly distinguished. The best known is Bedana (meaning seedless), a medium-sized fruit in which the seed is small and shriveled. Probably several distinct sorts are known by this name. McLean's, Dudhia, China, and Rose are other varietal names which appear in the lists of Indian nurserymen.

The Longan

(*Euphoria longana*, Lam.)

Opinions differ regarding the value of the longan. It is popular among the Chinese, but Americans who have tested longans produced in California and Florida have not as a rule considered them good. Frank N. Meyer says that they are improved by cooking, and that preserved longans are considered by some superior to preserved litchis, the flavor being thought more delicate.

According to Alphonse DeCandolle, the longan is a native of India, whence it has been introduced into the Malay Archipelago, southern China, and (recently) tropical America. It is a tree 30 to 40 feet high, resembling the litchi in habit and appearance. The leaves are compound, with two to five pairs of elliptic to lanceolate, glabrous, glossy, light green leaflets. The flowers are borne in terminal and axillary panicles, and are small and unattractive. The fruit is round, an inch or less in diameter, light brown in color, with a thin shell-like outer covering, and white flesh (aril) similar in character to that of the litchi but less sprightly in flavor. The single seed is dark brown and shining.

Meyer says: "The fruit, which is naturally brown, is generally artificially changed to a chrome-yellow. It is eaten fresh, canned, or dried. In the last condition one can obtain it at the Chinese New Year time in the most northern cities of the Empire. There are several varieties of longans, differing in size of fruit, productivity, and size of kernel. Their northern limit of growing seems to be, like that of the litchi, the region around Foochow."

Analysis of the longan by Alice R. Thompson has shown the ripe fruit to contain: Total solids 17.61 per cent, protein 1.41, total sugars 8.34, fat 0.45, and fiber, 0.63.

In French, the longan is commonly termed *œil de dragon* (dragon's eye). The Chinese name is spelled alternatively *longyen, long an, lung an, lingeng,* and so on. Botanical synonyms of *Euphoria Longana* are *Nephelium Longana,* Cambess., and *Dimocarpus Longan,* Lour.

In southern California and in southern Florida, the longan thrives and fruits abundantly if planted in situations not subject to severe frosts. It withstands lower temperatures than the litchi and is less exacting in its cultural requirements. P. D.

Barnhart, writing in the Pacific Garden, says of its culture in California: "We are of the opinion that the greatest success may only be obtained with it in the warmer foothill sections of the country, and that, too, beneath the sheltering arms of live oaks. It seems necessary to protect it from the direct sunlight and desiccating atmosphere of our summers, as well as from the frosts of winters. It requires an abundance of water during the summer months." It has been much more successful on the shallow soils of the Miami region in southern Florida than its relative the litchi.

Propagation is by seed, layering, and grafting, as with the litchi. Higgins remarks concerning the habits of the tree: "The statement has been made that it is a slower grower than the litchi, but this certainly does not hold true under Hawaiian conditions, where it is a robust tree far exceeding the litchi in vigor and rapidity of growth. As in the case of the litchi, seedlings frequently are very tardy coming into bearing." In southern China, where the longan is extensively grown, it is said to require more pruning than the litchi.

The fruit ripens somewhat later than that of the litchi, and is popular among the Chinese, quantities of it being sold in Hongkong and Canton during late summer. Doubtless some of the varieties cultivated in China are superior in quality of fruit to the seedlings which have been grown in the United States. It has been the general opinion of those who have tasted the American-grown longan that it is insipid and somewhat mawkish, although Barnhart considers it excellent.

The Rambutan (Plate XX)

(*Nephelium lappaceum*, L.)

In the Malay Archipelago are found several valuable tropical fruits which have not yet become extensively cultivated elsewhere. The rambutan is one of them. It is grown in nearly every garden in Singapore and Penang, and its fruit is one of the most delicious of the region. It resembles the litchi in character.

As seen in cultivation, the tree is 35 or 40 feet high, erect and stately in appearance. The compound leaves are composed of five to seven pairs of elliptic, obovate, or oblong leaflets, glabrate, about 4 inches long, shining and dark green above, paler beneath. The flower-panicles are axillary and terminal, loose and spreading in form, the flowers small, pubescent, the calyx campanulate, five- or six-cleft, the petals wanting. The fruits, which are produced in clusters of ten or twelve, are oval, about 2 inches in length, and covered with soft fleshy spines ½ inch long. They are crimson in color, sometimes greenish, yellowish, or orange-yellow. The outer covering, from which the spines arise, is thin and leathery, and is easily torn off, exposing the white, translucent, juicy flesh (aril) which adheres to the oblong, pointed, and flattened seed. The flavor is acidulous, somewhat suggesting that of the grape. It is usually relished by Europeans, though considered slightly inferior to its relative the litchi.

Apparently the rambutan is well distributed throughout the Malay Archipelago. H. F. Macmillan says: "It is curious that this fruit, which is so common in the low-country of Ceylon and in the Straits, appears to be scarcely known in India, Mauritius, Madagascar, etc." It has been introduced into the American tropics by the United States Department of Agriculture, but is not yet well established there.

The common name is taken from the Malayan word *rambut,* meaning hair, and has reference to the long soft spines with which the fruit is covered. *Rambustan, ramboetan,* and *rambotang* are forms sometimes used. The French spell it *ramboutan* and sometimes call the fruit *litchi chevelu* (hairy litchi).

The rambutan is eaten fresh. It has been found to contain about the same amount of sugar as the litchi and longan, as follows: Saccharose 7.8 per cent, dextrose 2.25, and levulose 1.25.

In climatic requirements the rambutan must be considered strictly tropical. It thrives in Ceylon up to elevations of 2000 feet, which means that it does not grow in the cooler parts of the island. It likes a moist hot climate and may not, therefore, succeed anywhere on the mainland of the United States, although there is a possibility that it might be grown in extreme southern Florida. It should be practicable to grow it in many parts of the American tropics.

Little is known regarding the culture of the tree. It succeeds on deep, rich, and moist soils, but its adaptability as regards soil and other conditions is not definitely understood. It is propagated by seed, and by air-layering in the same manner as the litchi; it has also been inarched successfully. Mature trees are productive, the bearing habits of the rambutan resembling those of the litchi. It is said that there are fifteen varieties, differing in color, size, and flavor, cultivated in the Malayan region, but they are not well known horticulturally.

The Pulasan (Fig. 43)

(*Nephelium mutabile*, Bl.)

In the markets of Singapore, the pulasan is sold as a variety of the rambutan. It is, however, a distinct species and is known elsewhere in the Malayan region under a different name.

The tree, which is considered to be indigenous in Java and Borneo, is not well known horticulturally. The leaves are compound, with two to four pairs of oblong to elliptic, acuminate leaflets commonly 5 to 10 inches long, glabrous and shining above, glaucous beneath. The fruit is larger than that of the rambutan, with a stouter stem, and is borne in closely-crowded clusters of three to five fruits, instead of loose clusters of a dozen or so. The pericarp or outer covering is thick, sometimes

$\frac{3}{8}$ inch, and the spines are short, blunt, and stout, much swollen near the base; whereas the pericarp of the rambutan is rarely more than $\frac{1}{4}$ inch thick, and the spines are much longer and taper uniformly toward the base. The flesh of the pulasan is less juicy than that of the rambutan, sweeter, and of less sprightly flavor. The size of the seeds is about the same in both species.

Other forms of the common name are *kapoelasan, capulasan,* and *pulassan.*

Like its congener the rambutan, the pulasan is probably suitable for cultivation only in moist tropical regions. It is not known to have been grown to fruiting age anywhere in tropical America, but there are many places where it should succeed. It is doubtful whether it will do so in Florida, and California is unquestionably too cool and dry for it.

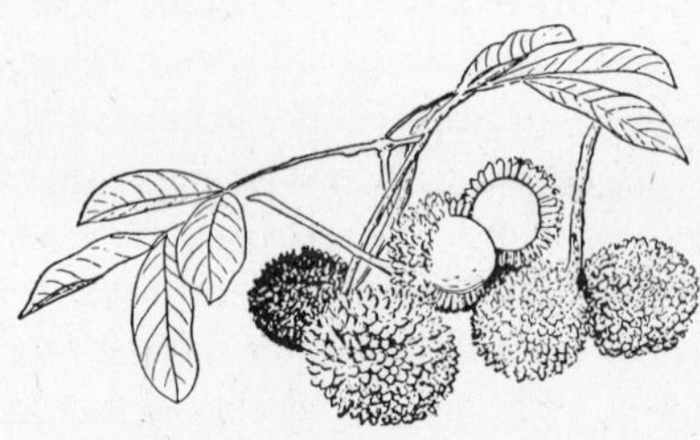

FIG. 43. The pulasan (*Nephelium mutabile*), a relative of the litchi which is cultivated in the Malayan Archipelago. The translucent, white, subacid pulp adheres closely to a large seed. (× $\frac{1}{4}$)

Harry H. Boyle says of the pulasan in Siam: "All the trees are propagated by marcottage (air-layering), budding and grafting being unknown arts in Siam. The flavor of some of the varieties is delicious and many trees produce seedless fruit."

THE AKEE (Plate XVIII)

(*Blighia sapida,* Koen.)

Like the oil palm (*Elæis guineensis*), now common on the coast of Brazil, the akee is an African plant which was brought to America in the days of the slave trade. According to William Harris, it reached Jamaica in 1778. It is now common in that island, and is cultivated on a limited scale in other parts of the West Indies, as well as on the mainland of tropical America. In the Orient it is rare. Its native home is in tropical West Africa.

On deep rich soils the tree becomes 35 or 40 feet in height. It is erect in habit, with an open crown and stiff branches. The leaves are abruptly pinnate, with three to five pairs of short-stalked, obovate-oblong leaflets, the upper ones 4 to 6 inches in length, the lowest pair much shorter. The small flowers are borne in short axillary racemes. The sepals and petals are five in number, the latter greenish white in color. The fruit is a curious-looking capsule, about 3 inches long, triangular in general outline, and straw-colored to magenta-red. When ripe it opens along three sutures, exposing three round shining seeds, with a whitish fleshy body at the base of each. The fleshy substance (technically the arillus), resembles in appearance the brain of a small animal. It is firm and oily in texture, and has a somewhat nutty flavor. When fried in butter it is a delicious morsel, and it is excellent boiled with salt fish. It has long been believed that the akee, unless cooked, is poisonous. J. J. Bowrey,[1] analytical chemist to the Government of Jamaica, found that:

"Unripe akees if eaten freely bring on vomiting. Decaying akees are decidedly unwholesome, and may even be very poisonous. This is true of many foods. Fresh ripe akees are good and harmless food, rather rich it is true, but to most persons quite wholesome. There may be individual idiosyncrasies with regard to akee, as there are to such usually harmless foods as mutton, duck, pork, mushrooms, etc. The red membrane of the akee, so commonly believed to be poisonous, is perfectly harmless. If the fruit be ripe and fresh, which can be known by its being open, the edible portion firm, and the red part bright in color, it may be considered a good and safe food. But if the fruit be not ripe, or if there are any signs of decay, such as mouldiness or softening of the edible portion, or a dingy color in the ordinary red part, the fruit should not be eaten."

[1] Kew Bull. 1892, p. 109.

The name akee came to America from Africa along with the fruit itself, and is generally used (sometimes as akee-apple) in the British colonies where the tree is grown. In Spanish-speaking countries the usual name is *seso vegetal*, or vegetable brains. *Cupania sapida*, Voigt., is a botanical synonym of *Blighia sapida*, Koen.

In tropical America the akee is grown most commonly in the hot moist lowlands. Since it has succeeded in southern Florida, however, the species cannot be considered strictly tropical in its requirements. When young it is susceptible to frost, but plants which have attained four or five years' growth have passed through temperatures of 26° above zero with very little injury. Several have been grown at Miami and Palm Beach, and the fruit which they have produced has been equal in every respect to that grown in the tropics. No large plants are known in California and it is doubtful whether the species will succeed anywhere in that state. It thrives on deep loamy soils with abundant moisture, but makes satisfactory (though slow) growth on the shallow sandy soils of southeastern Florida. It has come into bearing at Miami when about five years old.

Propagation is usually by seed, but vegetative means should prove successful. No horticultural varieties have been established.

The Mamoncillo

(*Melicocca bijuga*, L.)

Unlike its oriental relatives the litchi, the longan, and the rambutan, the mamoncillo is strictly an American plant. It is cultivated in the West Indies and on the neighboring mainland of South America, in which latter region it is considered to be indigenous. In Porto Rico and Cuba it is a popular fruit among the poorer classes.

In habit and foliage the species resembles the soapberry

(*Sapindus Saponaria*). The tree, which grows slowly, is erect, shapely, 30 to 40 or sometimes as much as 60 feet high. The leaves are compound, with two pairs of elliptic-lanceolate, acute, glabrous leaflets, the lower pair about half the size of the upper. The small flowers, which are produced in short panicles, are followed by clusters of smooth round fruits about the size of plums. The outer covering of these fruits is thick and leathery, and green on the surface; it incloses a large round seed surrounded by soft, yellowish, translucent, juicy pulp. The flavor is said to be usually sweet and pleasant, but in many varieties it is acid, especially if the fruit is not fully ripe.

The generic name Melicocca means honey-berry, and is intended to refer to the flavor of the fruit; but some of the mamoncillos grown in Cuba are frequently as sour as limes. Indeed, one of the common names for this fruit in southern Florida is Spanish-lime; it is also there called *genip*. *Mamoncillo* is the Cuban name. In Porto Rico it is known as *genipe*. In the French islands this same name (supposedly) is current, in the form *quenette* or *knepe*.

P. W. Reasoner says: "The fruit markets well in Key West, and there are a number of fine bearing trees in that place, and on the other islands. It is worthy of more attention all over south Florida." At Miami and Palm Beach it grows well, but some of the trees do not bear fruit. The mature plant withstands several degrees of frost without injury. It does not require rich soil, nor is it particularly exacting in other ways. So far as is known, it has never been grown to fruiting stage in California.

The mamoncillo has been propagated up to the present time exclusively by seed. It will probably lend itself, however, to the vegetative methods which are employed with its relatives. No horticultural varieties have been established.

CHAPTER XI

THE SAPOTACEOUS FRUITS

THE sapotaceous fruits are so named from the family Sapotaceæ, to which they belong, and which in turn is named from the old generic name Sapota (now represented in *Achras Sapota*, the sapodilla). The species are mostly tropical, although a few species of little economic importance are native in the United States north of the Florida Keys.

THE SAPODILLA (Plate XIX)

(*Achras Sapota*, L.)

Gonzalo Hernandez de Oviedo, who was one of the first Europeans to study the plants of the New World, called the sapodilla the best of all fruits. More recently, Thomas Firminger, an English horticulturist who lived in India, wrote of it that "a more luscious, cool, and agreeable fruit is not to be met with in this or perhaps any country in the world"; while the poetic French botanist, Michel Etienne Descourtilz, has characteristically described it as having "the sweet perfumes of honey, jasmine, and lily of the valley."

While it is scarcely possible to indorse the enthusiastic opinion of Oviedo, the sapodilla must be considered one of the best fruits of tropical America. It cannot vie, perhaps, with the pineapple or the cherimoya, but it is deservedly held in great esteem by the inhabitants of many tropical countries.

The tree is evergreen and stately, sometimes attaining a height of 50 to 75 feet, with a dense rounded or conical crown.

The wood is hard and durable; in fact, lintels believed to be made from it are found in the ruins of Tikal (Central America), dated 9.15.10.0.0 (Maya chronology) or 470 A.D. The branches often extend from the trunk horizontally. They are tough and pliable, which makes the sapodilla more resistant to cyclones and hurricanes than many other tropical fruit-trees. The bark contains a milky latex known commercially as chicle. This product is secured by tapping the trunk, and is exported in large quantities from southern Mexico and Central America to the United States, where it is used as the basis of chewing-gum. The leaves are entire or emarginate, ovate-elliptic to elliptic-lanceolate in outline, thick, stiff, shining, and 2 to 5 inches long. The small flowers are produced in the leaf-axils toward the ends of the branchlets; the calyx is composed of six small, ovate sepals, and the corolla is white, tubular, lobulate, with six stamens opposite the lobules. The ovary is ten- to twelve-celled, each cell containing one ovule. The fruit is variable in form, but commonly is round, oval, or conical, and 2 to 3½ inches in diameter. The thin skin is rusty brown and somewhat scurfy, giving the fruit a striking resemblance to an Irish potato. The flesh in the ripe fruit is yellow-brown, translucent, soft, sweet, and delicious, but when immature it contains tannin and a milky latex, so that it must not be eaten until fully ripe. The flavor has been likened to that of pears and brown sugar together; it is rich, slightly fragrant, and very pleasing to those who like sweet fruits. The seeds vary from none to ten or twelve and are hard, black, shining, obovate, flattened, and about ¾ inch long. They are easily separated from the flesh and give little trouble in eating the fruit.

The sapodilla is native to tropical America. Henry Pittier considers it indigenous in Mexico south of the Isthmus of Tehuantepec, in Guatemala, and possibly in Salvador and northern Honduras. It is particularly abundant in the lowlands of Tabasco, Chiapas, and the western part of Yucatan, throughout which region the wild trees are tapped for chicle gum. From its native home it has been carried around the world. It is grown on the western coast of India and in Bengal, and, according to H. F. Macmillan, was introduced into Ceylon about 1802, but it has not become widely cultivated in that island. One meets with the tree in some parts of Africa, and Gerrit P. Wilder says it is common in the Hawaiian Islands. Throughout tropical America, it is abundant from southern Brazil to Florida.

In California the sapodilla has not been a success. Occasional trees in favored locations have lived for several years, but they have never reached the fruiting stage. Frosts have eventually killed most of them, and even the coolness of California nights has proved unfavorable to their natural development. In Florida the plant's cultivation is limited to the east coast from Palm Beach (or perhaps farther north) southward to Key West, and on the west coast as far north as the Manatee River. Mature trees in that state have passed uninjured through temperatures of 28° above zero, according to P. W. Reasoner. On the Florida Keys the sapodilla is one of the favorite fruits.

The common name sapodilla, by which the fruit is known in Florida, is taken from the Spanish *zapotillo,* meaning small zapote. In Mexico the usual name is *chicozapote* (often abbreviated to *chico*); this is derived from the Nahuatl *tzicozapotl,* or gum zapotl. In Mexico and other Spanish-speaking countries it is also called níspero, a name which properly belongs to the European medlar, *Mespilus germanica.* The English have formed from this the term naseberry, which is current in the West Indies and India. In the latter country it is called in Marathi *chiku.* The Maya name *yá* is used in Yucatan. In southern Brazil one form of the fruit is called *sapotí,* another *sapota,* while at Pará the name is *sapotilha.* In German it is called *breiapfel,* and in French *sapotille.* The botanical synonymy is rather extensive: *Achras Sapota*, L., *Sapota Achras*, Mill., and *Sapota zapotilla*, Coville, are sometimes used.

The sapodilla is preëminently a dessert fruit. Rarely is it cooked or preserved in any way, although in Cuba and Brazil it is often made into a sherbet. According to Carl Wehmer [1] it contains about 14 per cent of sugar, of which 7.02 is saccharose, 3.7 dextrose, and 3.4 levulose. It also contains a small amount of acid and about 1 per cent of ash.

Although tropical in character, the sapodilla does not require a high degree of humidity nor entire freedom from frost. If liberally irrigated it can be grown in regions where the atmospheric humidity is low. The plant while young is injured by temperatures below freezing, but when mature it withstands 27° or 28° above zero. Although it prefers a rich sandy loam, it thrives on light clay and also on the shallow sandy soil underlaid with soft limestone which is found on the lower east coast of Florida. Indeed, its aptitude for rocky and forbidding situations on the Florida Keys is remarkable. It is said to grow well in India both on red sandy soil along the seashore and in the black alluvial land of the Dekkan.

It is the custom in India to plant sapodilla trees 15 to 20 feet apart. This is too close for the best results, particularly if the soil is rich and deep so that the tree grows to large size; 30 feet apart is probably close enough on good soils. V. N. Gokhale, writing in the Poona Agriculture College Magazine (1911), reports that in India the young plants are set in pits 1 foot wide and 2 to 3 feet deep in which a quantity of sheep-manure has been mixed with the soil, and that the mature trees are regularly supplied with manure two or three times a year.

Little attention has yet been given to pruning. Since the tree is of slow compact growth, it will probably require nothing more than the removal of an occasional unshapely branch. In

[1] Die Pflanzenstoffe.

southern Florida it thrives under the same cultural attention as citrus fruits.

The sapodilla is usually propagated by seed, but the variation among seedlings in productiveness as well as in quality, size, and shape of fruit necessitates some asexual means of propagation if the most desirable forms are to be perpetuated. Edward Simmonds has shown in Florida that the species can be budded in the same manner as the mango. Grafting and layering have been practiced in India.

Seeds, if kept dry, will retain their viability for several years. They should be sown in flats of light sandy soil, and covered to the depth of ½ inch. In warm weather germination takes place within a month. The young seedlings, after they have made their second leaves, may be potted off and carried along thus for a year or two, when they will be large enough to be set out in the open ground. Their growth is slow. If they are to be budded they should be planted in nursery rows which are 3 feet apart, and 18 inches apart in the row. In southern Florida, May has proved to be a good month for budding; in strictly tropical regions it can probably be done at any time of the year, provided the stock-plants are in active growth. Budwood should be chosen from young branchlets which have begun to lose their greenish color and assume a brownish tinge. It should be examined carefully to ascertain that the axillary buds or "eyes" are well developed. Shield-budding is the method employed, the details being practically the same as in budding the mango. After making the incision in the stock, the bud should be inserted promptly, since the latex soon collects around the wound and renders insertion difficult. Waxed tape should be used for wrapping. After three or four weeks, the stock may be headed back and the wrap loosened, leaving the eye exposed so that it may start into growth.

A. C. Hartless, superintendent of the Government Botanical Gardens at Saharanpur, India, has found that the sapodilla can be inarched and cleft-grafted on *Mimusops Kauki,* L. Propagated in this manner the tree is dwarfed and bears at an earlier age than when grown on its own roots; it is believed also to be more productive. V. N. Gokhale says that propagation in western India is by seeds and layering. Plants obtained from layers are believed to be more vigorous than those from seed. Eight to ten layers can be made each year from a bearing tree, choosing the branches close to the ground.

Seedling sapodillas rarely come into bearing until six to eight years of age, even when grown under favorable conditions. They usually fruit heavily, and often produce two crops a year, one being much lighter, however, than the other. Due to this habit, together with the natural variation in season among seedling trees, ripe sapodillas are to be found in the markets of tropical America almost throughout the year.

Experiments have shown that the fruit can be shipped successfully and with little care in packing. The skin is thin and delicate and the fully ripe fruit is injured very easily; but if picked while still hard or "tree ripe," it does not begin to soften for several days. Sapodillas have been shipped from the Florida Keys to New York, packed in tomato-crates which hold six small baskets, each basket carrying six good-sized fruits. For local consumption, or for shipping short distances, the common procedure in Florida is to pull the fruits from the trees and simply throw them into boxes or baskets, in which they are carried to market, the ripe ones being picked out daily.

The fruit-flies (Trypetidæ) are serious pests of the sapodilla in some regions, their larvæ infesting the ripe fruit and rendering it unfit for consumption. *Ceratitis capitata,* Wied., the Mediterranean fruit-fly, and *Anastrepha ludens,* Loew., the Mexican fruit-fly, are two of the most troublesome species. The tree is attacked by very few insect or fungous enemies.

Seedlings differ in productiveness, ripening season, and in size, shape, and character of their fruits. Those which are unusually choice or valuable should be propagated by budding, grafting, or layering, and established as named varieties. Occasionally a seedless kind is found, or one whose fruits are very large, weighing a pound or even more. Differences in flavor and quality of fruit are also noticeable. There are not as yet any named varieties known in the trade.

The Sapote (Fig. 44)

(*Calocarpum mammosum*, Pierre)

The sapote is one of the important fruits of the Central American lowlands. It furnishes to the Indians a nourishing and agreeable food, obtainable during a certain part of the year in considerable abundance. Cook and Collins remark: "It was this fruit that kept Cortes and his army alive on their famous march from Mexico City to Honduras."

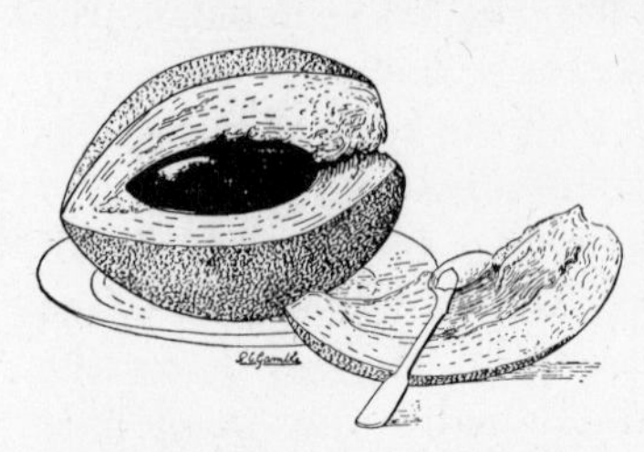

Fig. 44. The sapote (*Calocarpum mammosum*). (× $\frac{1}{3}$)

In the hot and humid lowlands the sapote becomes a large tree, often 65 feet high, with a thick trunk and stout branches. The Indians, when clearing the forest in order to plant coffee or other crops, usually spare the sapote trees they encounter, for they regard the fruit highly. The foliage is abundant, and light green in color. The leaves, which are clustered toward the ends of the stout branchlets, are obovate to oblanceolate in outline, broadest toward the apex, and 4 to 10 inches long. The small flowers are produced in great numbers along the branchlets. The sepals are eight to ten, imbricate, in several series; the corolla is tubular, whitish, with five lobes. The stamens are five and the ovary is hairy, five-celled, with one ovule in each cell. The fruit is elliptic or oval in form, commonly 3 to 6 inches long, russet-brown in color, the skin thick and woody and the surface somewhat scurfy. The flesh is firm, salmon-red to reddish brown in color, and finely granular in texture. The large elliptic seed can be lifted out of the fruit as easily as that of an avocado; it is hard, brown, and shining, except on the ventral surface, which is whitish and somewhat rough. To one unaccustomed to the exceedingly sweet fruits of the tropics, the flavor of the sapote is at first somewhat cloying because of its richness and lack of acidity. When made into a sherbet, as is done in Habana, it is sure to be relished at first trial. Inferior or improperly ripened sapotes will be found to have a pronounced squash-like flavor.

Pittier, whose studies of the sapotaceous fruits have done much to clear away the botanical confusion in which they have been involved, considers the sapote to be indigenous to Central America. Outside of its native area it is grown in the West Indies, in South America, and in the Philippines. In Cuba it is particularly abundant and the fruit highly esteemed. Though it has been planted in southeastern Florida it has never succeeded in that region. The limiting factor there seems to be unfavorable soil rather than temperature, while in California it has always succumbed to the cold, even when grown in the most protected situations.

In the British West Indies the sapote is called mammee-sapota, marmalade-plum, and marmalade-fruit. In the French West Indies it is known as *sapote* and *grosse sapote*. In Cuba it is called *mamey colorado* and, less commonly, *mamey zapote*. Throughout its native area, southern Mexico and Central America, it is known in Spanish as *zapote* (from the Nahuatl or Aztec name *tzapotl*) and this name is used also in Ecuador and Colombia. In the Philippines the term is *chico-mamey*. The more important botanical synonyms are: *Achras mammosa*, L., *Lucuma mammosa*, Gaertn., *Vitellaria mammosa*, Radlk., and *Achradelpha mammosa*, Cook. The name *mamey*, improperly applied to this fruit, results in its being confused with *Mammea americana*, L.

The Indians of Central America commonly eat the sapote out of hand, but it is occasionally made into a rich preserve and it may be employed in other ways. In Cuba it is used as a "filler" in making guava-cheese, and a thick jam, called *crema de mamey colorado,* is also prepared from it. The seed is an article of commerce in Central America, where the large kernel is roasted and used to mix with cacao in making chocolate.

The tree is tropical in its requirements. In Guatemala it is most abundant at elevations from sea-level to 2000 feet; at 3000 feet it is still quite common, but at 4000 feet it is rarely seen. At higher elevations it is injured by the cold and makes very slow growth. It thrives on heavy soils, such as the clays and clay-loams of Guatemala. It is believed in Florida that the plant does not like a soil which is rich in lime, and that for this reason it has failed to succeed at Miami and other points in the state where conditions otherwise seem to be favorable. P. W. Reasoner considered it to be as frost-resistant as the sapodilla.

Seedlings start bearing when seven or eight years old if grown under favorable conditions, and when of good size yield regularly and abundantly. The fruits are picked when mature, and laid away in a cool place to ripen, which requires about a week. If shipped as soon as picked from the tree, they can be sent to northern markets without difficulty. Sapotes from Cuba and Central America are often seen in the markets of Tampa and New Orleans. The season of ripening extends over a period of two or three months, usually beginning about August in the West Indies and Central America. Differences in elevation, and consequently in climate of course affect the time of ripening.

All of the sapote trees in tropical America are seedlings. Neither budding nor grafting has yet been used with this species, so far as is known. The seeds, which cannot be kept long, germinate more readily if the thick husk is removed before planting. They should be placed in sand or light soil, laid on their sides, and scarcely covered. When the young plants are six or eight inches high, they may be transferred to four- or five-inch pots. Their growth is rapid at first, but much slower after they have exhausted the food reserves stored in the large seed. It is probable that budding will prove as successful with the sapote as it has with the sapodilla. Seedlings differ greatly in the size, shape, and quality of their fruits. The best one should be propagated by some vegetative means.

The Green Sapote (Plate XX)

(*Calocarpum viride*, Pittier)

While greatly superior in flavor to its congener the sapote (*C. mammosum*), the green sapote is much more limited in its distribution. It is common in the Guatemalan highlands and is found also in Honduras and (rarely) in Costa Rica. Elsewhere it is not known, but it deserves to be cultivated throughout the tropics.

In habit and general appearance the tree greatly resembles the sapote, from which it can be distinguished (according to Pittier) "by the smaller leaves, downy and white beneath, the smaller and differently shaped sepals, the shorter staminodes and stamens, the latter with broadly ovate anthers, and above all the comparatively small, green, and thin-skinned fruit and the smaller, ovate seed." It is most abundant in northern Guatemala (the Alta Verapaz), where it grows usually at elevations of 4000 to 6000 feet. Unlike its relative the sapote, it does not thrive in the hot lowlands. The lower limit of its cultivation is approximately 3000 feet, the upper between 6000 and 7000 feet.

The fruit, which is known in Guatemala as *injerto* (Spanish) and *yash-tul* (Kekchi, green sapote), is much prized by the Indians of the Verapaz. The flavor is similar to that of the sapote, but more delicate, and the flesh is finer and smoother in texture. The largest fruits are nearly 5 inches long, turbinate to elliptic in outline, and brownish green to pale yellowish green in color; the skin thin, almost membranous, and easily broken. The flesh is pale red-brown in color, melting, sweet, and somewhat juicy. The seeds are commonly one or two, elliptic in form, and about 2 inches long. Usually the fruit is eaten fresh, but in some parts of Guatemala a preserve is prepared from it, similar to that made from the sapote.

The tree is productive, but has the disadvantage of not coming into bearing earlier than eight or ten years from seed. It is not systematically cultivated, but is met with in dooryards and around cultivated fields. The fruits are in great demand in the markets of Guatemalan towns. They ripen from October or November (depending on elevation) to February. When picked from the tree they are hard and can be carried long distances without injury, but after they have softened and are ready for eating they must be handled carefully, since the skin is thin and easily broken.

This species has been planted recently in California and Florida. It is more likely to succeed in the latter state than the sapote, since it is somewhat more frost-resistant. It is doubtful, however, whether it will survive temperatures below 27° or 28° above zero. Seed-propagation is the only method which has been employed up to the present time.

The Star-Apple (Fig. 45)

(*Chrysophyllum Cainito*, L.)

In Cuba, Jamaica, and several other tropical American countries, the star-apple is a common dooryard tree and its fruit is held in much the same estimation as the sapote, the sapodilla, and the sugar-apple. For its ornamental value alone it merits cultivation. Charles Kingsley, in his brief account of West Indian fruits, refers to the beauty of this plant. "And what is the next," he asks, after mentioning some of the trees seen on one of his rambles, "like an evergreen peach, shedding from the under side of every leaf a golden light, — call it not shade? A star-apple."

On the deep rich soils of Cuba the tree sometimes reaches 50 feet in height, although in southern Florida it rarely exceeds 30 feet. The leaves are oval or oblong, about 4 inches in length, deep green and glossy above, and golden-brown, with a sheen like that of satin, beneath. The flowers are small and inconspicuous, purplish white in color. The fruit is commonly round, sometimes oblate, and 2 to 4 inches in diameter. The surface is smooth, somewhat glossy, dull purple in some varieties, light green in others. On cutting the fruit transversely, it is found to be differentiated into two kinds of flesh; directly under the thin tenacious skin is a layer of soft, somewhat granular flesh, concolorous with the skin, and not very juicy; inclosed by this are eight translucent whitish segments in which the seeds are embedded. When the fruit is halved thus, transversely, these cut segments present a star-like appearance, whence the common name. Both kinds of flesh are sweet, entirely lacking in acidity, with the characteristic sapotaceous flavor. Normally there is one seed in each segment, but frequently several are aborted, leaving three to five in the fruit. They are ovate to elliptic in outline, flattened, $\frac{3}{4}$ inch long,

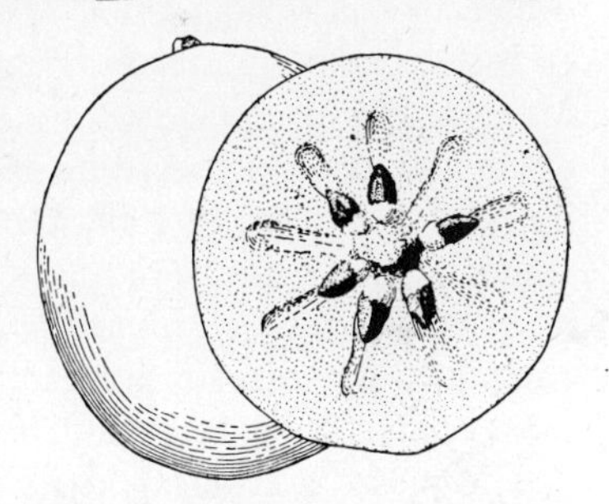

FIG. 45. The star-apple (*Chrysophyllum Cainito*), a popular fruit in Cuba. It is green or purple in color, and the flesh is melting, sweet, and pleasantly flavored. (× about $\frac{1}{2}$)

hard, brown, and glossy. The appearance of a halved star-apple is strikingly suggestive of that of the mangosteen.

The fruit is usually eaten fresh. In Jamaica it is sometimes made into preserves, and also (according to P. W. Reasoner) into a mixture somewhat cryptically called "matrimony," which is prepared by scooping out the inside pulp and adding it to a glass of sour orange juice. An analysis made in Hawaii by Alice R. Thompson shows the ripe fruit to contain: Total solids 11.47 per cent, ash 0.39, acids 0.12, protein 2.33, total sugars 4.40, fat 1.38, and fiber 0.85.

The tree is wild in the West Indies and in Central America. It is cultivated in the same area and also in South America, Mexico, Florida, and to a limited extent in Hawaii and a few other countries. According to H. F. Macmillan it was introduced into Ceylon in 1802, but it is not commonly grown anywhere in the Orient, so far as is known. In the English colonies it is known almost invariably as star-apple; in the French colonies (and sometimes in Cuba) it is called *caimite;* while in most Spanish-speaking countries the word is *caimito.*

The plant is tropical in its requirements. P. W. Reasoner notes: "When small, the tree is not apt to sprout up again if killed back by frost, and it is perhaps somewhat more tender than the sapodilla." Old trees are to be found at Miami and Palm Beach, Florida, which proves that the species is sufficiently hardy to grow in the southern part of that state. So far as is known, no plants have ever grown to fruiting size in California, although they have been planted in the most protected situations. The star-apple likes a humid atmosphere with relatively high temperatures throughout the year. Apparently it is not particular in regard to soil; it grows well both on the shallow sandy soils of southeastern Florida and on the deep clay loams of Cuba.

Propagation is usually by seed. Since there is much difference among seedlings, however, it will be desirable to employ some asexual means of propagation in order to perpetuate as varieties any choice kinds which originate. Budding will probably prove satisfactory. It is reported that cuttings can be grown, if they are made from well-ripened shoots and placed over strong moist heat. Seeds retain their viability for several months, are easily transported through the mails, and should be sown in light sandy loam.

Some trees yield heavy crops of fruit, while many others are shy bearers. The ripening season in the West Indies is April and May. The fruits are not good unless allowed to remain on the tree until fully ripe; if picked when immature they are astringent and contain a sticky white latex.

Two races are common, one green-fruited and the other purple-fruited. They are not known to differ in flavor or other characteristics except color.

The Canistel

(*Lucuma nervosa*, A. DC.)

Opinions differ regarding the value of the canistel. By some it is considered a delicious fruit; others find it too sweet and its musky flavor unpleasant. It is popular among residents of the Florida Keys and in Cuba. In the opinion of the author it is certainly not so good as the green sapote, the star-apple, or the abiu.

The tree, which reaches 15 to 25 feet in height, is commonly slender in habit, but sometimes broad and stiffly erect. It is of handsome appearance and for this reason is often planted in dooryards. The leaves are oblong-obovate to oblanceolate in outline, 4 to 8 inches long, glabrous, and bright green in color. The small flowers are produced upon the young wood in clusters of two to five. The fruit is round to ovoid in form, frequently pointed at the apex, orange-yellow and 2 to 4 inches long. The skin is membranous and the bright orange flesh soft and mealy in texture, resembling in appearance the yolk of a hard-boiled egg. The flavor is rich and so sweet as almost to be cloying, and is somewhat musky in character. The seeds, usually two or three in number (although the ovary is five-celled), are oval, about an inch long, hard, dark brown, and shining, except on the pale brown ventral surface.

So far as is known, the canistel is not cultivated commercially in any country, but it is grown as a garden tree in Cuba and southern Florida. The Cuban name *canistel* is presumably from the Maya *kanisté;* in Florida the names *ti-es* and egg-fruit are generally used. Botanically the species is often listed as *Lucuma rivicola* var. *angustifolia,* Miq.

The fruit, which in Florida matures from December to March, is eaten fresh. It is taken from the tree when mature and laid in the house to complete its ripening. Within three or four days it is soft and ready for eating.

The tree is fully as hardy as the sapodilla, and of similar cultural requirements. It grows in south Florida on the Keys and as far north as Palm Beach on the east coast and Punta Gorda on the west coast. P. W. Reasoner wrote in 1887: "Previous to the 'freeze' a specimen had been growing in Tampa for many years, which, after many discouragements by frost, finally produced fruit a few years ago." So far as known, the tree has never grown to fruiting size in California. In regard to soil it does not seem to be particular; it grows well on the heavy clay lands of Cuba and upon some of the poorest and most shallow soils of southern Florida. It shares with the sapodilla the ability to grow in apparently very unfavorable situations on the Florida Keys.

Propagation is usually by seed, but budding will doubtless prove successful. The husks should be removed from the seeds before they are planted. Though not a rapid grower, the tree comes into bearing when three to five years old.

While some trees produce fruit abundantly, others are poor bearers. As usual, there is much variation also in the size and quality of the fruits borne by different seedlings.

The Abiu (Fig. 46)

(*Pouteria Caimito*, Radlk.)

Although the abiu is one of the best of the sapotaceous fruits, it is not so widely cultivated as several other species. It greatly resembles the canistel in habit of growth and in foliage, but is easily distinguished from it by its light yellow fruit with white flesh. The tree reaches 15 or 20 feet in height. The leaves are obovate to lanceolate in outline, 4 to 8 inches long, acute, glabrous, and bright green. The fruit is ovate-elliptic (occasionally almost round) in form, 2 to 4 inches long, and bright yellow in color, with skin thick and tough. Surrounding the two or three large oblong seeds is the translucent flesh, which in flavor resembles the sapodilla but is of different texture. Until fully ripe it contains a milky latex which coagulates on exposure to the air and sticks to the lips in a troublesome manner.

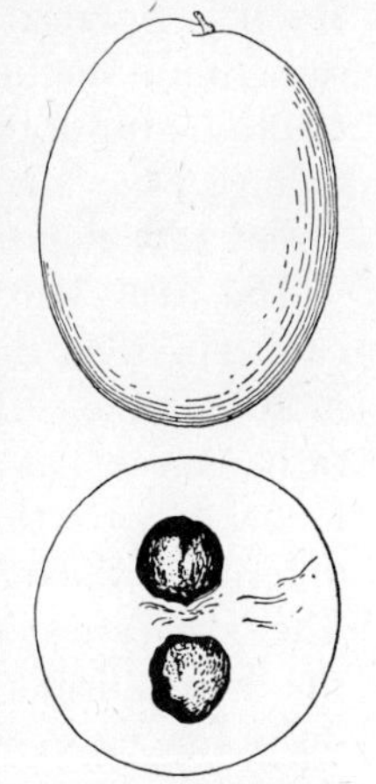

Fig. 46. The abiu (*Pouteria Caimito*), one of the best of the sapotaceous fruits. It is cultivated in Peru and Brazil, rarely elsewhere. The skin is yellow, the flesh whitish, melting, and sweet. (× ½)

Alphonse DeCandolle says of this species: "It has been transported from Peru, where it is cultivated, to Ega on the Amazon river, and to Pará, where it is commonly called *abi* or *abiu*. Ruiz and Pavon say it is wild in the warm regions of Peru, and at the foot of the Andes." Jacques Huber reports that its culture is now extensive at Pará and elsewhere in the Amazon Valley. It is a common fruit at Bahia and Rio de Janeiro, but outside of Brazil and Peru is little known. Recently it has been introduced into the United States, and should prove sufficiently hardy for cultivation in southern Florida, although probably it is too tender for any part of California. From its abundance in Pará it can be judged that it delights in a moist climate, uniformly warm throughout the year. It does well at Rio de Janeiro, however, where the weather is cool during part of the year.

Lucuma Caimito, Roem., is a synonym of *Pouteria Caimito,* Radlk. Because of its specific name *caimito,* this plant is sometimes confused with the West Indian *Chrysophyllum Cainito* (star-apple, see above), which is commonly known in Spanish as *caimito.* According to Pittier, its name in the Cauca Valley of Colombia is *caimo.*

The tree's cultural requirements are similar to those of the canistel. The Brazilians say that a soil rich in humus is the most suitable. Propagation is usually by seeds, but such vegetative means as budding and grafting should be successful. Huber reports that there are many seedling varieties at Pará, differing in the form and size of the leaves as well as in the fruits. The latter are sometimes round, sometimes elongated; in some the flesh is firm and in others soft and mucilaginous; some are insipid in flavor while others are very sweet and pleasant.

The Yellow Sapote

(*Lucuma salicifolia,* HBK.)

Both in foliage and fruit the yellow sapote closely resembles the canistel, but its fruit is, perhaps, slightly the better of the two. It is a small tree, attaining 25 feet in height, and usually of slender erect growth. The leaves are lanceolate, acute at the base and obtuse at the tip, 4 to 7 inches long, and light green in color. The small whitish or greenish flowers are solitary or in pairs in the leaf-axils. The fruit, commonly slender and

PLATE XVIII. Foliage and fruits of the akee (*Blighia sapida*).

extended into a long point at the apex, is 4 to 5 inches long, and orange-yellow in color. The skin is thin and delicate and the pulp soft and mealy, of the consistency and color of the yolk of a hard-boiled egg. In flavor it resembles the canistel. The seed is slender, nearly 2 inches long, light brown and glossy except on the whitish ventral surface.

The yellow sapote is most abundant in Mexico, but according to Pittier is found also in Panama and Costa Rica. The common names in Mexico are *zapote borracho* and *zapote amarillo.* The species is cultivated in that country from sea-level to elevations of 6000 feet. The fruit, which is eaten fresh, ripens in autumn and winter and is often seen in the markets.

While tropical in its requirements, the tree can be grown in regions which are subject to cool weather in part of the year. It is doubtful, however, whether it will stand more frost than its congener the canistel. In Mexico it grows on both light and heavy soils and in regions which are moist as well as in those which are comparatively dry. It has been propagated only by seed, but should lend itself to bud-propagation. As is common the fruits of different seedlings vary in form, size, and other characteristics.

The Lucmo

(*Lucuma obovata,* HBK.)

Pittier has recently called attention to this species, which has been cultivated in Peru since ancient times. It is a tree 25 to 35 feet high, with a dense rounded crown. The leaves, which are in bunches at the ends of the branchlets, are elliptic-ovate in outline, acute at the base, dark green above and paler or rusty below. The flowers are solitary or sometimes two or three together in the axils of the leaves. The fruit is round or ovate in form, about 3 inches long, green externally, with yellow flesh of mealy texture. The seeds are one to five in number (commonly two), flattened, and the size of chestnuts. The tree is thought to be a native of the maritime provinces of Chile and Peru. A few cultivated specimens have been seen in Costa Rica, but the species is not commonly grown outside its native region. It flowers and fruits throughout the year. The fruits must be stored in straw or chaff for several days after gathering before they are ready for eating. This species is believed to be represented by casts found in the graves of the ancient Peruvians. From the common name *lucmo* (sometimes *lucumo*) the generic name Lucuma is taken.

CHAPTER XII

THE KAKI AND ITS RELATIVES

THE genus Diospyros comprises about 200 species, mostly tropical and subtropical. One of them is the native persimmon (*D. virginiana*), which reaches as far north as Connecticut. The oriental kinds are becoming prominent fruits in the lower part of the eastern United States. Diospyros is the largest genus of the Ebony family, which is closely allied to Sapotaceæ. This genus and others furnish the ebony wood of commerce.

THE KAKI OR JAPANESE PERSIMMON (Plate XXI)
(*Diospyros Kaki*, L. f.)

The Japanese, who cultivate more than 800 varieties of the kaki, consider it one of their best fruits. The Chinese also value it highly and devote large areas to its production. Although it has been grown on a small scale in southern France for nearly a century, it is not believed to have reached the United States until the time of Commodore Perry's visit to Japan in 1856, and it was only in 1870 (or thereabouts) that grafted trees of superior varieties were first brought to this country.

Much attention has recently been devoted to the kaki, and it seems probable that it will assume an important position among the orchard-fruits of the cotton-belt and of California. If it does so, credit for its establishment on such a basis will be due largely to the United States Department of Agriculture as having introduced into this country the best Chinese and

Japanese sorts, and to H. H. Hume of Florida for his investigations of cultural problems. The name of Frank N. Meyer, late agricultural explorer for the Department of Agriculture, will be remembered by horticulturists in connection with the introduction of Chinese varieties.

The kaki is a deciduous tree growing up to 40 feet in height (though there are dwarf varieties which remain smaller than this), and having usually a round open crown. The leaves are ovate-elliptic, oblong-ovate, or even obovate in outline, acuminate at the apex, glabrous above and finely pubescent beneath, and 3 to 7 inches long. While it has usually been supposed that the kaki is diœcious, or rarely polygamous, Hume [1] has shown that a single tree may produce three kinds of flowers, perfect, staminate, and pistillate, in varying combinations. All of these are borne upon the current season's growth and open shortly after the shoots and leaves are developed. Staminate flowers are borne in three-flower cymes in the leaf-axils; the calyx and corolla are four-lobed and the latter has sixteen to twenty-four stamens inserted upon it in two rows. The pistillate flowers are solitary and axillary and have a large leaf-like calyx, a four-parted light yellow corolla, eight abortive stamens, and a flattened or globose, eight-celled ovary surmounted by a short four-parted style and much-branched stigma. Perfect flowers are intermediate in character between the staminate and the pistillate, and are most commonly associated with the former. Hume says: "Up to this time they have not been discovered on any varieties of the fixed pistillate-flowering type. In other words, it appears that the perfect flowers are a development from the staminate form and not from the pistillate form." It may be observed that the kaki corresponds in this respect to the papaya, in which perfect flowers are sometimes developed on trees which are normally staminate but never on those which are pistillate.

[1] Trans. St. Louis Acad. Sci., XXII, 5, 1913.

The fruit is oblate to slender conic in form, and from 1 to 3 inches in diameter. It has a thin membranous skin orange-yellow to reddish orange in color; soft (sometimes almost liquid) orange-colored pulp of sweet and pleasant flavor; and occasionally as many as eight elliptic, flattened, dark brown seeds, although there are frequently not more than half that number, and seedless fruits are of common occurrence.

The kaki was formerly thought a native of Japan, but it is now understood that it was originally confined to China, whence it was carried to Japan several centuries ago. Hume believes that the cultivated kakis may be derived from more than one wild species. This theory was suggested by the different reactions of certain varieties to the stimulus of pollination. After describing these reactions [1] he asks:

"Why is it that *D. kaki* presents these peculiar characteristics? Why is it, for instance, that Tsuru is always light fleshed whether the fruit contains seeds or not, while Yemon is light fleshed when seedless and dark fleshed when seedy? Is it not likely that *D. kaki* is not a true species but rather a mixture of two or more species, hybridized and grown under cultivation for centuries? Is it not possible that the present cultivated varieties known under the name of *D. kaki* are derived from two distinct species, one bearing dark fleshed fruit and the other light fleshed fruit? . . . In shape and peculiarities of fruit, color and characteristics of bark, size and shape of leaves, habit of growth and size of tree, they vary much more than any of our common fruits usually regarded as being derived from a single species."

From Japan the kaki has been carried around the world. Its cultivation in France has already been mentioned; it is limited principally to the Côte d'Azur (the Riviera) and Provence. On the opposite shore of the Mediterranean, in Algeria, it is grown to a limited extent. It has never been cultivated widely in

[1] Journal of Heredity, Sept., 1914.

India, but A. C. Hartless reports recently that certain varieties have proved successful at Dehra Dun and elsewhere, and grafted plants are being disseminated from the Botanical Garden at Saharanpur. In Queensland, where it is said to have been introduced less than forty years ago, it is meeting with favor but is not yet extensively grown. In the United States it has been planted chiefly in Florida, Louisiana, and California.

The name kaki, which is applied to this fruit in Japan, has become current in the United States and in southern France. Japanese persimmon and occasionally date-plum and Chinese date-plum are terms used in the United States, and *plaquemine* in France. The Chinese name is *shi tze.* Botanically the cultivated kakis are commonly grouped together under the name *Diospyros Kaki,* L. f., of which *D. chinensis,* Blume, *D. Schitse,* Bunge, and *D. Roxburghii,* Carr., are considered synonyms. French botanists have made botanical varieties or even species out of some of the forms which are elsewhere held to be mere horticultural varieties, *e.g., costata.*

In the United States the kaki is usually sold as a fresh fruit, to be eaten out of hand. In Japan certain varieties are used extensively for drying, the product somewhat resembling dried figs in character and being delicious. "The method of drying," writes George C. Roeding, "is simple. The skin is pared off and the fruits are suspended by the stems, tying them with string to a rope or stick and exposing them to the sun. They gradually lose their original form, turn quite dark and are covered with sugar crystals. . . . Fruit should be picked for drying when yellow and firm."

Methods of processing the mature fruit, so as to remove its astringency, are discussed on a later page. The chemical composition of five varieties is shown in the following table, from analyses made by H. C. Gore:

TABLE VI. COMPOSITION OF THE KAKI

Variety	Total Solids	Ash	Protein	Total Sugars	Tannin
	%	%	%	%	%
Hachiya . . .	25.06	0.49	0.64	17.71	0.88
Tane-nashi . . .	18.52	0.39	0.42	14.59	0.13
Triumph . . .	20.82	0.41	0.40	14.74	1.39
Tsuru	21.08	0.46	0.61	14.46	1.54
Zengi	21.83	0.49	0.73	14.72	0.41

Cultivation.

The kaki is distinctly a subtropical fruit and thus is not successful in the moist tropical lowlands, although there are many elevated valleys and plateaux in the tropics where it can be grown. Its culture in the United States is limited to regions which are suitable for the fig. Some varieties have survived temperatures as low as zero, while others are more tender. L. H. Bailey writes: "Many seedlings have been produced which seem to have increased frost-resisting powers. Instances are reported in which some of these trees have withstood the winters of east Tennessee. By successive sowing of seeds from these hardier seedlings we may look for a race of trees which will be adapted to the middle sections of the United States. There is a probability, also, that importations from the north of China and Japan may considerably extend the range northward in this country. Some varieties have succeeded in central Virginia and Kentucky."

Regarding the moisture requirements of the kaki, experience indicates that it does not need a high degree of atmospheric humidity if it is supplied with plenty of water at the root. T. Ikeda says of the trees in Japan: "They are very water-loving in habit and require a constant and sufficient supply of soil water." The behavior of the species in California has shown that it is entirely successful in a semi-arid climate, while experience in other regions indicates that it can be grown equally well in a region of reasonably heavy rainfall. In parts of India where the precipitation is extremely heavy it has not done well.

In soil requirements the kaki is not exacting. Emile Sauvaigo, one of the best French authorities, says: "It likes a deep, reasonably heavy, well-drained soil, and it does well on clays, when they are not too compact"; and Ikeda notes that the yield is larger, and the color and quality of the fruit better, when the trees are planted on heavy but well-drained loams. In California it has been observed that they make larger growth on heavy than on thin sandy soils, which would, of course, be expected. Satisfactory results are obtained in Florida on light sandy loams, particularly when they are moist; in fact, it seems difficult to give the plant too moist a situation, provided the drainage is good.

Florida nurserymen advise that the land on which kakis are to be planted be prepared in advance by growing a crop of cowpeas or velvet-beans and plowing them under to enrich the soil. Planting may be done in the lower South between November 15 and March 1, but preference is given to the period from December 1 to February 1. The trees should be spaced 18 or 20 feet apart (134 or 108 trees to the acre). As much as 24 feet is considered a desirable distance in California. The roots should not be allowed to dry out while the trees are being set. The tops should be cut back to 2 or 2½ feet on plants which have not large stems. Roeding says: "The tap-root should be cut back to 18 inches, and fresh cuts made on all the fibrous roots. After the trees are set, head them back to 18 inches. The first winter thin out the branches, not leaving more than four to form the head of the tree. Cut these back at least one-half. In the second, third, and fourth years pruning of the tree should be continued to fashion it into the typical goblet form."

Frequent and thorough cultivation of the grove during the spring and early summer is recommended for Florida. Cultivation should be discontinued about the middle of July and a cover-crop then planted. This may be cowpeas, velvet-beans, beggarweed, or a natural growth of weeds may be allowed to develop. Commercial fertilizers are used to advantage.

F. H. Burnette [1] writes as follows on this subject:

"Good clean culture is all that is required, the same that is given in any well-cared-for fruit orchard. In our heavy lands, or on soils similar in character to the soils of the bluff lands of Louisiana, sodding-over should never be allowed, if good crops are desired. Any good complete manure may be used. A good crop of cow-peas turned under every two or three years will be highly beneficial.

"During the first three years the growth of the tree should be watched in order to build a symmetrical, upright tree. This is not easy, for some of the varieties spread too much, and the leading upright branches are often overloaded and become broken easily, or are headed back by careless removal of the fruit. Ordinarily, after they begin to bear, there is little need of pruning. The tendency to overbear is so strong that new wood is not produced in abundance, and the tree becomes dwarf-like. Systematic thinning of the fruit is necessary to control this, as it will not do to leave the thinning to natural causes, and depend upon the tree to throw off all the fruit it cannot well take care of. The weakened condition from overbearing results in a sickly tree which readily becomes a prey to diseases and insects, and it requires a careful observer to train his tree and thin the fruit to the proper amount."

Propagation.

It has long been known, especially in Florida, that some varieties flower profusely but fail to develop any fruits. In other instances, though good crops are produced one season, yet the following year there is no fruit, even though climatic conditions may appear to be identical. This peculiar behavior was not understood until Hume showed that it was due to faulty pollination. In the Journal of Heredity for March, 1914, he writes:

[1] Bull. 99, La. Exp. Sta.

"It was not until 1909 that attention was called to the true cause of barrenness in *D. kaki*, and the year following the cause of sporadic fruitfulness was learned. It was known years before to a few that the flowers of *D. kaki* are of two kinds, pistillate and staminate, but that this fact had any practical bearing on the problem of unfruitfulness did not seem to occur to anyone. More recently the existence of perfect flowers, *i.e.*, those containing both stamens and pistils, was brought to light. These flowers have no practical bearing on the problem, as they are rare, and from some cause or other not yet clearly understood, their ovaries very seldom develop into mature fruit. Since 1909, the results of more than twenty thousand hand pollinations have fairly demonstrated that pollination will cause fruit to set and grow to maturity, when without it no fruit would be produced.

"The fruitfulness of certain trees or groups of trees in some seasons and not in others, even when pistillate flowers were present in goodly numbers each season, can now be explained by the fact that there are certain horticultural varieties of *D. kaki* which produce staminate flowers at irregular intervals. They may be found on certain trees one season and not the next. Many seasons may elapse before they appear again. It may even happen that never again are they produced, or they may be produced every other season. Many combinations of intervals or skips in the production of staminate flowers are possible and probable. A number of them have been observed and noted with references to particular trees. The staminate flowers, when they occur on these trees, are abundantly supplied with pollen and fertilize not only pistillate flowers on the same trees, but through the agency of insects the flowers of many trees surrounding them."

It was evident to Hume, therefore, that a variety was needed which could be depended on for the production of pollen to fertilize the flowers of trees which lacked the male element. The search for such a variety brought several to light, and one of them, the Gailey, is now recommended for planting as a pollinizer. By setting one of these trees to seven or eight of other kinds, productiveness is insured. Hume continues:

"It must be emphasized that the behavior of *D. kaki* in its relation to pollination, or of any other fruit for the matter of that, in any one locality, is no index to its behavior under any other set of conditions. Even though the conditions may appear to be the same, there are differences which we are too dull to detect or too ignorant to understand, but which nevertheless operate on the trees and influence the results. It is a matter of observation that under certain local seasonal and climatic conditions some varieties of *D. kaki* will set good crops of fruit without pollination (seedless of course) while under another set of conditions they do not do so. One season they may bloom freely and set all the fruit the trees should carry and with an equal amount of bloom in another season the same trees may bring no fruit to maturity.

"To sum up conditions as they are at present in the Lower South, and based on numerous observations extending over more than a decade, it is a fact that trees of all varieties of *D. kaki*, in good health and which bloom under normal weather conditions, can be depended upon to bear good crops if pollinated and it is equally true (a few varieties only excepted) that they will not do so if pollen is not provided. In the last two seasons it has been amply demonstrated that all that is necessary is to have staminate flowering trees in proximity to the pistillate ones and bees, wasps, flies and other insects will take care of the problem according to nature's own plan.

"What is the owner of an orchard already planted to do if he desires to place pollinizers in his orchard? It is quite easy to bud over branches here and there in properly placed trees. No preliminary cutting back is necessary, as the buds may be inserted where the bark is anywhere from one to three years old. The work should be done just as the leaves are coming out in the spring, using the ordinary method of shield-budding, and tying the buds in place with waxed cloth. The wraps should be left on about three weeks and as soon as the buds have taken, the branches should be cut back, leaving stubs five or six inches long to which the shoots from the buds may be tied as they grow out. These stubs should be removed at the end of one season's growth."

It may be mentioned that Tane-nashi, normally a seedless variety, fruits well without pollination, and it is thought that Tamopan may do the same.

The question of pollination is probably less important in semi-arid regions, such as California, than in the moist climate of Florida. The prospective grower should in any event use care in the selection of varieties, and satisfy himself as to the need of supplying pollinizers for them, before he undertakes to develop a commercial kaki orchard.

Horticultural varieties of the kaki are commonly propagated by budding and grafting. Several species of Diospyros are used as stock-plants.

The Chinese ring-bud or graft their plants upon the *ghae tsao* (*Diospyros Lotus*) and other species. The Japanese graft upon *D. Lotus,* on the *shibukaki* (an astringent variety of *D. kaki*), and occasionally on seedlings of the common sweet-fruited kaki. Ikeda states that stock-plants must be three years old and that grafting is done in early spring, using cions which have been stored for some days. Sauvaigo says that in southern France the kaki is grafted upon *D. Lotus, D. virginiana* (the common persimmon of the southern United States), and one or two other species. Crown-grafting and other methods are used, and the work is done in autumn or spring.

Hume considers that the best stock-plant for the southern United States is the common persimmon (*D. virginiana*), since it is more vigorous and produces a larger tree than other species. *D. Lotus* has been used in California but its value is not yet fully determined. Frank N. Meyer says of it: "As a stock, this persimmon may give to its grafted host a much longer life than the native American persimmon seems to be able to, for in China all the cultivated persimmons (kakis) grow much older than they do in America. Of some varieties there, one finds trees grafted on *D. Lotus* that are centuries old and still very productive."

Bailey writes: "The best method of propagating Japan persimmons is by collar-grafting upon seedlings of the native species (*Diospyros virginiana*), which are grown either by planting the seed in nursery rows or transplanting the young seedlings from seed-beds early in the spring. The seedlings can be budded in summer, and in favorable seasons a fair proportion of the buds will succeed. Thus propagated, the trees seem to be longer-lived than those imported from Japan. Inasmuch as the native stock is used, the range of adaptation as to soils and similar conditions is very great. As a stock, *Diospyros Lotus* is adapted to the drier parts of the West, where *D. virginiana* does not succeed."

Both cleft-grafting and whip-grafting are employed in Florida. Whip-grafting is considered best if the stock-plants are small. California nurserymen use the same methods and make a point of placing the graft as close to the root as possible.

Kaki trees begin bearing when three or four years old, and, proper attention being given the matter of pollination, produce heavy crops of fruit. Indeed, it is usually necessary in California to thin the fruit lest the trees injure themselves by overbearing. Pollination has been discussed on a previous page.

Picking and shipping.

If the fruit is to be shipped to distant markets, it should be gathered when fully grown but before it has begun to soften. Clippers or picking-shears should be used, and the fruit must be handled carefully, since it is easily bruised. Even when intended for home use it is preferable to gather it before it has begun to soften, and then ripen it in a dry warm room. Fruit treated in this manner is fully as good as that ripened on the tree.

Kakis should be packed for shipment as soon as picked. The six-basket carrier, commonly used for peaches, is employed in shipping them from Florida to northern markets. Each fruit is wrapped in thin paper.

Hume writes:

"Some of the varieties have dark flesh, others light flesh, still others a mixture of the two. The light and dark flesh differ radically in texture and consistency, as well as in appearance, and when found in the same fruit are never blended, but always distinct. The dark flesh is never astringent, the light flesh is astringent until it softens. The dark-fleshed fruit is crisp and meaty, like an apple, and is edible before it matures. Some of the entirely dark-fleshed kinds improve as they soften, like Hyakume and Yeddo-ichi; others are best when still hard, like Zengi. As they are good to eat before they are ripe, it is not so important that the dark-fleshed kinds be allowed to reach a certain stage before being offered to consumers unfamiliar with the fruit. The light-fleshed kinds, and those with mixed light and dark flesh, are very delicious when they reach the custard-light consistency of full ripeness. In some the astringency disappears as the fruit begins to soften, as with Yemon, and in a less degree with Okame, Tane-nashi; in others it persists until the fruit is fully ripe, as with Tsuru. The light-fleshed kinds should not be offered to consumers unacquainted with the fruit until in condition to be eaten. A person who has attempted to eat one of them when green and 'puckery' will not be quick to repeat the experiment. The 'puckery' substance in the immature persimmon is tannin. As the fruit ripens, the tannin forms into crystals which do not dissolve in the mouth, and in this way the astringency disappears."

Various methods are employed to remove the astringency of the light-fleshed kinds and render them fit for eating. The Japanese place them in tubs from which *saki* (rice beer) has recently been withdrawn; the tubs are then closed tightly, and after ten days the fruit is found to have lost its astringency and to be in condition for eating. George C. Roeding of California reports: "A new, simple process of alcohol inoculation has been practiced lately. Pierce the fruits at the bottom several times with a common needle dipped in alcohol, and pack them in a tight box or container lined with straw and with layers between the rows, keeping the box closed for ten days."

Several years ago H. C. Gore and his associates in the United States Department of Agriculture conducted extensive experiments looking toward the perfection of a method for processing kakis commercially. It was found that by placing the fruits in an air-tight drum or container and subjecting them to the influence of carbon-dioxide for a period of two to seven days, the astringency was entirely removed from certain varieties. With other kinds the method was not altogether successful. Since processing must always be tedious, it seems more satisfactory to plant only the sorts which do not require this treatment.

If the orchard comprises several varieties, ripe fruit may be picked in Florida from August to December or even later, and in California from September to December. Hume notes, regarding Florida: "The first persimmon to ripen is Zengi, in August; the whole crop does not come at this time, however, but continues to ripen for sixty days, the seedless ones being larger and later. . . . Early in September come the first Okames, continuing to ripen for a month. Hyakume ripens from September 15 to 30, the bulk of the crop ripening together, which is also true of Yemon, which ripens next. Some fruits of Triumph ripen in September, and it continues to ripen its fruits until December. At any time after the middle of October the whole crop of Triumph may be removed and ripened off the trees. Tane-nashi ripens with Yemon and Hachiya with Okame, Yeddo-ichi early in October, Costata later in the month, and Tsuru latest of all, often hanging on the trees until midwinter." Roeding gives the ripening season of the principal commercial varieties in California as follows: Tane-nashi in September, Hachiya in October, Hyakume in November, and Yemon in December.

Pests and diseases.

There are few insects or fungous diseases which need cause the American kaki-grower serious concern. The Mediterranean fruit-fly (*Ceratitis capitata* Wied.) attacks the fruit in Australia, but this insect has not yet made its appearance in the United States. A few scale insects are occasionally found in the kaki orchards of California and Florida, but the attacks of none have proved serious. Hume writes as follows with reference to Florida:

"The worst enemy of persimmon trees, and the only one worthy of note, is the flat-headed borer (*Dicera obscura*), a native insect. The adult is a hard, metallic beetle, about five-eighths inch in length. It lays its eggs in rough-barked places in the crotches of the tree, or in wounds made in pruning or resulting from injuries of any kind. The young borers hatched from these eggs bore through the bark, work between the bark and wood, later boring into the wood. The larvæ when well grown are about one inch long, white, with broad, flat heads and round bodies. That they are at work in a tree may be known by the discolored bark and by gum oozing from the trunk or branches. Cut away the bark with a sharp knife or chisel and destroy them. Paint the wounds thus made with good, thick, white-lead paint. Carefully paint all wounds when made, and scrape the rough-barked places on young trees. By careful attention to wounds on the trees, they may be prevented from entering, and the trees will live to a good old age."

Varieties.

Although 800 varieties are grown in Japan, Ikeda does not consider more than 90 to be valuable. In the United States the number offered by nurserymen is relatively small. The nomenclature of the horticultural varieties in Japan is somewhat confused, and doubtless nurserymen have multiplied the names. China possesses a considerable number of varieties, but relatively few of them are yet known in the United States.

Japanese writers classify kakis according as they are sweet or astringent. Hume points out that such a classification is not tenable, inasmuch as certain varieties fall in the sweet group when carrying seeds and in the astringent group when seedless. He writes in the Journal of Heredity for September, 1914:

"Based on the difference in flesh coloration under the influence of pollination, kaki may be divided into at least two groups, — first, those which show no change of color of flesh under the influence of pollination, and, second, those in which the flesh of the fruit is darkened under the influence of pollination. Since the change in color in the one case is directly due to pollination and in the other pollination has no effect whatever, we shall refer to those varieties which undergo no change in color as Pollination Constants and those which are light colored when seedless and dark colored when seedy we shall call Pollination Variants. Now, all varieties of *D. kaki* growing in this country or elsewhere may be referred to one or the other of these groups. If varieties which are constantly dark-fleshed whether seedy or seedless should be found, the group of Pollination Constants can then be divided into two groups of light- and dark-fleshed Pollination Constants. It is hardly probable that there are varieties which are dark-fleshed when seedless and light-fleshed when seedy, but if any such should be discovered a similar plan can be followed by dividing the group of Pollination Variants."

The varieties here described are grouped according to this classification. The number is limited to those which are well known in the United States, and are offered here by nurserymen. Regarding their relative merits, Hume says: "Tane-nashi, Triumph, Okame, Yemon, and Yeddo-ichi excel in quality, perhaps in the order named. Okame, on account of its long season, exquisite beauty, and superior quality, is the best for home use and the local market. Hachiya is valued for its immense size and showiness. For market, Tane-nashi and Yemon, of the light-fleshed kinds, and Hyakume and Yeddo-ichi, of the dark-fleshed kinds, are good shippers and desirable; Okame is also good." Fuyugaki, a variety recently introduced by the Department of Agriculture, now promises to excel all other kakis as a market fruit; it is never astringent (hence requires no processing), the appearance and quality of the fruit are both highly satisfactory, and the tree is very productive.

Group of pollination constants

Costata. — Form conical, pointed, somewhat four-angled in transverse section; size medium, length $2\frac{5}{8}$ inches, thickness $2\frac{1}{8}$ inches; surface salmon-yellow; flesh light yellow, dark-colored flesh or seeds seldom occurring; flavor astringent until the fruit is fully ripe, then sweet and pleasant. Ripening season very late.

Tree distinctive in appearance and a rapid erect grower. It does not produce staminate flowers in Florida. The fruit is remarkable for its good keeping qualities.

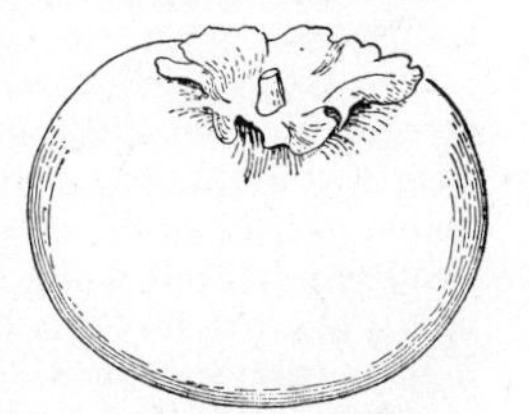

Fig. 47. The Fuyugaki kaki. (× about $\frac{1}{2}$)

Fuyugaki (Fig. 47). — Form oblate; size medium-large, length about 2 inches, thickness about $2\frac{3}{4}$ inches; base with sometimes four creases extending outward from the stem, the calyx reflexed in the ripe fruit; apex depressed, with smooth, regular, shallow basin; surface deep orange-red in color; skin thin, tough; flesh firm, meaty when ripe, deep carrot-orange in color, with minute, widely scattered dark specks; flavor sweet, with no astringency even in the unripe fruit; seeds $\frac{3}{4}$ inch long, few.

Recently introduced from Japan by the United States Department of Agriculture. Hume says: "It keeps well, and in quality is one of the best. We believe this variety will surpass all other Japan persimmons so far introduced as a market fruit. It can be placed on the market while still hard, and can be eaten without waiting for the fruit to soften."

Fig. 48. The Hachiya kaki. (× ½)

Hachiya (Fig. 48). — Form oblong-conical, with a short point at the apex; size very large, length 3¾ inches, thickness 3¼ inches; surface bright orange-red, with occasional dark spots and rings near the apex; flesh deep yellow, sometimes having a few dark streaks in it; flavor astringent until the fruit is fully ripe, then rich and sweet; seeds present. Ripens midseason to late.

Tree vigorous in growth, with a tendency to bear fruit in alternate years. It does not produce staminate flowers in Florida. The fruit is large and handsome. Said to be one of the principal varieties used in Japan for drying.

Ormond. — Form conical; size small to medium, length 2⅝ inches, thickness 1⅞ inches; base rounded, with the calyx reflexed; apex sharp, not creased, or only slightly so; surface smooth, bright orange-red, covered with a thin bloom; skin thin, tough; flesh orange-red, meaty, or jelly-like in the fully ripe fruit; seeds large, long, pointed. Ripening season late (December in Florida).

A fruit of good quality, and one which keeps well.

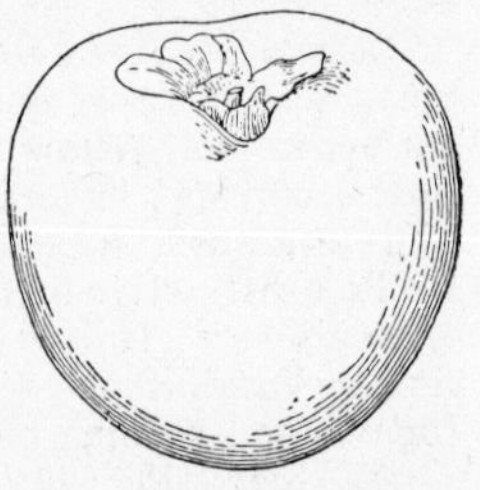

Fig. 49. The Tane-nashi kaki, one of the principal varieties used in Japan for the production of dried kakis, and now grown commercially in the United States. (× about ½)

Tamopan. — Form broadly oblate with a constriction around the middle; size large, weight sometimes 16 ounces, diameter 3 to 5 inches; surface smooth, orange-red in color; skin tough and rather thick; flesh meaty, light colored; flavor astringent until the fruit is fully ripe, then rich and sweet; seedless.

Introduced from China by the United States Department of Agriculture. The tree is a strong, upright grower.

Tane-nashi (Fig. 49). — Form roundish conical, very symmetrical; size large to very large, length 3⅓ inches, thickness 3⅜ inches; surface very smooth, light yellow to bright orange-red; flesh yellow, soft; flavor sweet and pleasant; seedless. Ripens early.

The tree is vigorous, prolific, and self-fertile, but it has shown a tendency in California to bear in alternate years. Extensively used in Japan for drying and considered a valuable market variety in the United States. Perhaps the most highly esteemed of the light-fleshed kinds.

Triumph. — Form oblate; size medium; surface yellowish to deep orange-red; skin thick; flesh yellowish red, translucent, soft and juicy; flavor astringent until the fruit is fully ripe, when it becomes sweet and pleasant; seedless or with as many as 5 to 8 seeds. Ripens in Florida from September to December.

The tree does not produce staminate flowers in Florida. A fruit of good quality, recommended for home use and for market.

Tsuru. — Form slender, pointed; size large, length $3\frac{3}{8}$ inches, thickness $2\frac{3}{8}$ inches; surface bright orange-red; flesh orange-yellow; flavor astringent until the fruit is fully ripe, when it becomes sweet and pleasant. Ripens very late.

Tree vigorous and productive, but does not produce staminate flowers in Florida.

Group of pollination variants

Gailey. — Form oblong-conical, sharp at the apex; size small; surface dull red, pebbled; flesh meaty, firm, and juicy; flavor pleasant.

This variety regularly produces staminate flowers every year, and is recommended for planting as a pollin zer n conjunction with the larger- and better-fruited sorts. One tree of Gailey should be planted to seven or eight of other varieties except Tane-nashi.

Hyakume. — Form roundish oblong to roundish oblate, always somewhat flattened at both ends; size large to very large, length $2\frac{3}{4}$ inches, thickness $3\frac{1}{8}$ inches; surface light buff-yellow, marked with rings and veins near the apex; flesh dark brown, crisp and meaty; flavor sweet, not astringent even while the fruit is still hard. Ripens midseason.

The tree is vigorous and productive, but never produces staminate flowers in Florida. One of the standard commercial varieties in California.

Okame. — Form roundish oblate, with well-defined quarter-marks, and the apex not depressed; size large, length $2\frac{3}{8}$ inches, thickness $3\frac{1}{8}$ inches; surface orange-yellow, changing to brilliant carmine, with a thin bloom which gives it a waxy translucent appearance; flesh light colored, brownish around the seeds, of which there are several; flavor astringent until the fruit begins to ripen, when it becomes sweet and pleasant. Ripens rather early.

The tree is vigorous in growth and a good bearer. It bears staminate flowers sporadically in Florida. The fruit is excellent in quality.

Yeddo-ichi. — Form oblate; size large, length $2\frac{1}{2}$ inches, thickness

3 inches; surface smooth or undulating, dark orange-red, and covered with a distinct bloom; flesh dark brown, tinged purplish; flavor sweet and rich, not astringent even while the fruit is still hard.

A fruit of excellent quality.

Yemon. — Form oblate, somewhat four-angled; size large, length $2\frac{1}{4}$ inches, thickness $3\frac{1}{4}$ inches; surface light yellow, changing to reddish and mottled with orange-yellow; flesh dull red-brown, except in occasional light-fleshed specimens; few-seeded or seedless; flavor sweet and pleasant after the fruit begins to soften.

A fruit of excellent quality.

Zengi. — Form round or roundish oblate; size small, length $1\frac{3}{4}$ inches, thickness $2\frac{1}{4}$ inches; surface yellowish red; flesh dark-colored; flavor sweet, even in the unripe fruit; seeds present. Ripens very early.

The tree is vigorous in growth and prolific in fruiting.

The Black Sapote (Plate XXII)

(*Diospyros Ebenaster*, Retz.)

Outside of Mexico the black sapote is little known; in that country, however, it is one of the popular fruits and is grown from sea-level up to elevations of 5000 or even 6000 feet. Unfortunately, the dark color of the flesh makes the fruit somewhat unattractive to those not familiar with it, but its large size, relative freedom from seeds, and its good quality make it a worthy tropical rival of the subtropical kaki or Japanese persimmon.

In the Mexican lowlands the black sapote, if grown on deep, rich, and moist soil, becomes a large and handsome tree, ultimately reaching 50 or 60 feet in height. In regions where the climate is cool or the soil is not favorable, it may not grow higher than 25 or 30 feet. The branchlets are dark colored, and the leaves elliptic or oblong in outline, usually obtuse at the apex, commonly 4 to 8 inches long, and bright green and shining. The flowers are small and white and resemble those of the kaki. They are polygamous, *i.e.*, some of them possess both stamens and pistils and others are staminate. The oblate fruit, which has a conspicuous green calyx around the stem

PLATE XIX. The sapodilla (*Achras Sapota*).

and is somewhat obscurely ribbed or lobed, is 2 to 5 inches in diameter and olive-green in color. The pulp which lies within its thin skin is soft, unctuous, dark chocolate brown in color, and of sweet flavor similar to that of the kaki but scarcely so pleasant. The seeds, one to ten in number (occasionally none), are oval, compressed, and about $\frac{3}{4}$ inch long.

William Philip Hiern, a recent monographer of the Ebenaceæ, following the botanist Manuel Blanco, considers the black sapote to be indigenous in the Philippine Islands. Other authorities, however, hold that its native home is in Mexico, and perhaps also in the West Indies. Many American plants were carried to the Philippines in the early days by the Spanish galleons which plied between Acapulco and Manila, and conversely, certain Philippine plants were brought to America. Elmer D. Merrill[1] observes regarding the black sapote: "Rarely cultivated, flowering in March; of local occurrence in the Philippines. Introduced from Mexico at an early date, and apparently formerly much more common than now." The existence of an Aztec name, *tliltzapotl* (if Manuel Urbina is correct in believing that this name was applied by the Aztecs to *Diospyros Ebenaster*) would argue an ancient cultivation in America, though it would not necessarily indicate that the species is indigenous here. But on the whole, the evidence seems to weigh heavily in favor of an American, as opposed to an Asiatic, origin.

At the present time, the black sapote is cultivated on a very limited scale in the West Indies and in Hawaii, and rarely in the East Indies. It has been planted at Miami, Florida, where it gives promise of being quite successful. It is sometimes injured by frost in that region, but danger from this source seems to be no greater than with the mango. Although many seedlings have been planted in California, they have failed to survive the winters, even when grown in the most

[1] Flora of Manila.

protected situations. The common name of the fruit in Porto Rico is *guayabote* or *guayabota;* in Hawaii it has been called black persimmon; while the usual terms in Mexico are *zapote negro* and *zapote prieto.*

The black sapote is eaten fresh. It is more highly esteemed by Europeans when the pulp is beaten with a small quantity of orange or lemon juice and served as a dessert. It should be chilled thoroughly before serving.

In its climatic requirements the species must be considered tropical, yet it will succeed in regions occasionally subject to temperatures of 28° or 30° above zero. Young plants, however, are killed by freezing temperatures, and for this reason it is necessary in Florida to protect them during the first few winters. In Mexico the species grows both in regions subject to heavy rainfall and those which are extremely dry, but in the latter it requires abundant irrigation. It is most commonly grown at elevations from 0 to 2000 feet, which indicates that it prefers a warm climate. It prefers a deep, moist, sandy loam, but has made fairly good growth in Florida on shallow sandy soil.

Like other fruits, the black sapote is grown in the tropics as a dooryard tree and is not often planted in orchard form. Little is known, therefore, regarding the cultural methods which will best suit it. Young trees are set in the open ground when one to two feet high, and should be spaced (if in the tropics and on deep soil) not closer than 40 feet, or 25 feet if in a subtropical climate (such as that of Florida) and on poor soil. Propagation is usually effected by means of the seeds, which retain their viability for several months if kept dry. They should be sown ½ inch deep in flats or pots of light loamy soil, and will germinate in about a month if the weather is warm. When three inches high, the plants may be transferred to three-inch pots. Their growth is slow and they require one to two years to reach suitable size for transplanting to the open ground.

P. J. Wester has found that the species may be propagated by shield-budding in much the same manner as the avocado and the mango. Using this method it is possible to perpetuate choice varieties which originate as chance seedlings. Wester says briefly: "Use mature, but not green and smooth, petioled budwood; cut the buds about an inch and a half long; insert the bud at a point where the stock is green or brown before it becomes rough."

Seedling trees do not come into bearing until they are five or six years of age. Even more time than this has been required in Florida. Mature trees usually bear regularly and heavily. The ripening season in the Mexican lowlands is July to September, somewhat later in the *tierra templada* or region which lies between 2500 and 4000 feet. If taken from the tree when mature and shipped immediately, the fruit may be sent to distant markets; but once it has softened (usually three to six days after it is picked), it is difficult to handle because of its thin delicate skin and the large mass of soft pulp.

No horticultural varieties have as yet been established. Seedlings differ noticeably in the size and character of their fruits, and it will be worth while to search out the best ones and propagate them by budding. Fruits $1\frac{1}{2}$ pounds in weight are seen at Tehuantepec, State of Oaxaca, Mexico.

THE MABOLO

(*Diospyros discolor*, Willd.)

Like the durian and the santol, the mabolo is a Malayan fruit little known outside its native area. It is a medium-sized tree with oblong-acute leaves 4 to 8 inches long, shining above and pubescent beneath. The fruit is round or oblate in form, about 3 inches in diameter, with a thin, velvety, dull red skin, and whitish, aromatic, rather dry flesh which adheres to the four to eight large seeds. P. J. Wester writes: "There is also a variety, rarer than the red, with yellowish to light brown fruits, the flesh of which is cream colored and sweeter, and less astringent. Trees bearing regular crops of seedless fruits are known in the Philippines. The main season of the mabolo extends from June to September, but scattered fruits are found at practically all seasons of the year. It is of medium vigorous growth and makes a desirable ornamental shade tree. It is indigenous to the Philippines and is fairly well introduced throughout the eastern tropics." It is not cultivated in the West Indies or elsewhere in tropical America, although a few trees may have been planted in botanic gardens and private collections.

CHAPTER XIII

THE POMEGRANATE AND THE JUJUBE

THE pomegranate and jujube are not closely related botanically, but the cultural requirements are similar. The pomegranate is the only genus of its family (the Punicaceæ), while the jujube (genus Zizyphus) is one of 40 or 50 genera of the Rhamnaceæ or Buckthorn family.

THE POMEGRANATE (Plate XXII)

(*Punica Granatum*, L.)

"Eat the pomegranate," sententiously said the prophet Muhammad, "for it purges the system of envy and hatred." Far earlier than in the days of Muhammad, however, was this fruit esteemed in the Orient. King Solomon possessed an orchard of pomegranate bushes; and when the Children of Israel, wandering in the Wilderness, sighed for the abandoned comforts of Egypt, the cooling pomegranates, along with figs and grapes, were remembered as longingly as the fleshpots.

It is with the grape and the fig that the pomegranate has been associated since the earliest times; but while in the East it still vies with them in popularity and importance, in America it occupies a minor position. Probably this is due: first, to the abundance here of other good fruits; and, secondly, to something in the character of the pomegranate which makes it particularly agreeable to inhabitants of hot arid regions. For this latter reason it might appeal in this country to a relatively small number; but even in the desert valleys of California and Arizona, where it should be most acceptable, Americans have not yet learned to appreciate it fully.

About 150 acres are now planted commercially to pomegranates in California. The total production in the United States, according to the census of 1910, is about 150,000 pounds.

If allowed to develop naturally, the pomegranate becomes a bush 15 to 20 feet in height. By training, it can be made to form a tree, usually branching close to the ground. It is semi-deciduous or deciduous in habit. The leaves are lanceolate to oblong (sometimes obovate) in form, obtuse, about 3 inches long, glossy, bright green, and glabrous. The handsome brilliant orange-red flowers are axillary, solitary, or in small clusters, and borne toward the ends of the branchlets. The calyx is tubular, persistent, five- to seven-lobed; the petals, five to seven in number, are lanceolate, inserted between the calyx-lobes. The ovary is embedded in the calyx-tube, and contains several locules in two series, one above the other.

The fruit is globose or somewhat flattened, obscurely six-sided, the size of an orange or sometimes larger, and crowned by the thick tubular calyx, giving an ornamental effect. It has a smooth leathery skin, which in the ripe fruit ranges from brownish yellow to red in color. Thin dissepiments divide the upper portion into several cells; below these, a diaphragm separates the lower half, which in its turn is divided into several cells. Each cell is filled with a large number of grains, crowded on thick spongy placentæ; these grains, which are many-sided and about ½ inch long, consist of a thin transparent vesicle containing reddish juicy pulp surrounding an elongated angular seed. The pulp is delightfully subacid in flavor.

Alphonse DeCandolle reached the conclusion that the "botanical, historical, and philological data agree in showing that the modern species is a native of Persia and some adjacent countries," an opinion which is generally accepted at the present day. The cultivation of the pomegranate, which began in prehistoric times, was extended, before the Christian era, westward to the Mediterranean region and eastward into China. At the present time it is a common fruit in India, Afghanistan, Persia, Arabia, and near the shores of the Mediterranean both in Europe and Africa, more particularly the latter. In America it is scattered from the southern United States to Chile and Argentina, probably reaching its greatest perfection in the arid regions of California, Arizona, and northern Mexico.

Throughout tropical America the plant is common in gardens and dooryards, but in many places it is grown more for its ornamental value than for its fruit. In humid climates the fruit is inferior in quality.

The ancient Semitic name *rimmon* has been adopted by the Arabs as *rumman*, and later the Portuguese *romã* or *roman* was formed from it. From the early Roman names *malum punicum* (apple of Carthage) and *granatum* have been taken the botanical name *Punica Granatum*, L., under which the species is known scientifically, and the common name *granada*, used throughout Spanish-speaking countries. From this same source, evidently, are the French *grenade* and the German *granatapfel*. Of the several names current in Hindustan *anar* is the commonest; *darimba* is the Sanskrit name. The Persians know the pomegranate-flower as *julnar*.

The fruit is peculiarly refreshing in character, hence is much eaten in hot countries. It is also used to prepare a cooling drink known as *grenadine;* but the beverage dispensed under this name in the Mediterranean region and tropical America commonly is colored and flavored artificially. The roots of the plant and the rind and seeds of the fruit are used medicinally in the Orient. The classical Arab lexicographers define the pomegranate as: "a certain fruit, the produce of a certain tree, well known; the sweet sort thereof relaxes the state of the bowels, and cough; the sour sort has the contrary effect; and that which is between sweet and sour is good for inflammation of the stomach, and pain of the heart. The pomegranate has six flavors, like the apple, and is commended for its delicacy, its quick dissolving, and its elegance."

In the United States the fruit has been more highly valued for its decorative effect than for other purposes. It is used on banquet-tables and as an adjunct to fruit salads. The principal chemical constituents of the pulp, as determined in Hawaii by Alice R. Thompson, are as follows: Total solids 17.52 per cent, ash 0.73, acids 0.13, protein 0.52, total sugars 16.07, fat 0.30, and fiber 0.32.

While the pomegranate can be grown throughout the tropics and subtropics, it produces good fruit only in semi-arid regions where high temperatures accompany the ripening season. In this respect it somewhat resembles the date-palm, although it is less exacting as regards heat than the latter and more frost-resistant. Like the palm, it requires plenty of water at the root, if good fruit is to be produced in abundance; nevertheless, it is able to withstand long periods of drought. Minimum temperatures of 15° or 18° above zero may not injure the plants severely. The sour varieties are said to be hardier than the sweet. No climate is too hot for the pomegranate, provided it receives ample water.

In regard to soil, the species is not exacting but it is considered to succeed best on deep, rather heavy, loams. It is on soils of this type that the excellent pomegranates of Mesopotamia are grown. A small amount of alkali is not injurious, nor does excessive moisture seem as detrimental to the pomegranate as to many other fruit-trees. George C. Roeding remarks: "I have used the pomegranate for a number of years in depressions in my vineyard where the ground was so damp for a good part of the year that grape vines invariably died. The pomegranate luxuriates in these spots."

When planted in orchard form, the bushes should be set 12 to 18 feet apart. Pomegranates are often planted in hedgerows, and under such conditions are ordinarily not more than 6 or 8 feet apart; but close planting and the permitting of development of suckers from the base of the plant naturally are detrimental to fruit-production.

Cultural practices in California have been described by Robert W. Hodgson, in "The Pomegranate."[1] The following extracts are taken from his publication:

"Pomegranate trees should be planted as early in the spring as the ground can readily be worked and is not too wet. However, as the pomegranate starts growth comparatively late in the spring, late planting is not accompanied by such disastrous effects as with the stone fruits. The best results seem to be obtained by planting in February and March.

"If the soil is in good condition, little care other than irrigation and two or three cultivations during the season is needed after planting. In older orchards the soil should be stirred at least once a month during the growing season.

"Some growers irrigate but little, while others apply as much as they give their citrus orchards. . . . If we set the water requirements of orange trees at fifty inches, including the rainfall, we may consider that the pomegranate requires thirty-five to forty inches. Some of this water comes as rain in the winter season. The rest is usually applied in two to five irrigations, distributed through the growing season. Some growers irrigate until July only. Others apply water once a month until September. The furrow system is used almost universally.

"To prune intelligently, one must consider the fruiting habit, and habit of growth of the tree. The pomegranate is a vigorous grower, sending up each year a number of shoots from the root which gives the plant a bush form unless otherwise trained. The fruit is borne terminally on short spurs produced on slow-growing mature wood. This wood bears for several years, but as the tree increases in size this wood loses its fruiting habit, which is assumed by the younger growth. Little or no fruit is produced in the interior of the tree.

"Bearing this in mind, it can clearly be seen that a heavy pruning, especially shortening in of the older wood, will greatly reduce the crop for the next two or three years.

"When the tree is planted it should be cut back to a whip at about 24 to 30 inches from the ground. As the buds put out and shoots are produced, these should be selected and thinned out to three or five or

[1] Bull. 276, Calif. Agr. Exp. Sta.

more scaffold branches which should be pinched back to make them stocky. These should be spaced some distance apart, — the lowest at least eight or ten inches from the ground, — and symmetrically arranged on the stem. The following winter the scaffold branches should be shortened to about three-fifths of their length. In the spring the new shoots arising from the scaffold branches (primary branches) should be restricted to two or three per limb. The main stem and frame limbs should be kept free from suckers at all times. The aim in pruning while the tree is young is to induce the formation of a stocky, compact framework. This should be accomplished by the end of the second or third year.

"After the framework is established all the growth is left and the tree comes into bearing. From this time to the age of 15 or 20 years, the tree increases slowly in size and yield. Pruning after the third year should be confined to a regular removal of all sucker growth arising from the root, and interfering branches as well as dead brush, and an annual thinning out or removal of some of the older branches. This should be done after the leaves fall in winter."

Propagation of the pomegranate is effected by means of seeds, cuttings, and layers. Seeds can be grown readily, but named varieties cannot be reproduced in this manner. Hodgson writes:

"The only method of propagation used commercially is by hardwood cuttings. These will grow in the open ground about as readily as willow cuttings. The stand obtained is very satisfactory and the method used very simple. In February or March hardwood cuttings ten to twelve inches long and one-quarter to a half inch in diameter are cut, usually from the shoots or suckers, and are planted in the open ground in nursery rows. These rows are ordinarily three feet apart and the cuttings spaced eight to ten inches in the row. The cuttings should be thrust almost their entire length into the earth, leaving only the top eye exposed. This eye is forced out and grows into the tree. Cuttings of this sort grow thriftily and are often ready for transplanting to the orchard or hedge by the following spring, although they are frequently left in the nursery row two seasons. Hardwood cuttings are sometimes cut in the fall and callused in sand over winter, then set out in early spring. This may result in a little earlier growth and consequently a larger tree that season, but is not necessary to insure striking root."

When grown under good cultural conditions, the plants come into bearing at three or four years of age. The yield is influenced by the character of soil and the method of pruning followed. On sandy soils light crops must be expected; and if suckers are allowed to develop unhindered, or if the mature plant is pruned of its fruit-bearing wood, little fruit can be produced. It is necessary to emphasize the importance of pruning in connection with pomegranate culture. A properly grown tree of mature size may yield 200 to 400 pounds of fruit annually, but one which has been subjected to incorrect pruning, or has a number of primary shoots growing from the base instead of a single trunk with laterals rising from it, will certainly give no such results.

Regarding the best methods of picking and handling the fruit, Hodgson says:

"On account of the common habit of splitting, the fruit of most varieties must be picked before fully mature. . . . Some trees will hold their fruit until winter and never show any splits.

"Fortunately, the pomegranate is one of those fruits which, after reaching a certain degree of maturity, continues to ripen in cold storage, where it will keep in excellent condition for five or six months. Not only does it ripen, but the quality is improved, the flavor becoming richer and more vinous. The rind shrinks and becomes thinner and tougher; the amount of rag decreases; and the seed coats appear to become more tender and edible. Several pickings should be made, the first about the first week in October, and two or three others at weekly intervals.

"Pomegranates are very securely attached to the fruiting wood by thick, strong stems, and should be clipped rather than pulled. . . . After sizing, the fruit is wrapped in tissue paper and packed. The commercial package used is the orange half-box. . . . The sizes run from 24 to 110 per box."

On this same subject Roeding[1] notes: "On account of its rather thick skin the fruit will withstand quite a lot of abuse. The one point to guard against is to pick the fruits before they are rained on, for when this occurs many of them will split, making them unfit for shipment. After they are gathered, the fruits, if stored in a cool, dry place, will keep for months; the

[1] Roeding's Fruit Growers' Guide.

skin loses its striking lustre, and the fruit shrinks some, but this in no way impairs the quality or flavor of the pulp."

The pomegranate has several enemies, both insect and fungous. In India, the larvæ of the anar butterfly (*Virachola isocrates* Fabr.) infest the fruit. A similar insect is the pomegranate butterfly of Egypt (*Virachola livia* Klug). Another lepidopterous pest, *Cryptoblades gnidiella* Miller, is also reported from Egypt. In California much damage has been caused by a disease known as heart-rot. "When an infected fruit is opened, the central cavity is found filled with a disgusting mass of decayed arils, black in color and disagreeable in odor. The decay usually shows no connection with the rind, being entirely surrounded by sound flesh. . . . Infection takes place in the blossom and progress of the fungus may be traced by a thread-like black line of decay from the stigma down through the stylar canal into the interior of the fruit." No remedy has been found for this disease up to the present time.

Several insects occasionally attack the tree in California, but none is said to be a serious pest. In Hawaii, the dreaded Mediterranean fruit-fly (*Ceratitis capitata* Wied.) is known to infest the fruit.

The varieties of the pomegranate are fairly numerous. Ibn-al-Awam, a Moor who wrote in the thirteenth century, described about ten kinds known in southern Spain at that time. At Baghdad, pomegranates are usually divided into three groups or classes, viz.: *ahmar* (red), *aswad* (black), and *halwa* (sweet). Several named varieties are known in Mesopotamia in a limited way, Salimi being considered the best. Ragawi, Halu, Aswad, and Amlasi are other forms.

The late Frank N. Meyer, describing the pomegranates of the Shantung Province of China, says: "There are dwarf varieties that grow only a few feet tall and bear but a few small scarlet fruits, while others grow from 15 to 30 feet tall and bear fruits one or more pounds in weight. There are varieties that have a white rind and are red inside and other kinds that are white both outside and inside."

Numerous varieties have been introduced into the United States from the Orient. Some of them are promising, but none is yet established in the trade. The following are the three principal varieties planted in those parts of the country where pomegranate culture is conducted commercially:

Wonderful. — Form oblate; size very large, the diameter sometimes 5 inches; base flattened; apex rounded, crowned with the prominent calyx; surface smooth, glossy, deep purple-red in color; rind medium thick, tough; flesh deep crimson in color, juicy, and of delicious vinous flavor; seeds not very hard.

Origin not definitely known; it was propagated at Porterville, California, in 1896, from a cutting obtained from Florida. Because of its vigor of growth, productiveness, and the excellent quality of its fruit, it has become the favorite commercial variety in California.

Paper-Shell. — Form globose; size large; surface glossy, pale yellow washed with pink; rind very thin; flesh bright red in color, juicy, and of pleasant flavor; seeds fairly tender.

Origin not definitely known; it was introduced into California from the southeastern United States. It is not so vigorous in growth as Wonderful, nor is the fruit so attractive, but it is productive and the fruit has good shipping qualities.

Spanish Ruby. — Form globose; size large; surface glossy, bright red in color; rind moderately thick; flesh crimson in color, juicy, of sweet aromatic flavor; seeds fairly tender.

Syns. *Purple, Purple Seeded.* A variety introduced into California from the southeastern United States. Commercially it is not important.

The Jubube

(Zizyphus spp.)

"The jujube," writes David Fairchild, "is one of the five principal fruits of China, and has been cultivated for at least 4000 years." It is only in the large-fruited Chinese varieties that the jujube is seen at its best. The inferior fruits which have been grown in southern Europe, Arabia, and northern India either represent a different species from those of China, or are varieties which have not been so highly improved by cultivation and selection.

Pliny recounted that the jujube was brought from Syria to Rome by the consul Sextus Papinius, towards the end of the reign of Augustus. It has, therefore, been known in southern Europe for more than 2000 years. It reached America some time during the nineteenth century, but only in the form of seedlings which yielded fruit of poor quality. With the introduction of the grafted Chinese varieties, obtained in 1906 and subsequent years by the United States Department of Agriculture, the jujube has become a fruit-tree worthy of the serious attention of horticulturists, and this it is now receiving throughout the southern and western parts of the United States.

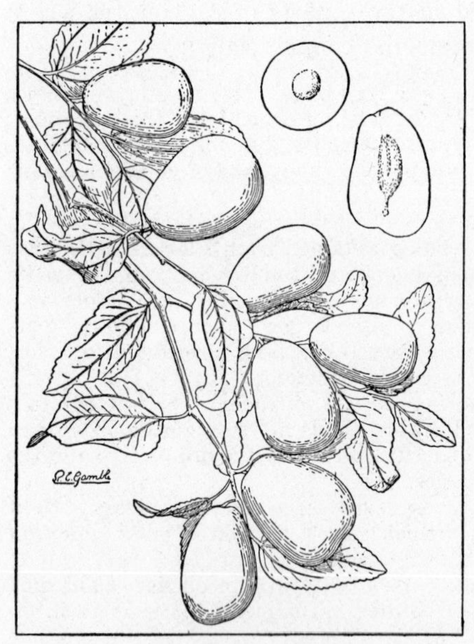

FIG. 50. The lang tsao, or "melting jujube" (*Ziziphus Jujuba*), from the Province of Shensi, China, now grown in California. The Chinese varieties of the jujube are better than those of other countries. (× ⅓)

The botany of this fruit is decidedly confused. Two species are cultivated in the Orient, differing but little from each other in botanical or horticultural characteristics. The Chinese jujube (Fig. 50) is considered to be *Zizyphus Jujuba*, Miller (*Z. vulgaris*, Lam., *Z. sativa*, Gaertn.), and the Indian jujube, *Zizyphus mauritiana*, Lam. (*Z. Jujuba*, Lam.). The principal difference between them seems to be that the leaves of the first-named are glabrous while those of the second are tomentose beneath. Further study will be required to show the proper classification of many cultivated forms.

The jujube is a small, somewhat spiny tree reaching a height of 25 or 30 feet. Its leaves are alternate, three-nerved, elliptic-ovate, ovate, or suborbicular in outline, commonly 1½ to 3 inches in length. The small greenish flowers are produced upon slender deciduous branchlets, or occasionally upon the old wood. The fruit is a small drupe, elliptic or oblong to spherical in form, from ½ to 2 inches in length, with a thin dark brown skin, and having whitish flesh of crisp or mealy texture and sweet agreeable flavor, inclosing a hard two-celled stone, elliptic to oblate in form and rough on the surface.

In searching botanical literature for data regarding the history and distribution of the jujube, it is impossible to determine, in many cases, whether *Z. Jujuba* or *Z. mauritiana* is the species discussed. One or the other (probably both in some instances) is cultivated in China, in the Philippines, through the Malayan region to India and Africa, and westward through Afghanistan, Persia, Arabia, and Asia Minor to the Mediterranean coast of France, Spain, and North Africa. In China the general name is *tsao;* in India *Z. Jujuba* is called the common jujube, *anab, unnab,* while *Z. mauritiana* is called the Indian jujube, *ber, bor,* and the like. In Arabia the common term for one species is *nabk.*

The late Frank N. Meyer, to whom we are indebted for many fine Chinese varieties of this fruit, observed, during his explorations in China, that the jujube could be used in several different ways. The fresh fruits of some varieties are excellent to eat out of hand. Dried, they resemble dates in character. Jujubes are sometimes boiled with millet and rice; they may be stewed or baked in the oven; they are used, raisin fashion, to make jujube-bread; and they are turned into *glacé* fruits by boiling them in honey and sugar sirup. Meyer particularly lauds the mi-tsao, or honey-jujube. "To prepare this," he says, "the Chinese take large, sound, dried fruits and boil them thoroughly in sugared water, after which they are taken out and dried in the sun or wind for a couple of days. When sufficiently dry they are given a slight boiling again and are partly dried. When dry enough to be handled, the skin is slightly slashed lengthwise with a few small knives tied together. Then the fruits are given a third boiling; now, however, in a stronger sugar water, and for the best grades of honey-jujube honey is added. When this process is finished they are spread out to dry, and when no longer sticky are ready to be sold."

A chemical analysis of the Chinese jujube made by the Bureau of Chemistry at Washington showed it to contain: Total solids 31.9 per cent, ash 0.73, acids 0.29, protein 1.44, total sugar 21.66 (sucrose 9.66, invert sugar 12.00), fat 0.21, hydrolyzable carbohydrates 2.47 and fiber 1.28.

Regarding the climatic and soil requirements of the jujube, Fairchild[1] writes:

"No weather appears to be too hot for it, and so far as resistance to cold is concerned, it has withstood temperatures of 22° F. without injury. Just how much lower winter temperatures it will withstand has not yet been determined. The range of territory, however, over which it is likely to prove a success as a fruit tree will probably be limited more by the length of the summer season than by the severity of the winter. The whole Southwest, with the exception of the elevated areas where cold summer nights occur . . . is a promising region in which to test the jujube. It enjoys brilliant sunshine, dry weather, and long, intensely hot summers, and although it will form good sized trees under other conditions, it appears to require these climatic factors to make it fruit early in life, regularly, and abundantly.

"As regards soil conditions, it appears to withstand slight amounts of alkali and to thrive with special vigor on the loess, or wind-drifted soil formations of China. . . . Under irrigation in northern California, and without irrigation in Central Texas, the trees have grown luxuriantly and fruited abundantly. In the warm humid region of Maryland, seedling trees have grown well, but fruited sparingly and irregularly. In Georgia, old seedling jujubes have fruited well."

[1] Journal of Heredity, Jan., 1918.

PLATE XX. *Upper*, the rambutan and other fruits; *lower*, a basket of green sapotes.

R. L. Beagles[1] says: "The jujube has endured a temperature of 13° F. at the Chico Station without any perceptible injury; it also withstands extreme heat, a temperature of 111° F. producing no apparent bad effects on trees and young grafted plants. . . . The tree starts into growth very late in the spring, which eliminates any danger from frost, and makes it a sure cropper. The fruits ripen in October and November."

Meyer[2] writes regarding the cultivation of the jujube in China:

> "In general, jujubes are grown in small groves or as single trees, but here and there one also meets regular orchards of them, covering perhaps 10 or 20 acres. In some localities the farmers plant them in rows through the fields. It seems that planted in this way, at a distance of five to ten feet apart, they produce the largest quantity and best quality of fruit. When in regular orchards the distance apart is from 15 to 25 feet, depending upon the variety and upon the personal preference of the planter.
>
> "The farmers, here and there, also have the practice of ringing their trees every year, claiming that thereby they considerably increase the crop. The jujube is about the only fruit tree around the roots of which the soil is not regularly cultivated, because the yield is found to be just as large without this work as with it."

Propagation is effected by seeds, grafting, root-cuttings, and one or two other means. Meyer reports regarding the methods employed in China: "As the varieties do not come true to seed, the trees are mostly propagated by the suckers which are nearly always found at their bases. Root cuttings can also be taken. Some varieties, however, do not readily produce suckers, and root cuttings are not successful. Then the Chinese resort to grafting the cions on wild stock. This grafting practice, however, seems to be confined to only a few localities, where the growers are men of considerable experience."

[1] California Citrograph, Oct., 1917.
[2] Bull. 204, Bur. Plant Industry.

The most satisfactory method of propagating the Chinese varieties in California has been whip-grafting. Seedling jujubes are used for stock-plants. These are easily grown, although the seeds (which are sown in drills in the open ground) are slow to germinate and it takes two years to produce a good plant. At one year of age many of them will be large enough to graft, but it is better to leave them until the second year.

J. E. Morrow, who has had experience in propagating the jujube at the United States Plant Introduction Field Station at Chico, California, notes that plants grafted in February sometimes grow to a height of three or four feet before the end of the year and mature a few fruits. He says further:

"The jujube root is one which does not like to be disturbed, and for quick results, and where climatic conditions will permit, I would advocate field-grafting on two-year-old roots. The cions are inserted close to the root, and covered with soil, which should not, however, be over one inch in depth above the top of the cion, so that when the ground settles after a hard rain the young plant will still be able to force its way through it.

"The argument in favor of bench-grafting is this: it may be done when the soil is too muddy or cold to permit outside work. The stock-plants are cut off just above the root, or the larger roots themselves are used as stocks. Upon these a cion about four inches long and of the diameter of a lead penicl is whip-grafted, and wrapped with raffia. A wedge-graft may be used if the stock is much larger than the cion. The grafts are then packed in boxes, between layers of moistened cedar or redwood sawdust or 'shingletow.' The box should be kept where temperature remains between 40° and 50°. In about a month calluses should have formed, and the grafts may be planted in the field. Grafting may be done in California any time in February or March, and the plants should go into the field not later than April 1. Cions may be cut between the first of December and the first of February, and stored until wanted for use."

The jujube is precocious and prolific in fruiting, and rarely fails to produce a good crop. Meyer observed in China that the plants begin to decline in vigor and productiveness after twenty-five or thirty years, and rarely live more than forty years. There are many varieties known in China, and not a few in other countries. Meyer has described ten of the best Chinese kinds in Bulletin 204, quoted above: most of these have been introduced into the United States and some have already produced fruit here. The best are considered to be the Yu, the Mu shing hong, and the Lang.

CHAPTER XIV

THE MANGOSTEEN AND ITS RELATIVES

Of the Guttiferæ or Garcinia family few plants are grown for fruit, and the mangosteen is the chief one. It is a tropical family of nearly 400 species and 30 to 40 genera. The family yields drugs, gums, and resins.

The Mangosteen (Plate XXIV)

(*Garcinia Mangostana*, L.)

Since the days when early voyagers returned to Europe with more or less fabulous stories of the wonders of the East, the mangosteen has received unstinted praise. It has been termed the "Queen of Fruits," "the finest fruit in the world," and Jacobus Bontius, who compared it to nectar and ambrosia, said that it surpassed the golden apples of the Hesperides and was "of all the fruits of the Indies by far the most delicious." Bontius was warranted in his enthusiasm. The combination of beautiful coloring with delicate enticing flavor entitles the mangosteen to rank above all other fruits of the Asiatic tropics. Indeed, it is doubtful whether the world possesses another tropical fruit which is its equal. It compares favorably with the most delicately flavored fruits of the Temperate Zone; Europeans and Americans who have been accustomed to the finely flavored peaches, nectarines, and pears of northern orchards find it delicious and unexceptionable, although they may criticize other tropical fruits as being insipid or mawkish.

Yet, strangely enough, this "prize of the Indies," admitted by all to be the finest fruit of the tropics, remains to this day extremely limited in its distribution, and known only to the favored few who have lived or traveled in the East Indies. David Fairchild, who has studied its requirements more exhaustively than any other man, is convinced "that the acclimatization of the mangosteen on the island of Porto Rico, and in many other parts of tropical America, is a possibility, and that the principal difficulties of its culture have probably arisen from an ignorance of the soil conditions demanded by the plant." Trees have fruited in Jamaica, Dominica, and Trinidad. There is a fruiting tree in Hawaii and a few others are scattered throughout the tropics in regions where it would have been said a few years ago that mangosteens could not be grown. There are grounds for the hope, therefore, that commercial production of this delectable fruit will not remain limited to a remote region in the eastern tropics.

The mangosteen is a small tree rarely over 30 feet high, with deep green foliage which glistens in the sunlight. The leaves are elliptic-oblong in form, acuminate at the tip, thick and leathery in texture, and 6 to 10 inches long. The flowers are polygamous; the staminate or male blossoms are borne in three- to nine-flowered terminal fascicles, and have orbicular sepals and broadly ovate, fleshy petals. The hermaphrodite flower is 2 inches broad, and is borne solitary or in pairs at the tips of the young branches. The sepals and petals resemble those of the male flower. The stamens are many, the ovary four- to eight-celled, with a sessile, eight-rayed stigma.

"This delicious fruit is about the size of a mandarin orange, round and slightly flattened at each end, with a smooth, thick rind, rich red-purple in color, with here and there a bright, hardened drop of the yellow juice which marks some injury to the rind when it was young. As these mangosteens are sold in the Dutch East Indies, — heaped up on fruit baskets, or made into long regular bundles with thin strips of braided bamboo, — they are as strikingly handsome as anything of the kind could well be, but it is only when the fruit is opened that its real beauty is seen. The rind is thick and tough, and in order to get at the pulp inside, it requires a circular cut with a sharp knife to lift the top half off like a cap, exposing the white segments, five, six, or seven in number, lying loose in the cup. The cut surface of the rind is of a moist delicate pink color and is studded with small yellow points formed by the drops of exuding juice. As one lifts out of this cup, one by one, the delicate segments, which are the size and shape of those of a mandarin orange, the light pink sides of the cup and the veins of white and yellow embedded in it are visible. The separate segments are between snow white and ivory in color, and are covered with a delicate network of fibers, and the side of each segment where it presses against its neighbor is translucent and slightly tinged with pale green. The texture of the mangosteen pulp much resembles that of a well-ripened plum, only it is so delicate that it melts in the mouth like a bit of ice-cream. The flavor is quite indescribably delicious. There is nothing to mar the perfection of this fruit, unless it be that the juice from the rind forms an indelible stain on a white napkin. Even the seeds are partly or wholly lacking, and when present are very thin and small." (Fairchild.)

Regarding the native home of the mangosteen, the classical Alphonse DeCandolle says: "The species is certainly wild in the forests of the Sunda Islands and of the Malay Peninsula. Among cultivated plants it is one of the most local, both in its origin, habitation, and in cultivation. It belongs, it is true, to one of those families in which the mean area of the species is most restricted."

The mangosteen is a common dooryard tree in the East Indies, particularly in Java and Sumatra. Much of the fruit sold in the markets comes from scattered trees. There are a few small orchards in Malacca and the Straits Settlements. The largest orchard in the world (containing, however, only 300 or 400 trees) is situated near Saigon, in Cochin-China. A few small orchards have been started in Ceylon, but mangosteens are not as abundant in that island as they are in the Malay Archipelago. So far as is known, the tree is not commonly grown anywhere in India, but there are said to be a few specimens in the Madras Presidency. Mangosteens grown in the Sulu Archipelago of the Philippines are often seen in the markets of Manila.

Concerning the behavior of this plant in the Hawaiian Islands, Gerrit P. Wilder says: "Many mangosteen trees have been brought to Hawaii, and have received intelligent care, but they have not thrived well, and have eventually died. Only two have ever produced fruit, one in the garden of Mr. Francis Gay of Kauai, which bears its fruit annually, and the other at Lahaina, Maui, in the garden formerly the property of Mr. Harry Turton."

Joseph Jones, curator of the Botanic Station at Dominica, in the British West Indies, writes in the Agricultural News (March 4, 1911):

"At the Point Mulatre estate, Dominica, two fine mangosteen trees, thirteen years old, are now fruiting for the first time. One specimen is bearing several dozen fruits, and the other a single fruit. There are known to be four bearing mangosteen trees in Dominica. As quite a number of estates possess a few young specimens of this interesting tree, it is probable that in the course of a few years the fruit will be fairly well known in the island, and may, in course of time, be available for export.

"One point in this connection is worthy of notice. The seedlings raised from trees established in the West Indies show much greater vigor, and thrive better, than did the original imported plants. This is probably due to acclimatization. With this increased vigor, and with great care in growing and selecting land and position, it may be possible to bring trees into fruit during their ninth or tenth year."

The Trinidad and Tobago Bulletin for January, 1914, says:

"In Government House Gardens there is a tree of the mangosteen which has now borne fruit more or less regularly for several years. There are also a few other fruiting trees in the Colony, *e.g.*, at Arima in the grounds of Mr. J. G. de Gannes and at Monte Cristo estate, the property of Mr. H. Monceaux.

"In addition to the old tree in the Government House Gardens there is another which has not yet borne fruit, and a group at St. Clair Experiment Station. The latter are now 11 years old and this month (January, 1914) one of them bore a single fruit for the first time. The age of this tree is definitely known as they were planted personally by Mr. J. C. Augustus, now the Curator of the Gardens. It will be of interest to know from others who have trees of any definite records of the age at which they begin to bear fruit in the Colony."

A number of trees have been planted in Cuba, the Canal Zone, and Porto Rico, but so far as known none of them is yet fruiting. In California and Florida there appears to be little hope for the mangosteen, since it is highly susceptible to frost-injury. If stock-plants are discovered which will impart hardiness, there is a possibility that it may yet be grown in the most protected situations in southern Florida.

The name mangosteen (in French *mangoustan*) is of Malayan origin. Yule and Burnell derive it "from Malay *manggusta* (Crawfurd), or *manggistan* (Favre), in Javanese manggis. . . . This delicious fruit is known throughout the Archipelago, and in Siam, by modifications of the same name." Botanically the species is *Garcinia Mangostana*, L.

The fruit is eaten fresh. The rind, or the entire fruit dried, is used medicinally in India. It contains tannin and a crystallizable substance known as mangostin. According to Carl Wehmer [1] the fresh fruit contains sugar as follows: Saccharose 10.8, dextrose 1, and levulose 1.2.

Cultivation.

Horticultural writers have asserted that the mangosteen can be grown only within four or five degrees of the equator. Experience has shown that such a statement is not warranted by facts. It is true that the tree is strictly tropical in its requirements and that its demands in regard to soil conditions are definite. There is no reason, however, to assume that it will not be possible to grow mangosteens successfully throughout the tropics wherever these conditions can be met. Fairchild considers that the unduly limited distribution of the

[1] Die Pflanzenstoffe.

tree is due to the difficulty which young plants have in establishing themselves, and he believes that a vast extension of mangosteen culture will take place when the root-system of this tree is thoroughly understood. "This may come about through the use of stocks which are less particular in their soil requirements. George Oliver's experiments have proved that the mangosteen can successfully be inarched upon a number of the related species of the same genus." Thus, on *Garcinia xanthochymus,* a vigorous and hardy species, it has done remarkably well. Since more than 150 species of Garcinia are known, there should be excellent possibilities of obtaining a stock-plant which will produce robust mangosteen trees on soils where they will not grow successfully on their own roots.

The mangosteen does not withstand frost, but the behavior of trees in Cuba and elsewhere shows that it is not injured by merely cool weather; that is, the constantly high temperatures of the equatorial belt are not essential to its success. Like the breadfruit and a few other strictly tropical species, it does not like temperatures below 40° or thereabouts. In Ceylon and Singapore the best orchards are on soils having a high clay-content, combined with plenty of coarse material and a small amount of silt, and where the water-table stands about six feet below the surface. "The impression is current," says Fairchild, "that the mangosteen requires a wet but well-drained soil and a very humid atmosphere. While the former statement appears to be true, the latter is not so, for the tree which has fruited on the island of Kauai (Hawaii) is in a dry but irrigated part of the island, with only six inches of rainfall, where it has to be irrigated twice a month."

The observations made by Fairchild during his studies of mangosteen culture in the Orient are of such importance that it is worth while to reproduce some of them here. He writes of his visit to W. H. Wright at Mirigama, Ceylon:

"His orchard consisted, at the time of my visit in 1902, of 23 trees and was then probably the largest in the colony. It was from eight to ten years old, having been planted with two-year-old trees which were sent him as a present from the Malay peninsula. The selection of a site for his orchard was a very happy one; a moist spot in his coconut plantation, a part of which had at one time been used as a rice field. The ground was so moist that open drains were cut through it to carry off the superfluous water and these are still kept in order. The soil of the squares on which the trees are growing is so moist and soft that, were it not for a layer of coconut husks, one's feet would sink in up to the ankle as he walks across them. The roots, under these circumstances, are bathed continually in fresh, not stagnant, moisture. Mr. Wright attributes his success in growing mangosteens to the fact that he has planted them on soil that never dries out, but has, at a few feet from the surface, a continual supply of fresh moisture. The water in his well, near by, is six feet from the surface of the ground. H. L. Daniel, who has been for 15 years trying to grow this fruit, and who, during that time, has planted over a hundred young trees, assures me that this is one of the secrets of the culture of this difficult fruit, and gives Mr. Wright credit for first finding it out.

"Another important detail relates to the matter of transplanting the young seedlings. Mr. Daniel plants the seeds in a small pot or coconut husk, and keeps them well watered and slightly shaded with a coarse matting of coconut leaves. He transplants them from this small pot to a larger one when the roots have filled it; and in removing he cuts off the tap-root if the latter is exposed. For two years these young plants are kept in pots and grow to a height of two to two and a half feet. It is useless to transplant them before they are at least two feet high, for the check given them, if too young, by the transplanting is so great that they refuse to grow.

"When transplanted, the plants are set in a hole three feet cube in size. Stiff soil is best but is not absolutely necessary, as they will grow in a light soil if the subsoil is a good paddy mud. From the first the young trees should be shaded with a matting of coconut leaves, which is suspended two feet or so above the top of the plant. This is to prevent wilting and subsequent death of the two red, partly developed leaves, which first appear from the seed, and which must be kept alive if the plant is to make a rapid growth. If these precautions of potting, shading, and selection of soil are followed, trees should come into bearing seven years from seed, producing a small crop of a hundred fruits or so. The subsequent treatment of the mangosteen orchard seems to be very simple, — no pruning of any kind is commonly practiced, although it might be advisable to prune; and little cultivating is done. A mulch of coconut husks about the base of the tree to keep the surface soil continually moist, and the application of a small amount of earth from the poultry-yard, sprinkled about underneath the trees each year, are the only attentions given them. Whether or not artificial fertilizers could be employed with profitable effect is a question that has not been answered."

In the same article [1] Fairchild writes of mangosteen culture in another region:

"In Singapore there are some small mangosteen orchards, that is, mangosteens mixed with other fruits. One which is easily accessible lies on the well-known road to the Botanic Gardens, some two miles from the Raffles Hotel. The land is low and wet and several drainage canals cut it up into large, square blocks. The soil is clay and evidently saturated with moisture. About each tree is a circular bit of cultivated soil, the rest being in grass, and scattered over the bare soil under the trees is a mulch of leaves and coconut husks. I do not know how old the orchard is, but it is presumably about 30 years of age. . . . Dr. Ridley, then Director of the Botanic Gardens in Singapore, remarked that though apparently in excellent condition this orchard was not productive. It was his belief that it needed pruning and his experience with a tree in Government House Gardens bears out his belief. He cut out the innermost branches from one of the lot of old mangosteen trees there, which had not borne well for years, and as a consequence it produced, the next year, an abundance of fruit. His opinion is that the trees should regularly be pruned of all the small inner branches."

Regarding the behavior of the mangosteen in Hawaii, Fairchild says: "Francis Gay, who planted the tree at Makaweli, Kauai, wrote that where the tree is growing the water is about six feet below the surface of the soil, that the tree is irrigated twice or three times a month, and that the rainfall of the region is six to seven inches a year. This tree of Mr. Gay's is about 25 years old, fruited first when ten years old and now bears only a few fruits per year. . . . It stands about 15 feet above sea-level in a spot well-protected from the winds by windbreaks and is growing on a sandy, alluvial soil."

[1] Journal of Heredity, Aug., 1915.

Propagation.

The work of George W. Oliver in the greenhouses of the Department of Agriculture at Washington has thrown much light on the requirements of young mangosteen plants, and on the best methods of propagation. The following extracts are taken from his report in Bulletin 202 of the Bureau of Plant Industry, "The Seedling-inarch and Nurse-plant Methods of Propagation."

"Few plants show the results of inattention on the part of the cultivator more plainly than the mangosteen. When once a plant becomes in the least sickly, there is little likelihood of its recovery on its own roots. The mangosteen does not take kindly to heavy soils; it prefers a well-drained soil containing a large proportion of decayed vegetable matter. When seedlings are removed from flats and put in pots some will die without apparent cause. An oversupply of water causing the soil to become in the least sour is certain to induce sickness much more quickly in the mangosteen than in other species of the genus. Therefore, great care is necessary in handling the plants, especially in the early stages of the seedlings.

"Unfortunately the mangosteen is not a strong-rooting plant, especially during the first year or two after germination. This peculiarity renders it particularly sensitive to dry weather and may account in part for the many failures to grow it successfully. Nearly all the other species of the genus have strong and abundant roots, even in the seedling stages. It therefore seems likely that the mangosteen will thrive better and under more widely varying soil and atmospheric conditions if the young plant is inarched to some species of the genus which has a good root system.

"The genus Garcinia is a large one, the Index Kewensis listing 228 species. Of these about 20 have been tried in the inarching experiments; and while the mangosteen unites with all of them, only a few can be recommended as promising stock-plants. Two other genera of the same family, Calophyllum and Platonia, have been tried. Two species of Calophyllum, *C. calaba* and *C. inophyllum*, are not satisfactory because the union between these and the mangosteen is imperfect. This is partly because the stems of the Calophyllums are softer than those of the seedling mangosteen and partly because the growth made by the former as they become older is much more rapid. *Platonia insignis* (see below), on the other hand, so far as the experimental work has progressed, is a very promising stock from one to three years after germination, and if it will grow under the conditions suitable for the mangosteen, it may turn out to be the best stock of all those tried. The most promising species of Garcinia for use as stock-plants for the mangosteen are *G. tinctoria*, *G. morella*, and *G. Livingstonei*, in the order named, the last a native of Portuguese East Africa. The two first named are from the Malay Peninsula."

Recent experiments have shown that *G. xanthochymus* is also promising. It is vigorous in growth, and adapted to many types of soil. Inarching the mangosteen is a simple process, essentially the same as inarching the mango. Oliver says further:

"None of the species of Garcinia used as stocks are difficult to raise from seeds, provided they are fresh. They are easiest to germinate when sown in soil composed largely of partially decomposed leaves mixed with a little loam and rough-grained sand. They should be potted as soon as the first leaves are well developed. All the Garcinias with the exception of *G. Mangostana* have magnificent root systems and they thrive under ordinary treatment in so far as soil watering and a considerable range of temperature are concerned.

"It is an important point to have the stock plants in an active stage of growth when the union is in progress, though the seedling mangosteens may be inarched while apparently dormant. Although the unions when both stock and cion are in a resting stage are fairly satisfactory, the difference in growth is easily observable when the stock plants are in active growth. To secure this active growth the stocks should be allowed to become dormant; then, when they are given larger pots, good drainage, and soil composed of rotted leaves, at least one-half, and the rest fibrous loam containing a little rough-grained sand, together with some rough charcoal and crushed bone, they will under high temperature respond with vigorous growth. When inarched in this condition the union is always satisfactory.

"All plants used as stocks have been from one to three years old. Within that period the age of the stocks seems to make little difference, especially when used as nurse stocks. Mangosteen seedlings seven months old united on nurse-stocks of three-year-old *Garcinia tinctoria* made very fine unions, and within six months after the union some of the mangosteen stems were almost as thick as those of the stocks."

P. J. Wester states that the mangosteen can be budded, and says: "Use mature, green and smooth, nonpetioled budwood;

cut the buds an inch and a half long; insert the buds in the stock at a point of the same appearance as the cion or at most where it is streaked with gray."

These methods of propagating the mangosteen are of recent inception but they promise to be of immense value in extending the area in which the tree can be grown, as well as in permitting the establishment of superior varieties, which is not possible when seed-propagation is the only means used. Fairchild writes: "When one considers that so far no selection of varieties of the mangosteen has been made, notwithstanding the fact that practically seedless fruits are of frequent occurrence, and further that the tree belongs to a large genus of fruit-bearing trees, at least fifteen of which are known to bear edible fruits, some of them as large as small melons, and that these are scattered in Australia, the Malay region, South China, Africa, Brazil, and Central America, it becomes evident that in the development and breeding of the mangosteen and in the discovery of a suitable stock for it, there lies a most promising field for horticultural research."

Season and enemies of the mangosteen.

Seedling trees may begin to bear fruit when seven or eight years old, but it is rare for them to do so before the ninth year. It is not yet known how many years will be required for an inarched or budded tree to come into bearing. In Ceylon the trees are said to bloom twice a year, once in August and again in January. The fruit from the first crop of flowers ripens in January, and that from the second in July and August. In Trinidad the fruiting seasons are said to be July and October. The January crop in Ceylon is a light one, not over 100 fruits to a tree, while in the August crop 500 to 600 fruits a tree may be expected.

As to marketing, Fairchild says: "Although the mangosteen is a very delicate fruit, it has an exceedingly tough, thick rind,

PLATE XXI. A young kaki tree in bearing.

and on this account it is likely to be a good shipper. Fruits which were sent in cold storage to Washington from Trinidad were excellent when eaten twenty-one days later, even though they had been out of cold storage over a week." Shipments are regularly made from the Straits Settlements to the markets of Calcutta. When the fruits decay, the rind hardens instead of becoming soft.

Little is known regarding the enemies of the mangosteen. W. N. C. Belgrave[1] reports a fungous parasite, *Zignoella garcineæ*, which causes the formation of cankers on the stems, working back from the young to the older branches. When the latter have been attacked, the foliage withers and eventually the entire tree dies. As a combative measure it is recommended to cut and burn trees which are attacked, in order to check the spread of the disease.

There are as yet no named varieties of the mangosteen in cultivation.

The Mamey

(*Mammea americana*, L.)

Christopher Columbus, after his first visit to Veragua in 1502, is said to have described the mamey as a fruit the size of a large lemon, with the flavor of the peach. Gonzalo Hernandez de Oviedo, about twenty years later, described it more fully and reported it as most excellent.

As a horticultural product, the mamey remains in very much the same position which it occupied at the time of the Discovery. It is a dooryard tree, nowhere cultivated on a commercial scale, but considered by the Indians a delicious fruit. Europeans who have settled in tropical America have learned that it yields a preserve which tastes remarkably like that made from the apricot.

[1] Agr. Bull. of the Federated Malay States, 3, 1915.

The tree, which is one of the most beautiful and conspicuous in the West Indies, reaches 60 feet in height. Its trunk sometimes attains a diameter of 3 or 4 feet, while the crown is of a deeper and richer green than that of most other trees. The leaves are oblong-obovate in form, rounded or blunt at the apex, 4 to 8 inches long, and thick and glossy. The white flowers, which are solitary or clustered in the axils of the young shoots, are fragrant and about an inch broad. The petals are four to six in number, the anthers numerous, and the stigma peltate. The fruit is oblate to round in form, and commonly 4 to 6 inches in diameter. It has a slightly roughened russet surface and a leathery skin about $\frac{1}{8}$ inch thick. Surrounding the one to four large seeds and often adhering to them is the bright yellow flesh, juicy but of firm texture. The flavor is subacid and pleasant, but the texture is so close that the fruit is commonly thought better when stewed.

The mamey is considered indigenous in the West Indies and the northern part of South America. Outside of its native region it is grown in Mexico and Central America, and occasionally in other regions, but it has not become common anywhere in the Orient, so far as is known. It is successfully cultivated in southern Florida as far north as Palm Beach. Though not common in this region, fine specimens are occasionally seen at Miami and other places. It is not grown in California, being too susceptible to frost for any part of that state.

Mamey, the name by which this fruit was known to the first Spanish settlers in the New World, is considered to have come from the aboriginal language of the island of Santo Domingo. From it have arisen the English common names mammee and mammee-apple, both widely used in the West Indies. The term *mamey de Santo Domingo* is sometimes used in Cuba and other Spanish-speaking countries to distinguish the species from the *mamey colorado* or *mamey zapote* (*Lucuma mammosa*). In southern Brazil it is known as *abricó do Pará* (Pará apricot). The most usual French name is *abricot de Saint Domingue.*

From the fragrant white flowers a liqueur is distilled in the French West Indies which is known as *eau de créole* or *crême de créole.* The wood is hard, durable, and well adapted to building purposes. It is beautifully grained and takes a high polish. The resinous gum obtained from the bark is used to extract chigoes from the feet.

The fruit is sometimes sliced and served with wine or with sugar and cream, but it is usually preferred by Europeans in the form of sauce, preserves, or jam. Mamey preserves are manufactured commercially in Cuba and a few other tropical countries.

The mamey is tropical in its requirements, and cannot be grown in regions which commonly experience more than two or three degrees of frost. Large trees were cut back to the trunks by a freeze of 26° above zero at Miami, Florida. While the best soil for it is a rich, well-drained, sandy loam, the tree has made good growth on the shallow sandy lands of southeastern Florida. Little attention has been given to its culture in any region. Seedlings do not come into bearing under six or seven years of age; when mature they usually bear regularly and abundantly. The ripening season in the West Indies is in the summer.

Propagation is usually by seeds, which germinate readily if planted in light sandy loam. Some asexual method should be employed to propagate desirable varieties originating as chance seedlings. Inarching, which succeeds with the mangosteen, should be applicable to this plant as well; budding may also prove to be successful, performed as with the mango. No named varieties have been established as yet. It will be worth while to search out the best existing seedlings in tropical America and propagate them.

The Bakurí

(*Platonia insignis,* Mart.)

In northern Brazil, particularly in the Amazon region, the bakurí occurs wild. It is scarcely known in cultivation, but the fruit gathered from trees in the forest is preserved in tins and sold commercially to a limited extent. The genus Platonia contains only one or two species.

The tree is described as large, with oblong, acute, leathery leaves. The flowers are solitary, terminal, rose-colored, and showy. J. Barbosa Rodrigues [1] says that the fruits are the size of oranges, bright yellow in color, with several seeds surrounded by white pulp. The flavor is acidulous, sprightly, and very pleasant.

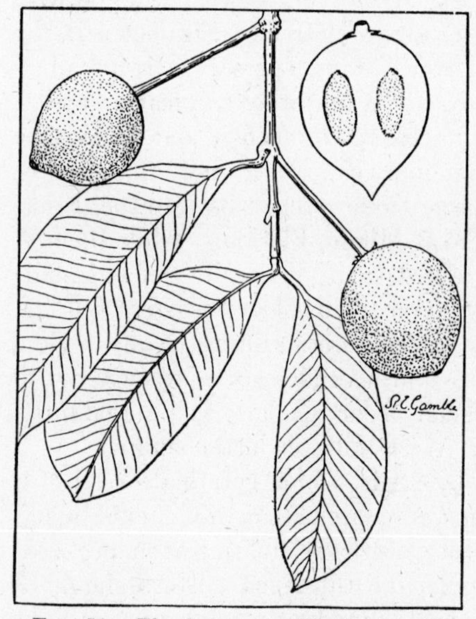

FIG. 51. The bakuparí (*Rheedia brasiliensis*), a Brazilian relative of the mangosteen. The skin is yellow, and the white pulp subacid and spicy in flavor. (× ½)

Jacques Huber of Pará, Brazil, writes: "The bakurí is a hardy tree with us and does not require careful cultivation. Cut down, it springs up easily from suckers which arise from the roots. In Marajo it is considered a weed, difficult to exterminate, especially in pastures near houses."

The bakurí is sometimes listed botanically as *Aristoclesia esculenta*, Stuntz. In Brazil it has been offered by nurserymen under the common name bakuparí, which properly belongs to a species of Rheedia.

The tree is probably strictly tropical in its requirements. It should repay horticultural attention.

[1] Hortus Fluminensis.

The Bakuparí (Fig. 51)

(*Rheedia brasiliensis*, Planch. & Triana)

This handsome tree is indigenous to the state of Rio de Janeiro in southeastern Brazil. It closely resembles its near relative the bakurí (*Platonia insignis*). The fruit is smaller in size than that of the latter species, and, while not considered so delicious, is highly prized by the natives, especially when prepared in the form of a *doce* or jam.

The tree, which is said to flower in December and ripen its fruit in January and February, is little known in cultivation. The fruit is ovate in form, sharp at the apex, and about 1½ inches long. It is orange-yellow in color and has a tough, leathery skin surrounding translucent snow-white pulp in which two oblong seeds are embedded. The flavor is subacid, suggesting that of the mangosteen.

Several other species of Rheedia produce edible fruits, but none of them is well known in cultivation. *R. edulis*, Planch. & Triana, is occasionally cultivated in Brazil under the name of *limão do matto* (wild lemon); it is a small, handsome tree with oblong glossy-green leaves and elliptic yellow fruits 2 inches long. The white pulp is highly acid. *R. macrophylla*, Mart., is said by Jacques Huber to be cultivated at Pará under the name of *bacury-pary*. Its fruits are said to resemble those of *Platonia insignis*, but are somewhat smaller and more acid.

CHAPTER XV

THE BREADFRUIT AND ITS RELATIVES

NOTWITHSTANDING their very different appearance, the breadfruits are of the same family (Moraceæ) as the mulberries, fig, and osage orange. The breadfruits, however, are tropical, whereas the fig is grown as a warm-temperate and subtropical fruit. The genus Artocarpus, comprising the breadfruit and its relatives, includes some 30 species.

THE BREADFRUIT (Figs. 52, 53)

(*Artocarpus communis*, Forst.)

Among the horticultural products brought to the attention of Europeans by the early voyagers to the East, few were considered of such interest and value as the breadfruit. The importance of its introduction into the British colonies in the West Indies was felt to be so great that His Majesty's government toward the end of the eighteenth century fitted out an expedition for the sole purpose of transporting the plants from Tahiti, in Polynesia, to Jamaica and other islands in the American tropics. On the failure of this expedition, due to the mutiny of the crew, a second and successful one was undertaken.

Contrary to expectations, the breadfruit did not prove of great value to the West Indian colonies. The banana is more productive and gives more prompt returns, and the negroes preferred to continue eating a fruit to which they were accustomed rather than trouble to cultivate the taste for a new one.

In Polynesia, however, the breadfruit still retains the important position which it occupied at the time the region was first visited by Europeans. There it is a staple food and

Fig. 52. The breadfruit (*Artocarpus communis*) is one of the staple foodstuffs of the Polynesians. It is cultivated on a limited scale in tropical America, where it was introduced toward the end of the eighteenth century. ($\times$ about $\frac{1}{7}$)

really entitled, by reason of its starchy character and the rôle which it plays in the native dietary, to the name which has been bestowed on it by the English.

The tree, when well grown, is one of the handsomest to be seen within the tropics. It reaches a height of 40 to 60 feet, and has large, ovate, leathery leaves which are entire at the base and three- to nine-lobed toward the upper end. Male and female flowers are produced in separate inflorescences on the same tree. The staminate or male flowers grow in dense, yellow, club-shaped catkins; the female, which are very numerous, are grouped together and form a large prickly head upon a spongy receptacle. The ripe fruit, which is composed of the matured ovaries of these female flowers, is round or oval in form, commonly 4 to 8 inches in diameter, green when immature but becoming brownish and at length yellow. The pulp is fibrous, pure white in the immature fruit and yellowish in the fully ripe one. The fruits are produced on the small branches of the tree upon short, thick stalks. Clusters of two or three are common.

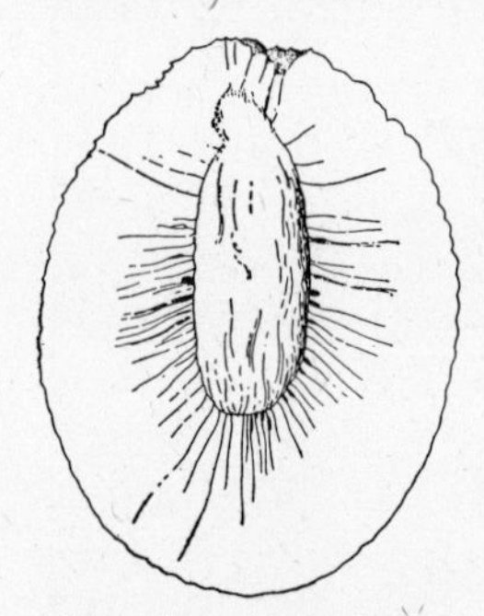

FIG. 53. The breadfruit, showing its internal structure. This is the seedless variety, generally cultivated in Polynesia; the other form has seeds as large as chestnuts, and is not highly valued. (× about $\frac{1}{4}$)

There are two classes of breadfruits, one seedless and the other carrying seeds. The former is propagated vegetatively, and is presumably the product of cultivation; the latter is often found in a wild state, and is not used in the same manner as the seedless kind. The seeds resemble chestnuts in size and appearance.

The breadfruit is believed to be a native of the Malayan Archipelago, where it has been cultivated since antiquity. From its native region it was carried to the islands of the Pacific in prehistoric times. Henry E. Baum,[1] who has written a lengthy history of this fruit, comments: "The open-boat journeys of the Polynesians in their peopling of the Pacific islands are marvelous from the point of view of seamanship alone. . . . Probably a hundred species of plants were introduced into Hawaii by the Polynesians, and

[1] Plant World, VI, 1903.

as a majority of their principal food-producing plants were propagated by cuttings alone, the difficulty in successfully carrying them across a wide expanse of ocean in open boats is obvious."

Spanish voyagers who visited the Solomon Islands in the sixteenth century encountered the breadfruit, and it is believed that it must also have been seen by the early Dutch and Portuguese sailors. In 1686 Captain William Dampier observed the plant at Guam and gave to the world an accurate description of the fruit and its uses. The famous Captain Cook, who explored the Pacific from 1768 until he met his death in the Sandwich Islands in 1779, is said to have suggested to the British the desirability of introducing the tree into the West Indies. The outcome was that notorious voyage under William Bligh, in the Bounty, which forms certainly the most dramatic incident in the history of plant introduction. The expedition sailed from England in 1787, and reached Tahiti, after a cruise of ten months, in 1788. A thousand breadfruit plants were obtained and placed on board ship in pots and tubs which had been provided for the purpose. Before the ship was out of the South Seas the crew, who had become enchanted with Tahitian life, mutinied and took charge of the ship, putting their commander and the eighteen men who remained loyal to him in a launch and setting them adrift. The mutinous crew sailed back to Tahiti, whence some of the members, accompanied by a number of Tahitians, migrated to Pitcairn's Island and established there an Utopian colony. After a trying voyage Bligh and his companions reached Tofoa, an island in the Tonga group, but they met with a hostile reception from the natives and were forced to continue their desperate pilgrimage. Fearing, because of their defenseless condition, to land on the Oceanic islands, they steered for the distant East Indies, which they were successful in reaching. "It appeared scarcely credible to ourselves," remarks Captain Bligh in his account of the voyage, "that, in an open boat so poorly provided, we should have been able to reach the coast of Timor in forty-one days after leaving Tofoa, having in that time run, by our log, a distance of 3618 miles; and that, notwithstanding our extreme distress, no one should have perished in the voyage."

Undaunted by the failure of the first attempt, a second was fitted out, again under the command of Bligh, who was promoted to the rank of Captain in the Royal Navy. This expedition, which sailed in 1792, secured 1200 breadfruit plants, as well as other valuable trees, and safely brought them to the West Indies.

The seeded breadfruit, which is much less valuable than the seedless variety, was introduced into the West Indies by the French ten years previous to Bligh's successful voyage.

At the present day the breadfruit is cosmopolitan in its distribution. Regarding its occurrence in Hawaii, Vaughan MacCaughey [1] says:

> "At the time of the coming of the first European explorers the breadfruit was plentiful around the native settlements and villages on all the islands: more plentiful than it has been at any subsequent period. It thrives in the humid regions of Kona and Hilo, on the island of Hawaii, and to-day there are many abandoned trees in these districts, marking the sites of once-populous Hawaiian villages. The extensive breadfruit groves of Lahaina, on Maui, were long famous for the excellence of their fruit. In humid valleys on Molokai, Oahu, and Kauai, the tree was also abundant, rearing its splendid dome of glossy foliage high above the surrounding vegetation.
>
> "It is distinctly a tree of the valleys and lowlands in Hawaii, and with the decadence of the Hawaiian population, and the utilization of fertile lowlands for sugar plantations, the majority of these fine old trees were sacrificed to make way for the white man's agriculture."

In some of the Polynesian Islands, the tree is of such ancient cultivation, and plays such an important part in the life of the people, that the natives are unable to conceive of a land where the breadfruit is not found.

[1] Torreya, March, 1917.

Westward from Polynesia and its native region (the Malay Archipelago), the breadfruit is grown in Ceylon and occasionally in India. In the American tropics it is nowhere an important product, but it is cultivated on a limited scale in the West Indies, the lowlands of Mexico and Central America, and on the South American mainland as far south as the state of São Paulo in Brazil.

There are probably no places on the mainland of the United States where it can be cultivated successfully. All parts of California unquestionably are too cold for it. Trees have been planted in extreme southern Florida, but so far as is known none has ever reached bearing stage, although there are fruiting specimens of the allied jackfruit in that state.

The seedless variety is invariably called breadfruit in English; the seeded variety sometimes breadnut. The Spanish name for the seedless form is *árbol del pan,* sometimes *masa pan;* the French *arbre à pain;* the Portuguese *arvore do pão* or *fruta pão;* the Italian *albero del pane;* and the German *brotbaum.* W. E. Safford [1] gives the following vernacular names: Seedless variety, — *lemae, lemai, lemay, rima* (Guam); *rima, colo, kolo* (Philippines); *'ulu* (Samoa, Hawaii); *uto* (Fiji). Seeded variety, — *dugdug, dogdog* (Guam); *tipolo, antipolo* (Philippines); *'ulu-ma'a* (Samoa); *uto-sore* (Fiji); *bulia* (Solomon Islands). Botanically the breadfruit is *Artocarpus communis,* Forst. The name *Artocarpus incisa,* L., is a synonym.

The methods of preparing breadfruit for eating are numerous. Safford writes: "It is eaten before it becomes ripe, while the pulp is still white and mealy, of a consistency intermediate between new bread and sweet potatoes. In Guam it was formerly cooked after the manner of most Pacific island aborigines, by means of heated stones in a hole in the earth, layers of stones, breadfruit, and green leaves alternating. It is still sometimes cooked in this way on ranches; but the

[1] Useful Plants of Guam.

usual way of cooking it is to boil it or to bake it in ovens; or it is cut in slices and fried like potatoes. The last method is the one usually preferred by foreigners. The fruit boiled or baked is rather tasteless by itself, but with salt and butter or with gravy it is a palatable as well as a nutritious article of diet."

Alice R. Thompson of Hawaii, who has published analyses of two varieties, says on the point of nutritive value: "The breadfruit is included in the table with bananas because it contains such high amounts of carbohydrates. In comparing it with the banana the hydrolyzable carbohydrates are seen to be much greater in amount. The breadfruit contains considerable amounts of starch even when ripe. The ash, fiber, and protein are high. The Samoan breadfruit was analyzed at a riper stage than the Hawaiian specimen, which may account for the larger proportion of starch to sugars in the former." Miss Thompson's [1] two analyses are as follows:

TABLE VII. COMPOSITION OF THE BREADFRUIT

Variety	Total Solids	Ash	Acids	Protein	Total Sugars	Fat	Fiber	Hydrolyzable Carbohydrates other than Sucrose
Hawaiian	41.82	.95	.04	1.57	9.49	.19	1.20	27.89
Samoan	26.89	1.15	.07	1.57	14.60	.51	.97	9.21

The above statements of uses and content apply solely to the seedless variety. In the seeded form the flesh or pulp is of little value, but the seeds, which are eaten roasted or boiled, are highly relished. They have something of the flavor of chestnuts.

The breadfruit tree is put to many uses in the Pacific islands.

[1] Report of the Hawaii Exp. Stat., 1914.

Cloth and a kind of glue or calking material are obtained from it, while the leaves are excellent fodder for live-stock.

In climatic requirements the tree is strictly tropical. MacCaughey sums up the necessary factors as: "A warm, humid climate throughout the year; copious precipitation; moist, fertile soil; and thorough drainage. The absence of any one of these conditions is a serious detriment to the normal growth of the plant, or may wholly prevent its fruiting. It is scarcely tolerant of shade, and in Hawaii large trees are almost invariably found growing in the open." It may be observed that in those parts of Central America where the breadfruit is cultivated it is found only in the lowlands, disappearing at elevations of about 2,000 feet. It is evident, therefore, that it is only successful in regions of uniformly warm climate.

Propagation of the seedless breadfruit is effected in the Pacific islands by means of sprouts from the roots. MacCaughey writes: "When growing in the soft moist soil which it prefers, the breadfruit roots shallowly and widely. Often a network of exserted roots is visible above the ground. This habit is of the greatest value in propagation. The wounding or bruising of the root at any given point stimulates the production of an offshoot, and young plants for transplanting are produced solely in this way. This mode of propagation is naturally very slow and laborious, as the young shoots grow slowly, and are very sensitive to injury."

P. J. Wester has developed in the Philippines a method which is more expeditious and satisfactory. Root-cuttings are used. The method is described by him as follows:

"A plant bed or frame should be filled with medium coarse river sand to a depth of 7 or 8 inches, — beach sand will do provided the salt has been thoroughly washed out. If sand is not procurable, sandy loam may be used.

"Larger cuttings may be made, but for the sake of convenience in handling and in order not to impose too severe a strain upon the tree that supplies the material, it is inadvisable to dig up roots for cuttings that are more than 2½ inches in diameter. Roots less than ½ inch in diameter should be discarded. Root cuttings 10 inches long have been very successful, but it is probable that a length of 8 inches would prove sufficient, and, if so, this would allow the propagation of a larger number of cuttings from a given amount of roots than if longer cuttings were made.

"Saw off the roots into the proper lengths and smooth the cuts with a sharp knife. Then make a trench and place the cuttings *diagonally* in the sand, leaving about 1½ to 2½ inches of the thickest end of each cutting projecting above the surface, pack the sand well, water, and subsequently treat like hardwood cuttings. When the cuttings are well rooted and have made a growth of eight to ten inches, transplant to the nursery. Great care should be exercised in not bruising, drying, or otherwise injuring the material from the digging of the roots to the insertion of the cuttings in the sand.

"The work should be done during the rainy season."

Seeds of the seeded breadfruit do not retain their vitality more than a few weeks, and should be planted promptly after they are removed from the fruit.

The varieties of the seedless breadfruit are numerous but imperfectly known. As many as twenty-five are said to occur in the Pacific islands, although MacCaughey states that only three are known in Hawaii. It is a curious circumstance that a tree as important as the breadfruit should have received so little scientific study; but exceedingly little is known regarding the cultural methods best suited to it and the relative merits of the different varieties propagated vegetatively. Concerning such matters as its place in Polynesian folklore, its history, and the uses of the fruit, however, there is an abundance of information in the accounts of early voyages as well as in the writings of modern authors.

The Jackfruit (Plate XXIII)

(*Artocarpus integrifolia*, Forst.)

"There is again another wonderful tree," wrote the pioneer traveler John de Marignolli in 1350, "called Chake-Baruke, as big as an oak. Its fruit is produced from the trunk, and not from the branches, and is something marvelous to see, being as big as a great lamb, or a child of three years old. It has a hard rind like that of our pine-cones, so that you have to cut it open with a hatchet; inside it has a pulp of surpassing flavor, with the sweetness of honey, and of the best Italian melon; and this also contains some 500 chestnuts of like flavor, which are capital eating when roasted."

Like other early travelers, Marignolli was inclined to exaggerate the merits of the new fruits with which he made acquaintance. The jackfruit is not generally considered first-class by Europeans. When preserved or dried it is better, but in tropical America the fruit is commonly not eaten except by the poorer classes. In the Orient, where it has been cultivated since ancient times, it seems to be held in greater esteem; H. F. Macmillan says that it "forms a very important article of food with the natives of the Eastern tropics." Both Theophrastus and Pliny, early writers who mentioned the jackfruit, give the same impression; Pliny describes it as the fruit "whereof the Indian Sages and Philosophers do ordinarily live."

The jackfruit is less exacting in its cultural requirements than its congener the breadfruit, and since it resists cool weather much better it is adapted to cultivation over a wider area.

The tree is large, stately, and handsome; under favorable conditions it may reach a height of 60 to 70 feet. The leaves are oblong, oval, or elliptic in form, 4 to 6 inches in length, leathery, glossy, and deep green in color. The flowers resemble in general those of the breadfruit, except that the pistillate or female blossoms are commonly produced directly on the bark of the trunk and larger limbs. The fruit is one of the largest in the world; some writers affirm that specimens have been known to weigh 80 pounds, although half this is a safer estimate. They vary from oval to oblong, and are sometimes 2 feet in length. The surface is studded with short hard points, and is pale green in the immature fruit, becoming greenish yellow and then brownish as ripening progresses. The fruit is divided inside into many small cavities each containing a seed surrounded by soft brownish pulp of pungent odor and aromatic flavor somewhat suggesting the banana. Thomas Firminger speaks rather discouragingly of this fruit. He says: "By those who can manage to eat it, it is considered most delicious, possessing the rich spicy flavor and scent of the melon, but to such a powerful degree as to be quite unbearable to persons of weak stomach, or to those unaccustomed to it."

The tree grows wild in the mountains of India and is ordinarily considered indigenous to that country. Alphonse DeCandolle believed that its cultivation probably did not antedate the Christian era. At the present day it is common in many parts of India, particularly in lower Bengal, and Macmillan observes that it has become semi-naturalized in Ceylon. In the Malayan region it is a common fruit-tree. The worthy Father Tavares states that it was introduced into Brazil by the Portuguese about the middle of the seventeenth century. It is now abundant in many parts of that country, particularly about Bahia. William Harris [1] gives the following account of its introduction into Jamaica:

"It was amongst the plants found on board the French ship bound from the Isle of Bourbon to Santo Domingo, which was captured by Captain Marshall of H. M. S. Flora, one of Lord Rodney's squadron, in June, 1782, and was sent to Mr. Hinton East's garden in Gordon Town. It was again introduced in the early part of 1793 when Captain Bligh of H. M. S. Providence brought it with other plants from the island of Timor in the Malay Archipelago. The tree is common all over the island, and is naturalized in the Cockpit country."

In Hawaii it is not abundant. It has never been a success in California, the climate having proved too cold for it. In

[1] Bull. Botanical Dept., 3, 1910.

southern Florida, however, there are several fruiting trees, but on the shallow soils of that region they do not grow to large size, and the fruits which have been produced were not of good quality. The species is probably too strictly tropical in its requirements to be entirely successful in any part of this country.

Concerning the origin of the name jackfruit, which is known to be an English adaptation of the Portuguese *jaca,* Yule and Burnell say: "Rheede rightly gives tsjaka (chakka) as the Malayalam name, and from this no doubt the Portuguese took *jaca* and handed it on to us." *Kanthal, kathal, panasa,* and *kantaka* are some of the vernacular names used in India. The French call it *jacque.* The orthography of the common English name might better be jakfruit, and indeed this spelling is employed by some writers, but the commoner form jackfruit will probably be hard to displace. *Artocarpus integra,* L., is a botanical synonym.

The fruit is eaten fresh, or it may be preserved in sirup, or dried like the fig. Thomas Firminger writes: "If the edible pulp of the fruit be taken out and boiled in some fresh milk, and then be strained off, the milk will, on becoming cold, form a thick jelly-like substance of the consistency of blanc-mange, of a fine orange color, and of melon-like flavor. Treated in this way the fruit affords a very agreeable dish for the table." Father Tavares has this warning: "It must be eaten when full ripe, and not at meal times; a cup of cool water should be taken immediately afterwards, never wine or other fermented drink, since these, when combined with the *jaca,* are poisonous." He adds that the seeds, boiled or roasted, are very pleasant and that they are used, pulverized, in making biscuits. The ripe fruits are often fed to cattle in Brazil. Alice R. Thompson of Hawaii has found the edible portion or pulp to contain: Total solids 23.20 per cent, ash 0.93, acids 0.27, protein 1.44, total sugars 15.15, fat 0.45, and fiber 1.3. The seed was found to contain: Total solids 50.82 per cent, ash 3.49, acids 0.16, protein 5.44, total sugars 1.87, fat 0.24, fiber 1.80, and hydrolyzable carbohydrates other than sucrose 23.53. Thus it will be seen that the pulp is rather high in protein and fiber and low in acids. The seeds have a high starch-content and very little sugar, while the protein-content is about 5 per cent.

The climatic requirements of the jackfruit consist in abundant precipitation and freedom from severe frosts. Probably it can be grown by the aid of irrigation in regions where there is little rainfall. Mature trees have passed through temperatures of about 27° above zero in southern Florida, but they were frozen to the large limbs. Though temperatures below freezing kill young trees and injure old ones, the jackfruit is not, like its congener the breadfruit, injured by cool weather several degrees above freezing. It prefers a rich, deep, and moist soil, but can be grown successfully on shallow and light soils such as some of those of southern Florida. In Brazil it grows well on clay and on sandy loam. Very little attention is given to cultural methods in the regions where the jackfruit is commonly grown. Like the breadfruit, it succeeds without much care from man, the sole necessity being abundant moisture.

Propagation is by seeds, which should be planted soon after their removal from the fruit. The method of propagation by means of root-cuttings or suckers, which is practiced with the seedless breadfruit, is said not to be successful with this species.

According to Paul Hubert, young trees come into bearing when five years of age. It is doubtful, however, whether they can be depended on to fruit so early. Thomas Firminger writes: "The jackfruit is not borne, like most other fruits are, from the ends of branches, but upon stout footstalks projecting from the main trunk and thickest branches of the tree. In no other way, indeed, could its ponderous weight be sustained. The situation of the fruit, moreover, is said to vary with the age of the tree; being first borne on the branches, then on the trunk, and in old trees on the roots. Those borne on the roots, which discover themselves by the cracking of the earth above them, are held in the highest estimation." When grown in a cool climate the fruits are of inferior quality. The ripening season extends over several months.

Paul Hubert states that *Batocera rubra* L. attacks the tree in some regions. This insect, which is a cerambycid beetle, causes much damage to fig trees in India by boring in their trunks, and probably works on the jackfruit in the same manner. The larva of a moth, *Perina nuda* F., is said by H. Maxwell-Lefroy to feed on the jackfruit throughout India.

"Of this tree," says the excellent Rheede, "they reckon more than thirty varieties, distinguished by the quality of their fruits, but all may be reduced to two kinds; the fruit of one kind is distinguished by plump and succulent pulp of excellent flavor, being the Varaka; that of the other, filled with softer and more flabby pulp of inferior flavor, being the Tsjakapa." This classification is borne out by more modern writers. Thomas Firminger speaks of the hard and soft kinds, and the same two forms are known in Brazil. H. F. Macmillan gives the following résumé of the subject:

"Jak-fruit occurs in several varieties, the two most distinct in Ceylon being: (1) 'Waraka,' distinguished by a firm fruit, which the natives recognize by the sound when flicked with the fingers; (2) 'Vela,' characterized by its softer rind, through which the finger may be thrust when approaching ripeness, the pulp being less sweet than that of the former variety. Of these there are several subvarieties, as 'Kuru-waraka' (with small and almost round fruit), and 'Peni-waraka' ('honey jak'), which has a sweetish pulp. A variety called 'Johore jak,' with hairy leaves and a small oblong fruit with a most overpowering odor, is greatly esteemed by those who eat the fruit."

Since these "varieties" are propagated by seed, they should properly be termed races. Of true horticultural varieties, propagated vegetatively, there are none.

THE MARANG

(*Artocarpus odoratissima,* Blanco)

The marang has been brought recently to the attention of horticulturists by P. J. Wester, who considers it a fruit of unusual promise. It resembles the jackfruit and the seeded breadfruit in appearance, but is superior in quality to either of these. The tree, which grows wild in the southern Philippine Islands and the Sulu Archipelago, is medium-sized, with large, dark green entire or three-lobed leaves 18 to 24 inches long. Wester describes the fruit as roundish oblong in form, about 6 inches in length, with the surface thickly studded with soft greenish yellow spines $\frac{1}{3}$ inch long. The rind is thick and fleshy, the flesh white, sweet, and juicy, aromatic and of pleasant flavor; it is separated into segments (about the size of a grape) which cling to the core, and each segment contains a whitish seed nearly $\frac{1}{2}$ inch long. "When the fruit is ripe, by passing a knife around and through the rind, with a little care the halves may be separated from the flesh, leaving this like a bunch of white grapes." In the Philippines it ripens in August.

The tree is strictly tropical in its requirements and probably will not succeed in regions where the temperature falls below 32° or 35° above zero. It likes a moist atmosphere and abundant rainfall. The marang has been introduced into the United States, but does not promise well either in Florida or in California.

PLATE XXII. A basket of pomegranates; right upper corner, the black sapote.

CHAPTER XVI

MISCELLANEOUS FRUITS

HAVING discussed in the different chapters the fruits that are more or less closely related botanically and culturally, we may now put the remaining kinds together in a single final fascicle. Most of these fruits are of very minor importance horticulturally. Here the reader will find accounts of the durian, santol, langsat, carambola, bilimbi, tamarind, carissa, ramontchi, umkokolo, ketembilla, white sapote, tuna, pitaya, tree tomato, genipa.

THE DURIAN (Plate XXIV)

(*Durio zibethinus*, Murr.)

Except for the fact that a few trees have been planted in the West Indies and elsewhere, and that P. J. Wester has shown that it can readily be budded (thus paving the way for its improvement), the durian occupies the same position to-day which it held when first observed by Europeans in the fifteenth century, — that of a semi-cultivated fruit of great importance to the inhabitants of the Malayan region.

Its tardy dissemination has probably been due to the perishable nature of its seeds, making it difficult to carry the species from one part of the tropics to another. It must be admitted, also, that the fruit is not one which has invariably met with a favorable reception from Europeans. Because of its strong disagreeable odor many do not like it, but others become extremely fond of it.

In its native home the durian becomes a large tree. It has obovate-oblong leaves 6 to 7 inches long, leathery in texture, shining on the upper surface and scaly on the lower. The flowers, which are produced in cymes, have a bell-shaped five-lobed calyx and five oblong petals. The fruit is oval in form, 6 to 8 inches long, covered externally with short woody protuberances. It is five-valved, and within each compartment are several seeds surrounded by clear pale brown custard-like pulp of strong gaseous odor and rich bland taste. The following description by a distinguished durian-eater, Alfred Russel Wallace,[1] gives an excellent idea of this remarkable fruit:

"The banks of the Sarawak River are everywhere covered with fruit trees, which supply the Dyaks with a great deal of their food. The Mangosteen, Lansat, Rambutan, Jack, Jambou, and Blimbing, are all abundant; but most abundant and most esteemed is the Durian, a fruit about which very little is known in England, but which both by natives and Europeans in the Malay Archipelago is reckoned superior to all others. The old traveller Linschott, writing in 1599, says: — 'It is of such an excellent taste that it surpasses in flavor all the other fruits of the world, according to those who have tasted it.' And Doctor Paludanus adds: — 'This fruit is of a hot and humid nature. To those not used to it, it seems at first to smell like rotten onions, but immediately they have tasted it they prefer it to all other food. The natives give it honorable titles, exalt it, and make verses on it.' When brought into a house the smell is often so offensive that some persons can never bear to taste it. This was my own case when I first tried it in Malacca, but in Borneo I found a ripe fruit on the ground, and, eating it out of doors, I at once became a confirmed Durian eater.

"The Durian grows on a large and lofty forest tree, somewhat resembling an elm in its general character, but with a more smooth and scaly bark. The fruit is round or slightly oval, about the size of a large coconut, of a green color, and covered all over with short stout spines, the bases of which touch each other, and are consequently somewhat hexagonal, while the points are very strong and sharp. It is so completely armed, that if the stalk is broken off it is a difficult matter to lift one from the ground. The outer rind is so thick and tough, that from whatever height it may fall it is never broken. From the base to the apex five very faint lines may be traced, over which

[1] The Malay Archipelago.

the spines arch a little; these are the sutures of the carpels, and show where the fruit may be divided with a heavy knife and a strong hand. The five cells are satiny white within, and are each filled with an oval mass of cream-colored pulp, imbedded in which are two or three seeds about the size of chestnuts. This pulp is the eatable part, and its consistence and flavor are indescribable. A rich butter-like custard highly flavored with almonds gives the best general idea of it, but intermingled with it come wafts of flavor that call to mind cream-cheese, onion-sauce, brown sherry, and other incongruities. Then there is a rich glutinous smoothness in the pulp which nothing else possesses, but which adds to its delicacy. It is neither acid, nor sweet, nor juicy, yet one feels the want of none of these qualities, for it is perfect as it is. In fact to eat Durians is a new sensation, worth a voyage to the East to experience.

"When the fruit is ripe it falls of itself, and the only way to eat Durians in perfection is to get them as they fall; and the smell is then less overpowering. When unripe, it makes a very good vegetable if cooked, and it is also eaten by the Dyaks raw. In a good season large quantities are preserved salted, in jars and bamboos, and kept the year round, when it acquires a most disgusting odor to Europeans but the Dyaks appreciate it highly as a relish with their rice. There are in the forest two varieties of wild Durians with much smaller fruits, one of them orange-colored inside; and these are probably the origin of the large and fine Durians, which are never found wild. It would not, perhaps, be correct to say that the Durian is the best of all fruits, because it cannot supply the place of the subacid, juicy kinds, such as the orange, grape, mango, and mangosteen, whose refreshing and cooling qualities are so wholesome and grateful; but as producing a food of the most exquisite flavor it is unsurpassed. If I had to fix on two only, as representing the perfection of the two classes, I should certainly choose the Durian and the Orange as the king and queen of fruits.

"The Durian is, however, sometimes dangerous. When the fruit begins to ripen it falls daily and almost hourly, and accidents not infrequently happen to persons walking or working under the trees. When a Durian strikes a man in its fall, it produces a dreadful wound, the strong spines tearing open the flesh, while the blow itself is very heavy; but from this very circumstance death rarely ensues, the copious effusion of blood preventing the inflammation which might otherwise take place. A Dyak chief informed me that he had been struck down by a Durian falling on his head, which he thought would certainly have caused his death, yet he recovered in a very short time."

The area in which the durian is indigenous has not been determined with certainty. The species is generally believed to be native to Borneo and other islands of the Malay Archipelago, but Sir Joseph Hooker considered that its distribution as an indigenous plant probably did not extend to the Malay peninsula. He thought that *Durio malaccensis,* Planch., which grows in Malacca and Burma, might be the wild form of the durian.

The region in which the tree is commonly found extends from the northern Federated Malay States through the Dutch East Indies and up into the Philippines as far as Mindanao. A single tree is known to have fruited in Hawaii, and another in Dominica, British West Indies. The species is seen occasionally in Ceylon and other tropical countries, but outside of the Malayan region its cultivation is limited mainly to botanic gardens.

The name *durian* (or *dorian*) is the only one by which this fruit is known to Europeans. Yule and Burnell say: "Malay *duren,* Molucca form *durivan,* from *duri,* a thorn or prickle (and *an,* the common substantival ending; Mr. Skeat gives the standard Malay as *duriyan* or *durian*)." Various spellings of the word are found in the early writers.

An analysis made in the Philippines by W. E. Pratt shows the fruit to contain: Total solids 44.5 per cent, ash 1.24, acids 0.1, protein 2.3, invert sugar 4.8, sucrose 7.9, and starch 11.0. In the Philippine Journal of Science, November, 1912, O. W. Barrett writes: "The chemical body which is responsible for the very pronounced odor is probably one of the sulfur compounds with some base perhaps related to that in butyric acid; it is not an oil nor a sugar, not a true starch nor an inulin, but according to Dr. W. E. Pratt it is a substance new to the organic chemist. The pulp contains a compound which, it is believed, is related to erythrodextrin, but seems to exist, if such, in a new form in this fruit."

In its climatic requirements the durian is tropical, probably strictly so. The few experiments made indicate that it will not succeed anywhere on the mainland of the United States. It is limited to regions free from frost, and delights in a deep rich soil and abundant moisture. There are many places in the West Indies and elsewhere in tropical America where it should be quite at home. In the Malayan islands, where it is commonly grown, the tree receives little cultural attention, hence nothing is known regarding pruning, irrigation, or other matters which usually give the northern horticulturists much concern. Propagation is ordinarily by seeds, which do not keep long after they are removed from the fruit. It has been learned that they can be shipped successfully from the eastern to the western tropics if they are packed in a mixture of charcoal and coconut fiber, slightly moistened.

The method of budding practiced by Wester, to which reference has been made, differs very little from shield-budding as applied to the avocado and mango. By means of this method of propagation it will be possible to perpetuate superior seedlings, and the number of years required for the tree to come into bearing should be lessened. Wester recommends that the budwood be well beyond the tender stage, but not so old that it is brittle. The petioles should be removed some time before the budwood is to be used, and the petiole-scars given time to heal over; if this is not done, decay may attack the buds. The inverted T-incision is preferred.

No horticultural varieties have yet been established, but several seedling races or forms are known to exist. Barrett says: "In passing we should not forget that there are durians and *durians;* some are said to be without a strong odor, while to our certain knowledge some of the Borneo varieties are not at all objectionable. Borneo has at least six and probably ten varieties; some of these have only one or two seeds and are comparatively small fruits, while others are fully as large as our largest Jolo or Lake Lanao (Mindanao) forms; the pulp of some is nearly white, while that of others is pale salmon or even orange in color."

The Santol

(*Sandoricum Koetjape,* Merr.)

Few writers recommend the santol as a fruit worthy of extensive cultivation. It is known chiefly in the Malayan region, where it is indigenous. The tree is medium sized, attaining to 50 feet in height. The leaves are trifoliate with elliptic to oblong-ovate, acuminate leaflets 4 to 6 inches in length. The greenish flowers are borne in axillary panicles and are followed by globose or oblate fruits about 2 inches in diameter, brownish yellow and velvety on the surface, with a thick tough rind inclosing five segments of whitish translucent pulp which adheres to the large seeds.

"The santol," writes P. J. Wester, "is one of the most widely distributed fruits in the Philippines. The tree is hardy, of vigorous and rapid growth, and succeeds well even where the dry season is prolonged. The fruit is produced in great abundance, in fact in such profusion that large quantities annually rot on the ground during the ripening season, which extends principally from July to October. It should be stated that the waste of the fruit is due principally to its poor quality; in fact, from the European point of view most of the santols are barely edible. However, now and then trees are found whose fruit is of most excellent flavor, and when a fruit shall have been found that also has the feature of being seedless or semi-seedless, like the mangosteen, it is believed the now practically unknown santol will become one of the most popular of the tropical fruits."

Sandoricum indicum, Cav., is a botanical synonym.

The Langsat (Fig. 54)

(*Lansium domesticum,* Jack)

While it cannot be said to rival the mangosteen, the langsat is considered one of the best fruits of the Malayan region.

Like the mangosteen it differs from many other tropical fruits in being juicy and of aromatic subacid flavor, instead of richly sweet. It is doubtless to this characteristic that it owes its popularity among European travelers and residents in the East.

Like several other excellent Malayan fruits, the langsat has not yet become generally cultivated outside of the Asiatic tropics. Its introduction into the western hemisphere has been accomplished, but it is only found as yet in a few botanic gardens and private collections of rare plants.

The tree is erect, symmetrical in form, usually somewhat slender, and attains a height of 35 to 40 feet. Its pinnate leaves are composed of five to seven elliptic-oblong to obovate acuminate leaflets 4 to 8 inches in length. The small subsessile flowers are borne on racemes or spikes arising from the larger branches. The fruit varies in form and character, but is generally oval or round, 1 to 2 inches in diameter, velvety and straw-colored, with a thick leathery skin inclosing five segments of white, translucent, juicy, aromatic flesh, and one to three large seeds. The tree is cultivated in many islands of the Malay Archipelago and in the Philippines. Regarding its importance in the latter region, P. J. Wester writes: "The lanzone is extensively grown for the Manila market in Laguna Province, east of Santa Cruz, and is also cultivated to a considerable extent in Misamis, Zamboanga, the Sulu Archipelago, and around Argao in Cebu." As indicated by Wester's note the common name in the Philippines is *lanzone* (often spelled *lanzón*). In the Malay Archipelago the forms *lansa* and *lanseh* are sometimes seen, and also the name *ayer-ayer*.

FIG. 54. The duku, a variety of the langsat (*Lansium domesticum*) which grows in the Malayan Archipelago. (× 1/7)

While it is most commonly eaten out of hand, the culinary uses of the fruit are several. The edible portion is said to contain 1.13 per cent of protein, 1 of acid, and 4.9 of sugar.

In its climatic requirements the plant is distinctly tropical. Wester says: "The lanzone is of vigorous growth and succeeds best under somewhat the same climatic conditions as the mangosteen. It will not grow where there is a pronounced or prolonged dry season, and in the Philippines it is usually grown in half-shade interplanted with the coconut." Experiments indicate that it is not suitable for cultivation in Florida or California, the climate of both states probably being too cold for it. In Cuba and the Isle of Pines it has shown more promise.

Little is known regarding cultural methods, since the langsat usually occurs in the Malayan region as a dooryard tree, or along roadsides, where it receives no cultural attention. Propagation is commonly by seeds, which should be planted as soon as possible after they are removed from the fruit; but Wester has shown that cleft- and side-grafting are successful, and one or the other of these methods should be used to propagate choice varieties, and to insure early fruiting. Wester says: "The cion should be well matured but not of old growth, 2½ to 3½ inches long, ¼ to ⅜ inch in diameter, and inserted in the stock 2½ to 4 inches above ground; when at that height it is ¼ to ⅜ inch in diameter; cover all wounds with grafting wax. Shield-budding has been done but the percentage of successful buds is small."

The langsat occurs in two distinct forms, one termed *langsat* and the other *duku* or *doekoe*. The typical langsat is borne in clusters of five or six up to twenty or thirty, and the individual fruits are round or oval in form, about an inch long, with a comparatively thin skin. The duku is produced in small clusters of two to five fruits, and is round, from 1 to 2 inches in diameter, with a thicker, darker-colored skin more leathery than that of the langsat.

THE CARAMBOLA (Fig. 55)

(*Averrhoa Carambola*, L.)

"There is another fruit called Carambola," wrote the Dutch traveler Linschoten in 1598, "which hath 8 corners, as bigge as a smal aple, sower in eating, like unripe plums, and most used to make Conserues." The Chinese and the Hindus eat the carambola when green as a vegetable, when ripe as a dessert. It is widely distributed in the tropics, but in America it is not so highly esteemed as in the Orient.

FIG. 55. A flowering and fruiting branch of carambola (*Averrhoa Carambola*), an Asiatic fruit sometimes cultivated in tropical America. (× ½)

The tree is small, handsome, and grows up to 30 feet in height. It has compound leaves composed of two to five pairs of ovate or ovate-lanceolate leaflets, rounded at the base and acute to acuminate at the apex, 1½ to 3 inches long, glabrous, light green above and glaucous below. The small white or purplish flowers are borne in short racemes from the bark of the young and old branches. The petals are five; the stamens ten, but five are without anthers. The fruit is oval or elliptic in outline, translucent yellow or pale golden brown in color, 3 to 5 inches long, and three-, four-, or five-ribbed longitudinally, so that a cross-section is star-shaped. "It contains a clear watery pulp," writes W. E. Safford, "astringent when green and tasting like sorrel or green gooseberries, but pleasantly acid when ripe, or even sweet, with an agreeable fruity flavor, and a strong perfume like that of the quince."

While the native home of this species is not definitely known, it is believed to be indigenous to the Malayan region, whence it was early brought to America. It is now cultivated in southern China, and from there westward to India. Safford states that it grows in Guam, but is not common. It also grows in the Philippines and in Hawaii. In America it is most abundant in Brazil, where it was doubtless introduced by the Portuguese. It does not grow in California, but succeeds in southern Florida. E. N. Reasoner has a handsome specimen in his tropical fruit shed at Oneco, near Bradentown, a place which would be too cold for the species were it not given some protection during the winter. It is rare on the lower east coast of Florida.

The name *carambola* is said to have come from Malabar, and was early adopted by the Portuguese. In upper India the fruit is called *kamranga* or *kamrakh*. The presence of a Sanskrit name, *karmara*, and the accounts of early writers, indicate that the plant was known in India before the time of European colonization. The Chinese are said to call the fruit *yongt'o* or foreign peach. In the Philippines it is termed *balimbing* as well as *carambola;* in Guam *bilimbines*.

The fruits, when fully ripe, are eaten out of hand, or they may be stewed. When slightly unripe they are used for jelly and pickles. Like the bilimbi, the carambola contains potassium oxalate, and for this reason the unripe fruit is used in dyeing and to remove iron-rust. In southern China carambolas are preserved in tin and exported to other countries. An analysis made in Hawaii by Alice R. Thompson shows the ripe fruit of the sweet variety to contain: Total solids 8.22 per cent, ash 0.42, acids 0.78, protein 0.71, total sugars 3.40, fat 0.75, and fiber 1.23.

In its climatic requirements the tree may be considered tropical. It withstands very little frost and when young is injured by temperatures above the freezing-point. It prefers a warm moist climate and a deep rich soil, but it can be grown successfully on sandy soils and heavy clays, and in northern India it thrives where the climate is dry. Cold is the limiting factor in California and Florida; in the latter state it may succeed from Palm Beach southwards, but plants have often failed to grow at Miami. When young the carambola is delicate and requires careful attention.

Safford states that the tree is long-lived and a constant bearer, producing, in Guam, several crops a year. Father Tavares writes of its behavior in Brazil: "During the entire year it loads itself with successive crops of flowers and fruits, except for a short period when it is devoid of foliage."

Propagation is readily effected by means of seeds, and P. J. Wester has shown that budding is successful. He states that budwood should be beyond the tender stage, but not so old that it is brittle. It should not be used if the petioles have fallen. The buds should be cut an inch in length, and inserted in inverted T-incisions, the operation of budding being essentially the same as with the avocado.

No horticultural varieties of the carambola are yet established. Sweet and sour seedling forms are sometimes recognized.

The Bilimbi

(*Averrhoa Bilimbi*, L.)

Like its congener the carambola, this tree is probably a native of the Malayan region, but it is known only as a cultivated species. The fruit is too highly acid to be eaten out of hand; it may be pickled in the same manner as the cucumber, which it resembles in appearance; it may be preserved in sirup; or it may be used as a relish with meat or fish.

The tree, which grows to about 30 feet in height, may be distinguished readily from the carambola by its larger leaves, which have five to seventeen pairs of leaflets in place of two to five. The crimson flowers have ten stamens, all perfect. The fruit, known in different regions as *bilambu, balimbing, blimbing, blimbee,* and *camias,* is cylindrical or obscurely five-angled, 2 to 4 inches long, greenish yellow and translucent when ripe, with soft juicy flesh containing a few small flattened seeds.

The requirements of the tree are much the same as those of the carambola. It is usually propagated by seeds. P. J. Wester reported that attempts to bud it were not successful. No horticultural varieties are grown.

The Tamarind (Fig. 56)

(*Tamarindus indica,* L.)

In addition to the usefulness of its fruit, the tamarind has the advantage of being one of the best ornamental trees of the tropics. It is particularly valued in semi-arid regions, where it grows luxuriantly if supplied with water at the root. From India to Brazil, its huge dome-shaped head of graceful foliage enlightens many a dreary scene.

The fruit became known in Europe in the Middle Ages. Marco Polo mentioned it in 1298, but it was not until Garcia d'Orta correctly described it in 1563 that its true source was known; it was thought at first to be produced by an Indian palm. The New England sea-captains who traded with the West Indies in the eighteenth and nineteenth centuries frequently brought the preserved fruit to Boston from Jamaica and other islands, but in recent years it has become scarcely known in the United States. In Arabia and India, however, it is a product of considerable importance.

When grown on deep rich soils the tree may attain to 80 feet in height, with a trunk 25 feet in circumference. The small pale green leaves are abruptly pinnate, with ten to twenty pairs of opposite, oblong, obtuse leaflets, soft and about $\frac{1}{2}$ inch long. The pale yellow flowers, which are borne in small lax racemes, are about 1 inch broad. The petals are five, but the lower two are reduced to bristles. The fruit is a pod, cinnamon-brown in color, 3 to 8 inches long, flattened, and $\frac{1}{2}$ to 1 inch in breadth. Within its brittle covering are several obovate compressed seeds surrounded by brown pulp of acid taste.

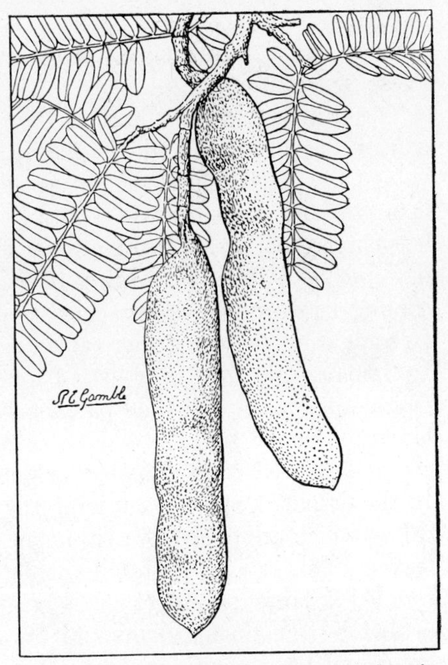

FIG. 56. The tamarind (*Tamarindus indica*), a leguminous fruit-tree whose brown pods contain an acid pulp used in cooking, and to prepare refreshing drinks. $(\times \frac{1}{2})$

The tamarind is believed to be indigenous to tropical Africa and (according to some authors) southern Asia. It has long been cultivated in India and it was early introduced into tropical America. It succeeds in southern Florida and has been grown in that state as far north as Manatee, where a large tree was killed by the freeze of 1884.

It is not sufficiently hardy to be grown in any part of California.

Yule and Burnell say: "The origin of the name is curious. It is Arabic, *tamar-u'l-Hind,* 'date of India,' or perhaps rather in Persian form, *tamar-i-Hindi.* It is possible that the original name may have been *thamar,* 'fruit' of India, rather than *tamar,* 'date.'" In French it is *tamarin,* in Spanish and Portuguese *tamarindo.*

The fruit is widely utilized in the Orient as an ingredient of chutnies and curries and for pickling fish. In medicine, it is valued by the Hindus as a refrigerant, digestive, carminative, laxative, and antiscorbutic. Owing to its possession of the last-named quality, it is sometimes used by seamen in place of lime-juice. With the addition of sugar and water it yields a cooling drink or *refresco,* especially well known in Latin America. In some countries tamarinds are an important article of export. In Jamaica the fruit is prepared for shipment by stripping it of its outer shell, and then packing it in casks, with alternate layers of coarse sugar. When the cask is nearly full, boiling sirup is poured over all, after which the cask is headed up. In the Orient the pulp containing the seed is pressed into large cakes, which are packed for shipment in sacks made from palm leaves. This product is a familiar sight in the bazaars, where it is retailed in large quantities; it is greatly esteemed as an article of diet by the East Indians and the Arabs. Large quantities are shipped from India to Arabia.

The pulp contains sugar together with acetic, tartaric, and citric acids, the acids being combined, for the most part, with potash. In East Indian tamarinds citric acid is said to be present in about 4 per cent and tartaric about 9 per cent. The following analysis has been made in Hawaii by Alice R. Thompson: Total solids 69.51 per cent, ash 1.82, acids 11.32, protein 3.43, total sugars 21.32, fat 0.85, and fiber 5.61. Commenting on this analysis, Miss Thompson says: "The tamarind is of interest because of its high acid and sugar content. It is supposed to contain more acid and sugar than any other fruit. The analysis reported by Pratt and Del Rosario shows the green tamarind to contain little sugar, but the sugar increases very greatly on ripening."

The tree delights in a deep alluvial soil and abundant rainfall. Lacking the latter, it will make good growth if liberally irrigated. The largest specimens are found in tropical regions where the soil is rich and deep. On the shallow soils of southeastern Florida the species does not attain to great size. When small it is very susceptible to frost, but when mature it will probably withstand temperatures of 28° or 30° above zero without serious injury. It is usually given little cultural attention, and is not grown as an orchard tree.

Propagation is commonly by means of seeds. These can be transported without difficulty, since they retain their viability for many months if kept dry. They are best sprouted by planting them ½ inch deep in light sandy loam. The young plants are delicate and must be handled carefully to prevent damping-off. P. J. Wester has found that the species can be shield-budded in much the same manner as the avocado and mango. He says: "Use petioled, well-matured, brownish or grayish budwood; cut the buds one inch long; age of stock at point of insertion of bud unimportant."

Seedling trees are slow to come into bearing. A mature tree is said to produce, in India, about 350 pounds of fruit a year.

Little is known of the insect pests which attack the tamarind. H. Maxwell-Lefroy mentions two, *Caryoborus gonagra* F., and *Charaxes fabius* Fabr., the latter a large black, yellow-spotted butterfly whose larva feeds on the leaves. Both these insects occur in India.

Thomas Firminger speaks of three varieties of tamarind which are grown in India, but does not know whether they can be depended on to come true from seed. M. T. Masters, in the

"Treasury of Botany," states that the East Indian variety has long pods, with six to twelve seeds, while the West Indian variety has shorter pods, containing one to four seeds. Seedlings undoubtedly show considerable variation in the size and quality of their fruit, which accounts for the different "varieties" which have been noted by many writers. Since none of these has yet been propagated vegetatively, they are of little horticultural importance.

The Carissa (Fig. 57)

(*Carissa grandiflora,* A. DC)

For its ornamental value as well as its edible fruits the carissa deserves to be cultivated throughout the tropics. Within the last few years it has become fairly common in southern Florida, and it has been found to succeed in southern California.

The plant is a large, much-branched and spreading shrub, reaching 15 or 18 feet in height. It is armed with stout branched thorns, and the dense foliage is deep glossy green in color. The leaves are ovate-acute, mucronate, thick and leathery, and 1 to 2 inches long. The flowers, which are borne in small terminal cymes, are star-shaped, fragrant, and about 2 inches broad. The plant blooms most profusely in early spring, but produces a few flowers throughout the year. The fruits, most of which ripen in summer, are ovoid or elliptic in form, 1 to 2 inches long, with a thin skin inclosing the firm granular reddish pulp, toward the center of which are several papery almost circular seeds. David Fairchild, who studied this plant in Natal (its native home), writes of it: "On the markets of Durban the long, brilliant red fruit of the amatungula is commonly sold; in fact, during January and February it is one of the commonest fruits to be seen in the stalls. Though variable in size and shape, it has generally an elongated form, with a distinct point, and the diameter of a good-sized Damson plum. The thin red skin covers a pink flesh with a milky juice, which in flavor is sweet but lacks character, although much praised by European residents for use in making fruit salads."

The name under which this fruit is known in Natal is *amatungula.* In the United States it is called Natal-plum as well as carissa. The botanical name *Arduina grandiflora,* E. Mey., is a synonym of *Carissa grandiflora.*

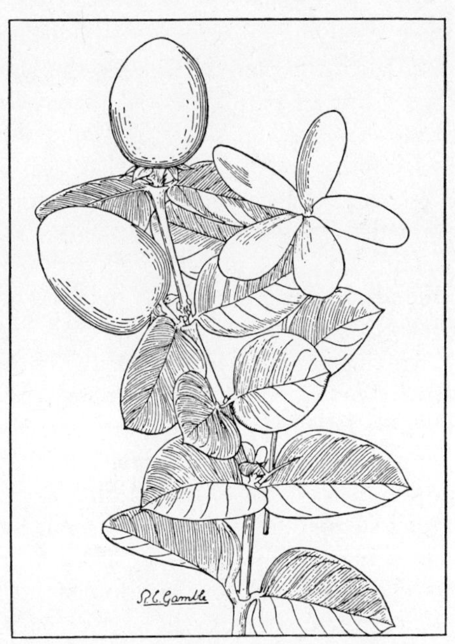

FIG. 57. The carissa (*Carissa grandiflora*) is a handsome shrub from South Africa, with fragrant white flowers and scarlet fruits whose flavor suggests raspberries. (× ½)

In Florida, the carissa is not generally relished when eaten out of hand. When stewed it yields a sauce which greatly resembles that made from cranberries. It is also used for jelly and preserves. According to an analysis made in Hawaii by Alice R. Thompson, its chief chemical constituents are: Total solids 21.55 per cent, ash 0.43, acids 1.19, protein 0.56, total sugars 12.00, fat 1.03, and fiber 0.91.

The plant is not exacting in its climatic requirements. It will grow in warm, moist tropical regions, and in the dry subtropics wherever the temperature rarely falls below 26° or 28° above zero. In California it is sometimes injured by frost, but in southern Florida this is rarely the case. It succeeds on soils of varying types, red clay, sandy loam, and light sand. It is somewhat drought-resistant.

The carissa is particularly valued as a hedge plant. It withstands shearing admirably and its growth is compact and low. "To make an amatungula hedge," writes Fairchild, "is a very simple matter. The seeds are sown in a seed-bed, and when the young plants are six inches high they are transplanted to the place chosen for the hedge and set a foot apart, alternately in parallel rows, distant from one another a foot or more. As the plants grow they are trimmed into the desired hedge form, and the oftener they are trimmed the thicker they interweave their tough, thorny branches, making an impenetrable barrier for stock of all kinds., When in flower the white jasmine-like blossoms show off strikingly against the dark background of foliage; and the red fruit which follows is quite as pretty."

Cuttings, when planted directly after removal from the parent bush, do not form roots readily unless grown over bottom heat; but a method has been devised by Edward Simmonds at Miami, Florida, whereby nearly every one will grow. This consists in notching young branchlets while still attached to the plant, making a cut halfway through the stem 3 or 4 inches from the tip. The branchlet is then bent downward and allowed to hang limply until the end of the second month, when a callus will have formed on the cut portion, and the cutting may be removed and placed in sand under a lath shade, requiring another month to strike roots.

The carissa is also propagated by layering, and it is not difficult to bud, using the common method of shield-budding, essentially the same as practiced with the avocado. Late spring is the best time to do the work.

It has been noted in Florida and more particularly in California that many carissa plants are unproductive. This matter has never been fully investigated, but the preliminary studies

PLATE XXIII. The jackfruit (*Artocarpus integrifolia*), the largest tropical fruit.

of Allen M. Groves at Miami, Florida, suggest that the difficulty may be due to lack of the necessary insects to effect cross-pollination. It has been observed, however, that occasional plants uniformly bear heavily, and the vegetative propagation of such eliminates all danger of unproductiveness.

There are as yet no named varieties in the trade.

Another species of carissa, and one which is sometimes confused with *C. grandiflora,* is *C. Arduina,* Lam. (*C. bispinosa,* Desf., *Arduina bispinosa,* L.). This can be distinguished from *C. grandiflora* by the smaller size of the flowers, which are only $\frac{1}{2}$ inch broad in place of nearly 2 inches, with the corolla-segments much shorter than the tube; and by the oblong-obtuse fruit, which is only $\frac{1}{2}$ inch in length and contains one or two lanceolate seeds, instead of fifteen or twenty circular ones. The species is not commonly cultivated in the United States, but is said to be used as an ornamental plant in Cape Town, South Africa.

The karanda (*Carissa Carandas,* K. Sch.), a species common in India, has been introduced into the United States, but is not often planted either in California or Florida. It is distinguished from *C. grandiflora* and *C. Arduina* by the corolla-lobes being twisted to the right instead of to the left in the bud; by the oblong or elliptic-oblong leaves with rounded or obtuse tips; and by the spines being simple in place of bifurcate. Its fruits are less than an inch long, and contain three or four seeds. They are used in India for pickles and preserves.

The Ramontchi (Fig. 58)

(*Flacourtia Ramontchi,* L'Hérit.)

While it must be listed among the minor fruits, the ramontchi (more commonly known in the West Indies as Governor's-plum) is not devoid of interest and merit. It is an excellent hedge plant, and its plum-like fruits, which are produced in great abundance, make good jam and preserves.

If allowed to develop to maximum size, the plant may become a large shrub or small tree about 25 feet high. It is armed with long slender thorns. The leaves are broadly ovate in outline, 2 to 3 inches long, acuminate, and commonly serrate. The staminate and pistillate flowers are normally produced on separate plants, as in the papaya; it is, therefore, necessary to

FIG. 58. The ramontchi (*Flacourtia Ramontchi*), often called governor's-plum, comes from Madagascar. Its maroon-colored fruits, of subacid flavor, are valued principally for making preserves. (× $\frac{3}{5}$)

plant trees of both sexes in order to have fruit. The flowers are small and inconspicuous, the fruits round, about an inch in diameter, and deep maroon colored when fully ripe, having a thin skin surrounding soft juicy pulp and several small thin seeds. The flavor is sweet and agreeable in some varieties, acid and somewhat strong in others.

The ramontchi is considered a native of southern Asia and Madagascar. It is now widely scattered throughout the tropics, but is not extensively cultivated in any region. It can be grown in southern Florida as far north as Fort Pierce. In California it has never been very successful. With protection during the first winters it may be possible to grow it in the mildest sections of the latter state. It withstands light frosts after it has attained a few years' growth, and is not exacting in its cultural requirements. It grows on soils of various types, and in moist climates as well as in those which are rather dry.

Propagation is usually effected by means of seeds. When multiplied in this manner, however, many more male plants are produced than are required for the pollination of the females, and it is not possible to perpetuate choice varieties. Vegetative propagation, most likely by means of budding, will have to be applied to this species before its cultivation can be made altogether satisfactory.

The Umkokolo (Fig. 59)

(*Dovyalis caffra*, Warb.)

While its scented fruit is not of great value for eating out of hand, the umkokolo, often called in English kei-apple, is a useful and interesting plant. It is unexcelled for hedges in regions where the temperature does not commonly fall below 20° above zero.

The native home of the species is on the Kei River in South Africa. It is a tall vigorously-growing shrub, with rich green foliage and long, stiff, sharp thorns. The leaves are oblong-obovate, about 2 inches in length, often in small clusters at the bases of the thorns. Staminate and pistillate flowers are produced on separate plants, and both are without petals. The fruit is oblate or nearly round, bright golden yellow, and about 1 inch in diameter. The thin skin incloses a yellow, melting, juicy pulp and five to fifteen flattened pointed seeds. The flavor is aromatic, highly acid unless the fruit is fully ripe.

Because of this, the fruit is most commonly used to make jam and preserves.

Outside of its native region the umkokolo has been planted to a limited extent along the shores of the Mediterranean in France, Algeria, and Italy; in northwestern Australia; and in Florida and California. In Florida it is said to have succumbed to the cold during the severe winter of 1894–1895, and in California it has been killed by temperatures of 16° above zero. The usual winter temperatures in the southern parts of both states, however, are too high to injure it, and the species can be grown safely as far north as the Lake region in central Florida and favored sections of the San Joaquin Valley in California.

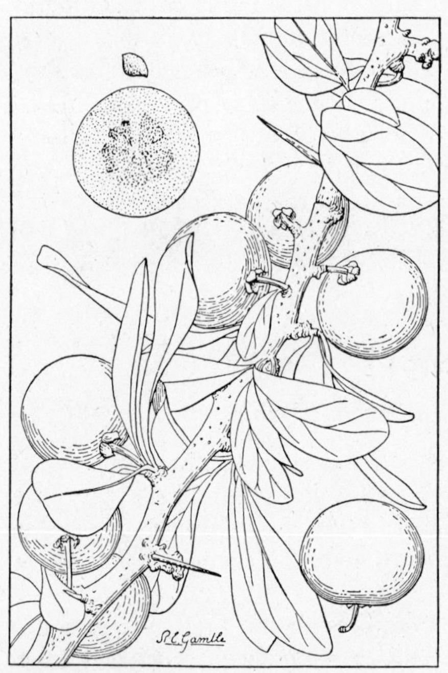

Fig. 59. The umkokolo or kei-apple (*Dovyalis caffra*) is a large thorny shrub from South Africa, excellent for hedges. (× ½)

Botanically the umkokolo is a Dovyalis (latterly written Dorylis), and it is sometimes listed as *Aberia caffra*, Harv. & Sond. *Umkokolo* is one of the vernacular names of its native region in South Africa. The name kei-apple is often spelled incorrectly kai-apple.

The plant is not exacting in its cultural requirements, and is decidedly drought-resistant. It is most successful in a subtropical climate, and on a soil rich in humus.

It is considered one of the best hedge plants in South Africa, since its long sharp thorns make it impenetrable. To form a hedge the bushes should be set 3 to 5 feet apart, and should be trimmed on both sides once a year. For the production of fruit, they should not be set closer than 12 to 15 feet, and both staminate and pistillate plants must be present. One of the former (male) is considered to be sufficient for twenty to thirty of the latter (female). If sufficient seedling plants are grown so that there are sure to be some of both sexes, satisfactory results will be obtained; otherwise, it is best to propagate staminate and pistillate plants by layering or some other vegetative means, and to plant no more staminates than will be required to furnish pollen.

In the Mediterranean region and in the United States, the plants flower in April and May and ripen their fruit from August to October. Seeds may be sown in pans or flats of light sandy loam. Plants propagated in this manner will begin to bear when four or five years old. Propagation by layering is practiced in Queensland, and the species will probably lend itself to shield-budding, since P. J. Wester has shown that another member of the same genus can be propagated readily in this way. The ripe fruit is sometimes attacked by the Mediterranean fruit-fly (*Ceratitis capitata* Wied.).

The Ketembilla

(*Dovyalis hebecarpa,* Warb.)

The ketembilla is a better fruit than its congener the umkokolo, but the plant is somewhat more limited in its distribution. From its native home in Ceylon it has been brought to the Western Hemisphere, where it may now be found in a few gardens in Florida, Cuba, and California; elsewhere it is little known. Since it is more tropical in its requirements than the umkokolo, it is not suitable for cultivation in the Mediterranean basin, except perhaps in the most favored situations.

In growth and habit the plant is less robust than its congener, although it reaches about the same ultimate height, 15 to 20 feet. The branches are slender, often drooping under the weight of their fruit, and the thorns are long and sharp, but not so formidable as those of the umkokolo. The leaves are lanceolate or oval in outline, acute, entire or subserrate, and 2 to 4 inches long. The fruit is of the same size and form as that of the umkokolo, but maroon-purple in color and more velvety on the surface. The purplish pulp is sweet and luscious, with a flavor resembling that of the English gooseberry, a fruit which the ketembilla suggests so strongly in appearance and character as to give rise to the common name Ceylon-gooseberry. *Aberia Gardnerii*, Clos., is a botanical synonym.

The plant does not withstand drought as well as the umkokolo, and is injured by temperatures considerably above 20°. While it succeeds in southern Florida, the climate of most parts of southern California has usually proved too cold for it. It likes plenty of moisture, both in the atmosphere and in the soil, and under proper conditions bears enormous crops of its attractive fruits.

The distribution of the sexes is the same in this species as in the umkokolo, and it is, therefore, necessary to insure the proximity of staminate and pistillate plants if fruit is desired. It has been reported that isolated plants of both species are sometimes fruitful, which suggests that they may in occasional instances produce perfect flowers and not require cross-pollination. If plants of such character are found, they should be propagated by budding or grafting, since they would be of considerable value. P. J. Wester reports that shield-budding is successful. He says: "Use petioled, preferably spineless, not too old budwood with tomentum still present; cut buds an inch to an inch and a quarter long; age of stock at point of insertion of buds unimportant." Propagation by seeds is easily effected, as with the umkokolo.

The White Sapote (Fig. 60)

(*Casimiroa edulis*, La Llave)

In the highlands of Mexico and Central America, where it is believed to be indigenous, the white sapote ranks among the principal cultivated fruits. Outside of this region it is not well known, although it has, in recent years, attracted attention in California and Florida.

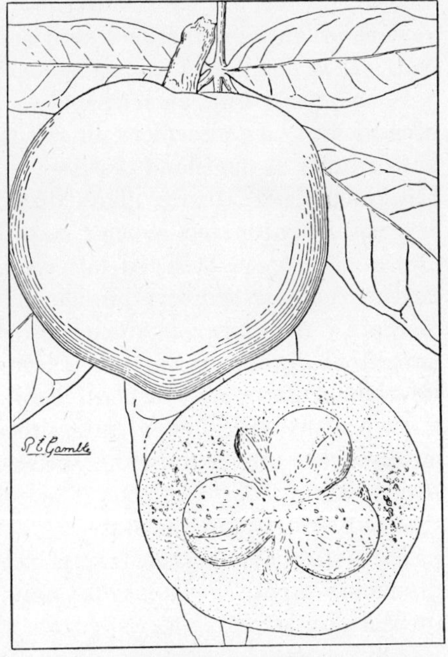

Fig. 60. The white sapote (*Casimiroa edulis*), a common fruit of the Mexican and Central American highlands, is now grown in California and Florida. (× ½)

The Aztecs of ancient Mexico used the term *tzapotl* to designate soft sweet fruits such as the sapodilla and its allies. The lack of acidity and the heavy sweetness of these fruits makes them less acceptable to palates accustomed to apples and peaches than the mangosteen and certain other tropical fruits. They are, however, liked by many northerners, and natives of tropical regions consider them perfect.

The white sapote is a medium-sized erect or spreading tree, having palmately compound leaves, small inconspicuous flowers, and yellowish green fruits the size of an orange. The fruits have a thin membranous skin, yellowish flesh of soft melting texture and sweet or slightly bitter flavor, and one to five large oval or elliptic seeds.

In its native region the white sapote is a fruit of the highlands. Throughout Mexico and Guatemala it is found at elevations of 2000 to 3000 feet, and occasionally as high as 9000 feet. It is not grown in regions subject to heavy rainfall.

It has borne fruit at La Mortola, in southern Italy, and is occasionally seen elsewhere on the Riviera. It is said also to have fruited in the island of Jersey. Although introduced into California from Mexico about 1810, it has not yet become extensively cultivated in that state, and large trees are rare. One of the oldest, believed to have been planted more than a century ago, is growing on De La Guerra Street in Santa Barbara. A number of younger trees, most of them propagated by F. Franceschi and distributed about 1895, are fruiting in various parts of southern California; although some of these produce small bitter fruits, others bear large ones of delicious flavor. In Florida the species has not been cultivated so long as in California, but it has proved quite successful in the southern part of the state.

The Aztec name for this fruit is *cochiztzapotl*, meaning "sleep-producing sapote." It is commonly known in Mexico at the present day as *zapote blanco* (white sapote). In Guatemala it is called *matasano*.

The fruit is usually eaten fresh, but attempts have been made in Central America to prepare a sweet preserve from it on a commercial scale. Some of the early writers considered the white sapote unwholesome, and stated that it would induce sleep if indulged in too freely, but recent experience does not indicate that there are grounds for such beliefs. Francisco

Hernandez observed that the seed, if eaten raw, was poisonous to animals and men. An analysis of the fruit made at the University of California shows it to contain: Water 72.64 per cent, ash 0.44, protein 0.64, total sugars 20.64 (invert sugar 8.44, sucrose 12.20), fat 0.46, crude fiber 1.26, and starch and the like 3.92.

In its climatic requirements the tree is distinctively subtropical. It is not altogether successful in Central America below 3000 feet, and it thrives at elevations of 5000 to 6000 feet. It is even found in places which are too high (*i.e.*, too cold) for the avocado. It prefers a well-drained sandy loam, but may be grown on clay if the drainage is good, and in Florida it has done well on shallow sandy soils underlaid with soft limestone. It is drought-resistant, but succeeds much better in dry regions if irrigated like the orange.

While young, the tree should be watered liberally to encourage growth, and when it is about three feet high it should have the terminal bud removed, in order to induce branching; three or four laterals will develop, and these in turn, after they have grown to a length of one or two feet, should have the terminal buds removed. Unless this is done, the tree may grow ten or twelve feet high before it branches.

Seeds should be planted as soon as possible after their removal from the fruit in flats of light porous soil, or singly in three-inch pots, covering them to the depth of an inch. If the weather is warm, or artificial heat is provided, germination will take place within three or four weeks. The young plants should be grown in pots until two or three feet high, when they may be set out in the open ground.

Seedlings do not come into bearing until seven or eight years old, and many produce fruit of inferior quality. For this reason propagation should be effected by some vegetative means. Shield-budding is successfully practiced, the method being essentially the same as with the avocado. Stock plants should be selected from young vigorous seedlings whose stems are about $\frac{3}{8}$ inch in diameter at the base. Budwood is taken from the ends of the branches, but of fairly well matured wood which has acquired an ash-gray color. The buds are cut about $1\frac{1}{2}$ inches long, leaving any wood that may adhere to them, and are inserted in T-incisions, after which they are bound firmly in place with waxed tape. At the end of two to four weeks, depending on the weather, they may be unwrapped and then rewrapped loosely, leaving the bud exposed so that it may start into growth, at the same time lopping back the stock to a point three or four inches above the bud. In the tropics budding can probably be done at almost any season of the year; in California spring and summer, when the stock plants are in most active growth, are the best times.

Seedling variation results in some trees being very productive, while others bear little fruit. No budded trees have yet come into bearing. The ripening season in Guatemala is April and May; in Florida it is May; in Mexico it extends from May to July; and in California it begins in September and ends in November. Because of its thin skin and delicate texture, the fruit does not ship well, unless picked while still hard and dispatched so as to reach its destination before it has had time to soften.

Several horticultural varieties have been described, but none has been propagated or planted extensively. Harvey and Maechtlen are two which have been offered by the trade in California; Parroquia and Gillespie have been described, but not propagated.

The Tuna

(Opuntia spp.)

Several species of Opuntia, notably *O. Ficus-indica,* Mill., and *O. megacantha,* S. D., are extensively grown in tropical and subtropical countries for their fruits, commonly known as tunas, prickly-pears, or Indian figs.

Among the aboriginal inhabitants of tropical America, the tuna (using this term in a comprehensive sense) has long been held in high esteem. It was early introduced into southern California by the Franciscan monks, and is now found abundantly in many places, particularly around the old missions. From America it was carried to Spain by the early voyagers, and from that country it spread along the Mediterranean littoral and finally to many other regions. It is now cultivated and naturalized throughout the tropics and subtropics.

The edible-fruited opuntias are erect or spreading plants, growing from 10 to 25 feet in height. They have large flattened branches made up of more or less rounded joints, which in popular language are called leaves. Usually these joints bear long sharp spines, although in some species they are almost spineless. The flowers, which are produced toward the upper part of the joints, are yellow or red and rather showy. The oblong to pear-shaped fruits, commonly 2 to 3 inches in length and green to deep maroon in color, contain soft, whitish, translucent pulp intermixed with numerous large bony seeds. The pulp is juicy with a pleasant, although not pronounced, flavor. The principal objection to the tunas is the great quantity of hard seeds which they contain.

O. Ficus-indica has fewer spines and somewhat differently colored fruit from *O. megacantha;* both these species are cultivated in the southwestern United States as well as in Mexico, the Mediterranean region, and elsewhere. Several other species produce edible fruits, but their cultivation is not extensive.

A considerable quantity of tunas is shipped annually to the United States from Sicily, and an important trade could be developed betweeen the United States and Mexico.

Because of its rather high nutritive value, the tuna forms

an important article of diet in many regions. It is eaten fresh, dried, or prepared in various ways. Griffiths and Hare have discussed this subject fully in "The Tuna as Food for Man."[1] The ripe fruit contains: Total solids 19.66 per cent, ash 0.40, acids 0.18, protein 0.98, total sugars 13.42, fat 0.23, and fiber 2.79.

J. W. Toumey, writing in Bailey's "Standard Cyclopedia of Horticulture," says: "It has been ascertained that some of the best varieties are capable of producing on lean, sandy or rocky soil, ill-suited for growing ordinary crops, as much as 18,000 pounds of fruit to the acre. When it is considered that this is equal to 2500 pounds of sugar, as well as other valuable food constituents, it may be readily seen that the food value from the standpoint of nutrition is considerable."

Little cultural attention is usually given to the opuntias in the regions where they are grown for their fruit. To quote Toumey again: "Plantations are usually made on dry slopes of hills, as the plants do not thrive where there is much moisture or on heavy clay soils. Joints, cut or broken from the plants, are used instead of seeds, and are planted at distances of 6 to 8 feet in furrows from 6 to 15 feet apart. No tillage is practiced, as they grow rapidly, and in a few years smother out all other growth. Before planting, the cuttings are exposed in half sunlight from seven to fifteen days, that they may partially wither, in order to facilitate rooting.

"An important advantage in the culture of these plants is the regularity of the yearly crop. They begin to bear in about three years after planting, and continue in bearing for many years."

Numerous varieties or forms, usually local in their distribution, are distinguished in Mexico and elsewhere. In spite of the attention given in recent years to the improvement of this fruit by breeding, still further advances must be made before varieties are obtained which will become popular as table-fruits among North Americans.

[1] Bul. 116, Bur. Plant Industry.

THE PITAYA (Fig. 61)

(Hylocereus, Lemaireocereus, and Cereus)

The fruits of many cacti are known in tropical America by the name pitaya, also spelled *pitahaya*, *pitajaya*, *pitajuia*, *pitalla*, and *pithaya*. These belong to several genera, formerly classified under the genus Cereus, but the best fruits are obtained from the genera Hylocereus and Lemaireocereus. Pitayas are commonly larger than tunas, and by some are considered superior to the latter in quality, but their use is less extensive.

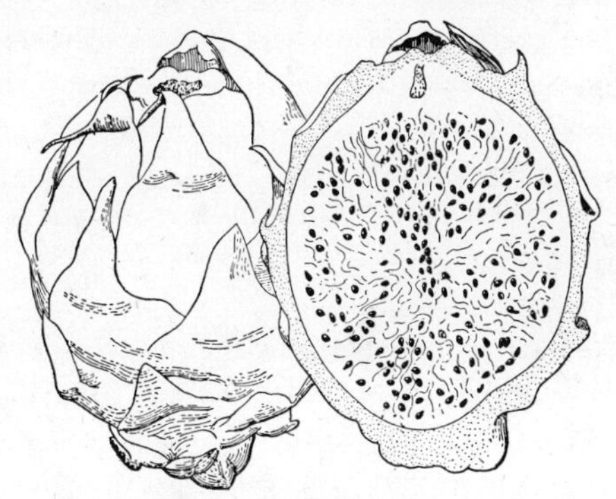

FIG. 61. The pitaya (*Hylocereus undatus*) is widely cultivated in the American tropics. Its bright rose-colored fruits contain white translucent pulp of pleasant taste; they are produced by a climbing cactus which bears handsome night-blooming flowers. (× ⅓)

The genus Hylocereus has several species which produce good fruits. The widely cultivated plant which usually passes under the name of *Cereus triangularis* is properly *Hylocereus undatus*, Brit. and Rose; the true *C. triangularis* is found in Jamaica, but rarely elsewhere. All of these plants are climbing in habit, and have three-angled stems. They produce large, showy, night-blooming flowers, and oblong or oval fruits, bright pink to red in color, sometimes more than 3 inches in length, with large leaf-like scales on the surface. The flesh is white and juicy and is filled with numerous minute seeds. In southern Mexico the fruits are used in various ways: they may be eaten out of hand; employed in making cooling drinks and sherbets; and for preserves.

Somewhat distinct are the pitayas furnished by several species of Lemaireocereus, notably *L. griseus*, Brit. & Rose, and *L. queretarensis*, Brit. & Rose, and their allies. These are common wild plants in Mexico and elsewhere, and *L. griseus* is often cultivated. The fruits are globose, about 2½ inches in diameter, and covered with many small clusters of spines. These are brushed off the red fully ripe fruit, leaving it in condition to be eaten. The flesh is dark red to purple, sweet and delicious in flavor.

The propagation and culture of these plants resembles that of the tunas; the Hylocereus group, however, is much better adapted to a moist tropical climate than the latter.

The Tree-tomato (Fig. 62)

(*Cyphomandra betacea*, Sendt.)

Several food-plants which were cultivated by the agricultural Indians of ancient Peru have become of economic importance to the modern world, one of them, the potato, of immense value. The tree-tomato, a bush fruit which was planted in their gardens high upon the mountain-sides, is now grown in the hill-regions of India and Ceylon, as well as in several other countries.

In its native home, where it forms a miniature tree 5 or 6 feet high, O. F. Cook says the plant is cultivated at elevations of 6000 to 10,000 feet. In California it grows 8 or 10 feet high. It has large cordate-ovate, subacuminate leaves, small pinkish flowers, and oblong fruits produced in clusters of three or more. In length these fruits are about 2 inches; in color and in general character they resemble tomatoes, to which they are, of course, closely related. "It has become thoroughly established in many hill gardens," writes H. F. Macmillan of the tree-tomato in Ceylon, "and is commonly grown about Nuwara Eliya for market. The egg-shaped and smooth-skinned fruit, produced in great abundance and in hanging clusters at the ends of the branches, is in season almost throughout the year, but chiefly from March to May. At first greenish purple, it changes in ripening to reddish yellow. The subacid and succulent fruits are refreshing and agreeable when eaten raw, but their chief use is for stewing; they may also be made into jam or preserve. The tree is a quick grower, and commences to bear fruit when about two years old, remaining productive for several years."

FIG. 62. The tree-tomato (*Cyphomandra betacea*), a fruit produced by a half-woody shrub from South America, closely resembles the tomato in character, and is useful in the same ways as the latter. (× ½)

It has been found in California that the species withstands several degrees of frost. It may be killed back to the large limbs by a temperature of 26° to 28° above zero, but it promptly recovers. In Mexico and Central America, where it is known as *tomate,* it is cultivated by the Indians at elevations of 4000 to 8000 feet. It likes a rich loamy soil and grows best when abundantly irrigated. It does not require a high degree of atmospheric humidity.

Propagation is effected by means of seeds which germinate readily and develop rapidly into strong plants.

The Genipa

(*Genipa americana*, L.)

In parts of Brazil and in Porto Rico the genipa is a popular fruit. Elsewhere it is of little importance. Outside of its native area, which is considered to be northeastern South America and the West Indies, it is indeed scarcely known.

When well grown the tree is stately and handsome in appearance. It reaches a height of 60 feet or more, and has a straight, slender trunk branching 10 or 15 feet above the ground. The leaves are oblong-obovate in form, entire or dentate, dark green in color, and about a foot long. The flowers, which in Brazil are produced in November, are small, and light yellow in color. The fruits are the size of an orange, broadly oval to spherical in form, and russet-brown. After being picked they are not ready to be eaten until they have softened and are bordering on decay. Beneath the membranous skin is a thin layer of granular flesh, and within this a mass of soft brownish pulp in which numerous small compressed seeds are embedded. The flavor is characteristic and very pronounced; it may be likened to that of dried apples, but is stronger, and the aroma is more penetrating.

The genipa, known in Brazil as *genipapo*, in Porto Rico as *jagua*, and in the British West Indies as *genipap* and marmalade-box, is eaten fresh, and used to prepare an alcoholic beverage known as *licor de genipapo*. A refreshing drink known as *genipapada* is also made from it, and, when green it furnishes a dye used by some of the Brazilian Indians in tattooing.

In its climatic requirements the tree is tropical. It is not known to have been grown in California or Florida, although it might succeed in the southern part of the latter state. It prefers a humid atmosphere and a deep rich loamy soil containing plenty of moisture. Propagation is usually by seeds, which are easily germinated. P. J. Wester, who has experimented with the tree in the Philippines, finds that it can be shield-budded in the same manner as the avocado. He says: "Use mature, bluish-green, smooth, non-petioled budwood; cut the buds about an inch and a half long; age of stock at point of insertion of bud unimportant." By utilizing this method of propagation it will be possible to perpetuate choice varieties which originate as chance seedlings.

BIBLIOGRAPHY

Much of the literature on tropical fruits exists in the form of bulletins and brief articles in the horticultural press. Reference has been made in the text of this work to the most important.

The more extensive works containing information on the history, cultivation, varieties, pests and diseases of tropical fruits are listed below. An asterisk is placed before those which will be found particularly useful by the tropical horticulturist or fruit-grower.

Ballou, H. A., Insect Pests of the Lesser Antilles. Pamphlet No. 71 of the Imperial Department of Agriculture for the West Indies. Barbados. 1912.

Benson, Albert H., Fruits of Queensland. Department of Agriculture and Stock, Brisbane. 1911.

*Capus, G., and Bois, D., Les Produits Coloniaux, Origine, Production, Commerce. Librairie Armand Colin, Paris. 1912.

Clute, Robert L., Practical Lessons in Tropical Agriculture, Book 1. The World Book Co., Yonkers-on-Hudson, N.Y. and Manila, P. I. 1914.

*Cook, Melville T., The Diseases of Tropical Plants. Macmillan and Co., Ltd., London. 1913.

*Cook, O. F., and Collins, G. N., Economic Plants of Porto Rico. Contributions from the United States National Herbarium, vol. VIII, pt. 2. Government Printing Office, Washington. 1903.

DeCandolle, Alphonse, Origin of Cultivated Plants. 2d ed. International Scientific Series. Kegan Paul, Trench, Trübner and Co., Ltd., London. 1909.

*Essig, E. O., Injurious and Beneficial Insects of California, 2d ed. State Commission of Horticulture, Sacramento, California. 1915.

*Fenzi, E. O., Frutti Tropicali e Semitropicali (esclusi gli Agrumi). Instituto Agricolo Coloniale Italiano, Firenze. 1915.

Firminger, Thomas A. C., Manual of Gardening for Bengal and Upper India. Barnham, Hill and Co., Calcutta. 1869.

Haldane, R. C., Subtropical Cultivations and Climates, a Handy Book for Planters, Colonists, and Settlers. William Blackwood and Sons, Edinburgh and London. 1886.

*Hubert, Paul, Fruits des Pays Chauds, tome 1. H. Dunod et E. Pinat, Paris. 1912.

Ikeda, T., The Fruit Culture in Japan. Seibido, Tokyo. Without date.

Jumelle, Henri, Les Cultures Coloniales, Légumes et Fruits. 2d ed. Librairie J. B. Baillière et fils, Paris. 1913.

*Macmillan, H. F., A Handbook of Tropical Gardening and Planting, with special reference to Ceylon. 2d ed. H. W. Cave and Co., Colombo, Ceylon. 1914.

Maxwell-Lefroy, H., Indian Insect Life, a Manual of the Insects of the Plains. Thacker, Spink and Co., Calcutta and Simla. 1909.

Pierce, W. Dwight, A Manual of Dangerous Insects Likely to be Introduced into the United States through Importations. Office of the Secretary, United States Department of Agriculture. Government Printing Office, Washington. 1917.

*Pittier, H., Ensayo sobre las Plantas Usuales de Costa Rica. H. L. and J. B. McQueen, Inc., Washington, D. C. 1908.

*Popenoe, Paul B., Date Growing in the Old World and the New. West India Gardens, Altadena, California. 1913.

Reasoner, P. W., and Klee, W. G., Report on the Condition of Tropical and Semi-tropical Fruits in the United States in 1887. Bulletin 1, Division of Pomology, United States Department of Agriculture, Washington. 1891.

Ribera Gomez, D. Emilio, Manual sobre Árboles Frutales, Escrito especialmente para America. Garnier Hermanos, Paris. No date.

Riviere, Ch., and Lecq, H., Cultures du Midi, de l'Algérie, et de la Tunisie. J. B. Baillière et fils, Paris. 1906.

*Roeding, George C., Roeding's Fruit Growers' Guide. Published by the author, Fresno, California. 1919.

*Safford, W. E., The Useful Plants of the Island of Guam, with an Introductory Account of the Physical Features and Natural History of the Island, of the Character and History of its People, and of their Agriculture. Contributions from the United States National Herbarium, vol. IX. Government Printing Office, Washington. 1905.

Plate XXIV. *Upper*, the mangosteen; *lower*, the durian.

*SAUVAIGO, ÉMILE, Les Cultures sur le Littoral de la Mediterranée; Provence, Ligurie, Algérie. 2d ed. Librairie J. B. Baillière et fils, Paris. 1913.

VON MUELLER, FERDINAND, Select Extra-Tropical Plants Readily Eligible for Industrial Culture or Naturalization. American edition, revised and enlarged. George S. Davis, Detroit, Michigan. 1884.

*WATT, SIR GEORGE, The Commercial Products of India, being an abridgment of "The Dictionary of the Economic Products of India." E. P. Dutton and Co., New York. 1908.

*WESTER, P. J., Plant Propagation in the Tropics. Bulletin 32 of the Bureau of Agriculture, Manila, P. I. 1916.

WICKSON, EDWARD J., The California Fruits and How to Grow Them. The Pacific Rural Press, San Francisco, California. 1910.

WILCOX, E. V., Tropical Agriculture, the Climate, Soils, Cultural Methods, Crops, Live Stock, Commercial Importance and Opportunities of the Tropics. D. Appleton and Co., New York and London. 1916.

WILDER, GERRIT P., Fruits of the Hawaiian Islands. Revised ed. The Hawaiian Gazette Co., Ltd., Honolulu. 1911.

*WOODROW, G. MARSHALL, Gardening in India, 5th ed. Printed at the Education Society's Press, Byculla, Bombay. 1889.

WOODROW, G. MARSHALL, The Mango: Its Culture and Varieties. Alexander Gardener, Paisley. 1904.

YULE, SIR HENRY, and BURNELL, ARTHUR COKE, Hobson-Jobson; being a glossary of Anglo-Indian colloquial words and phrases, etymological, historical, geographical and discursive. New edition, edited by William Crooke. J. Murray, London. 1903.

INDEX

B

C

T

U

V

W

X

Y

Z